AF632415

Identification of PRAWNS/SHRIMPS OF INDIA and their Culture

Identification of PRAWNS/SHRIMPS OF INDIA and their Culture

Dr. A.D. DHOLAKIA
M.Sc., LL.B. (Sp), Ph. D.
(Retd.) Research Officer and Head
Fisheries Research Station,
Junagadh Agricultural University, Sikka
Gujarat, India

2010
DAYA PUBLISHING HOUSE
Delhi - 110 035

© 2010 ANSHUMAN D. DHOLAKIA (b. 1947–)
ISBN 81-7035-652-0
ISBN 978-81-7035-652-3

All rights reserved, including the right to translate or to reproduce this book or parts thereof except for brief quotations in critical reviews.

Published by : **Daya Publishing House**
1123/74, Deva Ram Park
Tri Nagar, Delhi - 110 035
Phone: (011) 27383999
Fax: (011) 23260116
e-mail : dayabooks@vsnl.com
website : www.dayabooks.com

Showroom : 4760-61/23, Ansari Road, Darya Ganj,
New Delhi - 110 002
Phone: 23245578, 23244987

Laser Typesetting : **Classic Computer Services**
Delhi - 110 035

Printed at : **Chawla Offset Printers**
Delhi - 110 052

PRINTED IN INDIA

Acknowledgement

I dedicate this book jointly to my wife Smt Sangita to whom I received utmost encouragement and appreciation. But for the help received from her to keep myself free from any domestic chore whatsoever, the writing of the book would not have been possible, and to Late Prof. N. D. Chhaya who was my guide in my Ph. D. Thesis "Marine prawn Fisheries and its culture in Saurashtra with special reference to *Penaeus merguiencies* deMan". I also express my gratitude to my late mother Smt. Jayshriben and father Shri Dwijendraray.

I am highly thankful to Shri A. U. Buch, Deputy Commissioner of Fisheries (Retired), Department of Fisheries, Government of Gujarat, Gandhinagar, for writing Foreword of this book. I am also highly thankful to Dr. D. C. Bhatt, Head Marine Sciences, Bhavnagar University, Bhavnagar for Recommending this book for the use of students and teachers.

Prof. Anshuman D. Dholakia

Acknowledgement

[illegible]

[illegible]

[illegible]

Foreword

I have seen the contents of this book and found that Prof. Dholakia has put tremendous and useful effort in identification of prawn and shrimp species of India. Early identification of the commercial species from the Mysis and Postlarvae will help to separate unwanted species from the culture batch. For precious identification at these stages, he has given drawing of each part of the body. For identification of adults he has given drawings and colour photographs wherever it is possible.

In this book he has given identification of 78 Marine water prawn and 17 freshwater shrimps totaling 95 prawn and shrimp species. Identification and culture activities were explained using 535 drawings, 87 colour photographs and 30 tables. Besides identification he has covered culture aspects also for this requirement of different parameters like Site selection, design and construction of culture pond, type of soil, stocking, feed nutrition requirement, feed preparation, method and quantity of feed distribution are explained and also possibilities of diseases observed in commercial cultured prawn/shrimp are given.

I know Dr. Dholakia since 1977. He has done his Ph. D. in marine Prawn Culture. He has done excellent work when he was in-charge of prawn culture laboratory at Okha. He has published many research papers. Looking to his experience, I am sure that this Book will be helpful to B. F. Sc. and M. F. Sc. students, teachers and research workers of Fisheries Faculty of respective colleges, technical officers/staff of fisheries, as well as to supervisors of prawn/shrimp culture farms. I wish him all success in his life.

A.U. Buch

Deputy Commissioner of Fisheries, (Retired)
Department of Fisheries, Govt. of Gujarat.
Gandhinagar, Gujarat

Residence

"Ashirwad" A/67-C Plot No 1006/2
Opp. S. T. Depot. Sector–7 C
Gandhinagar – 382007, Gujarat

Preface

Prawns are recognized as a major delicacy for the table and in many parts of the world prawn fisheries have developed extensively during the last forty years. Traditionally, prawning was a shallow-water fishery; prawns being trawled from coastal and estuarine waters; but with the increased demand for prawns, trawling and trapping have now been extended down to 400 fathoms.

The commercial prawns of India can be grouped in to Penaeid and non-penaeid. Penaeid form a little over 50 per cent of the total marine prawn catch, the rest being non-penaeid.

It is necessary to identify them and only commercially important and high demand prawn can be cultured for profitable business. In this book I have given identification of 78 Marine water prawn and 17 freshwater shrimps. Identification of prawn and shrimp of commercial importance is given from Mysis stage to Post larvae stages. Identification using carapace and how to identify prawn/shrimp in field are also given. I have tried to give justice to this subject with the help of 535 drawings, 87 colour photographs and 30 tables. In this book I have tried to cover identification of 95 prawn and shrimp species. The subject is justified using more than 90 references.

Different parameters like Site selection, design and construction of culture pond, type of soil, stocking, feed nutrition requirement, feed preparation, method and quantity of feed distribution, and some major diseases in commercial cultured prawn/shrimp are given. Culture method in details of marine prawn as well as freshwater shrimp is given. In this book the requirement of syllabus for B.F.Sc and M.F.Sc. approved by I.C.A.R. is also looked in to with respect to the subject of this book.

Details of selection of prawn/shrimp species for culture considering local condition are also discussed.

Looking to high demand of culture prawn due to eco-friendly and unpolluted environmental condition, business people would like to culture prawn/shrimp. This book will be highly helpful to them as well as to fishery students, teachers and research workers.

Prof. Anshuman D. Dholakia
201, Shashwat Apartment,
Nr. Vaibhav Laxmi Temple
B/H Drive-in-Cinema
Memnagar, Ahmedabad – 38 0052

Recommendation

I have seen the contents of this book. Looking to the high demand and taste of Prawn/Shrimps, it is necessary to know about them. Prof. Dholakia has tried to identify about 95 prawn/shrimp species in this book. He has also given the culture systems for both Marine as well as freshwater prawn/Shrimp.

In present days when day by day landings are decreasing, it is necessary to culture prawn/shrimp. It is my experience that many students and people are facing difficulty in identifying proper prawn/shrimp.

This is very important to identify prawn/shrimp properly before starting their culture. Efforts made by Prof. Dholakia is to be admired. I recommend using this book as reference book by students and teachers who are in this field may get proper guidance.

Prof. D.C. Bhatt
Head, (Marine Sciences) (Retd.)
Bhavnagar University, Bhavnagar

Contents

Identification of Prawns/ Shrimps Covered in this Book

Sl.No.	Family	Species	Page No.
	PENAEIDAE		
1.	***Penaeus***	*Penaeus monodon* Fabricius	105, 112, 126, 301
2.		*Penaeus semisulcatus* de Hann.	105, 113, 129, 301
3.		*Penaeus indicus* Milne Edward. New name: *Penaeus* (Fenneropenaeus) *indicus* H. Milne Edwards, 1837	106, 113, 130, 300
4.		*Penaeus merguiensis* de Man	106, 113, 132, 301
5.		*Penaeus japonicus* Bate New Name (*Marsupenaeus japonicus*)	105, 112, 137
6.		*Penaeus latisulcatus* Kishionuye New Name: *Penaeus* (Melicertus) *latisulcatus* Kishinouye, 1896	105, 144
7.		*Penaeus penicillatus* Alcock	106, 135,
8.		Penaeus canilatus	106
9.		*Penaeus canaliculatus*	112, 146
10.		*Penaeus esculentus*	142
11.		*Penaeus vannamei* New Name: *Litopenaeus vannamei*	139
12.		*Penaeus longistylus*	143

Sl.No.	Family	Species	Page No.
72.		*Acetes sibogae* Hansen	111, 191
73.		*Acetes erythraeus* Nobili, 1905	111, 191
74.		Acetes cochinensis Rao	111
75.		Hippolysmata ensirostris (Kemp)	192
76.	**Penaeidae: Sub family Aristeinae**	*Aristeomorpha woodmasoni* Calman	104, 192
77.		*Aristeus alcocki Ramdan*	104, 192
78.		*Aristeus virilis*	104
79.		*Aristeus semidentatus* Bate, 1881	193
80.		*Atypopenaeus stenodactylus* (Stimpson, 1860)	105, 193
81.	**Palaemo-noidea**	*Macrobrachium rosenbergii* (de Man)	195, 197
82.		*Macrobrachium malcomsonii* (H. Milne Edw)	195, 203
83.		*Macrobrachium nipponense* (De Haan, 1849)	204
84.		*Macrobrachium lamarrei*	195, 207
85.		*Macrobrachium villosimanus*	195
86.		*Macrobrachium kulsiense*	211
87.		*Macrobranchium veliense*	210
88.		*Macrobrachium nobilii* (Henderson and Matthai, 1910)	211
89.		*Macrobranchium rude* (Heller)	195, 212
90.		*Macrobranchium idella* (Heller)	195, 212
91.		*Macrobranchium equidens* (Dena)	213
92.		*Macrobranchium scabriculum* (Heller)	196,
93.		*Macrobrachium mirabile*	196, 213
94.		*Macrobrachium javanicum*	195
95.		*Palaemon serenus*	122, 213
96.		*Palaemon tenuipes*	120, 215
97.		*Palaemon styleferus*	
98.		*Palaemon serratus*	122, 215
99.		Palaemon affinis M. Edwards, 1837	120, 122, 216
100.		*Palaemon longirostris*	216
101.		*Palemon natator*	215

Part I
IDENTIFICATION

Chapter 1
Introduction

Crustaceans are landed in all the maritime States of India, but the amount of landings very from State to State. The landings of East Coast of India form only about 17 per cent of the total crustacean landings, while the balance of about 83 per cent is landed on the West Coast of India. Among the States, Maharashtra ranks first by contributing about 48 per cent of the total crustacean landings in India followed by Kerala which contributes on an average 28 per cent of the average annual production of crustaceans. In fact, the major crustacean fishery of India are today located in the two States, Maharashtra and Kerala.

The prawns constitute a large group of Crustaceans varying in size from microscopic to 35 cm long. Although nearly 2500 species are known, less than 300 are economically harvestable; and of these about 100 comprise much of the annual world prawn and shrimp catches. (Dholakia, A. D. 2004)

Prawns are recognized as a major delicacy for the table and in many parts of the world prawn fisheries have developed extensively during the last forty years. Traditionally, prawning was a shallow-water fishery; prawns being trawled from coastal and estuarine waters; but with the increased demand for prawns, trawling and trapping have now been extended down to 400 fathoms.

Prawns are widely distributed, occurring in marine, brackish and fresh waters from the equator to the Polar Regions. They are found from nearly shallow waters to depth of nearly 5700 m. However, most of the commercial prawns are taken on the continental shelf at depth of less than 100 m.

The commercial prawns of India can be grouped in to Penaeid and non-penaeid. Penaeid form a little over 50 per cent of the total marine prawn catch, the rest being non-penaeid. (Dholakia, A. D. 1993)

There was a despot about the use of word Prawn and Shrimp. The term "Prawn" used in India is identical to term "Shrimp" used in Western countries. As decided in

Indo-Pacific Fisheries Council meeting held at Tokyo in 1955, only Penaeid, Pandalids and Palmonidis varieties will be properly known as "Prawn" while other small varieties will be identified as "Shrimp"

Prawns and Shrimps are Decapod Crustacea

Before deciding if prawns and shrimps are different, it has to be agreed that they are at least a bit similar. All prawns and shrimps are crustaceans, which are mostly aquatic animals with a hard skin (an exoskeleton) over a segmented body. Crustaceans belong to the subphylum Crustacea. They are like insects, which also have an exoskeleton, but differ in usually having many pairs of legs, instead of three pairs. The Decapoda, the group of Crustacea to which all prawns and shrimps (and lobsters and crabs) belong, have five pairs of legs on the main part of the body, plus five pairs of swimmerets on the abdomen or tail. It is the muscular tail that is edible. The classification of the Decapoda is very complex, even to a carcinologist (a scientist who studies Crustacea).

Prawns (Figure 1.2)

The crustaceans that Australians call prawns belong to one decapod family, Penaeidae. Adults grow to about 200 mm long. In Asia prawns are raised in coastal farms.

Penaeids live close to the seafloor in shallow water, burrowing in the mud during the day and moving only at night, when they can be caught by trawl nets. Prawns reproduce by dispersing their eggs freely into the water, where the young prawns hatch and swim into estuaries to grow up.

There are about 70 species of prawns in Australia, but only 10 are of economic significance: banana prawn, Endeavour prawn, tiger prawn, king prawn, red-spot prawn and school prawn are some of the names used for different species or groups of species.

Shrimps (Figure 1.1)

Shrimps, on the other hand, are members of the Caridea, another group of Decapoda comprising many families. Most carideans are not edible, or they are too small (rarely more than 40 mm long). Carideans produce eggs that are carried by the adult female, attached to the swimmerets under the tail.

Telling Shrimps and Prawns Apart

An important difference between penaeids and carideans, besides the way they reproduce, is in the way the segments of the abdomen (tail) overlap. In penaeids (prawns) the sides of all segments overlap the segment behind, like roof tiles. In carideans (shrimps) the sides of the second segment overlap both the one before and the one after (see drawing). And in prawns the first three of the five pairs of legs on the body have small pincers, while in shrimps only two pairs are claw-like. In some shrimps one or other of the first two pairs of legs is bigger than the other whereas in prawns all the legs are similar lengths.

A shrimp in the USA is a prawn in Australia!

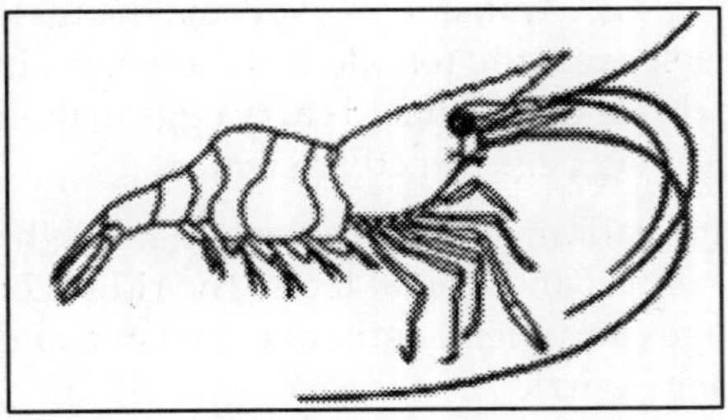

Figure 1.1: A Typical Shrimp

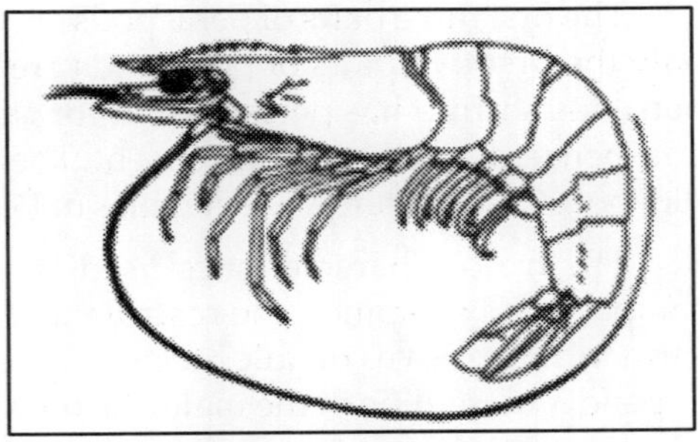

Figure 1.2: A Typical Prawn

(*Source*: Museum Victoria, Information Sheet No. 10295, August, 2006)

But that is not the end of the story. There is a large fishery for penaeids in the southern USA, especially in the Gulf of Mexico, and Americans call them shrimp! Penaeid and Non-penaeid varieties can easily be identified in field as under.

Commercial shrimp and prawns of India can be grouped into Penaeid and Non-penaeid. These two groups of shrimps and prawns are having some important distinguishing features by which it can be easily separated.

Penaeid (Prawn)

Exoskeleton of second segment of the body does not cover overlap on first segment but overlaps only on third segment.

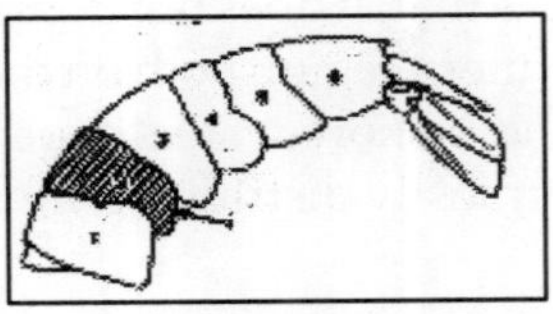

The pleurae of the exoskeleton overlap only of the third segment in penaeids

Figure 1.3: Overlapping of Abdominal Segment

Non-penaeid (Shrimp)

Exoskeleton of second segment of the body overlaps on both first and third segment.

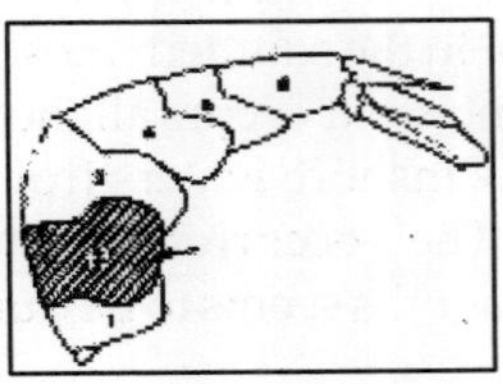

The pleurae on either side of the exoskeleton of the second abdominal segment overlap the pleurae of the first and third segment in non-penaeids

Figure 1.4: Overlapping of Abdominal Segment

The first three pairs of peraepods are chelate in penaeids. While in non-penaeids only the first two pairs of peraepods are chelate. For transferring sperms the male penaeids shrimp has petasma and for storing sperms the female has thelycum. In non-penaeids such organ are absent. The females of non-penaeid carry eggs in their pleopods as a cluster, while females of Penaeids lay eggs directly in water.

The major characteristics used in identification of shrimps are mainly the carapace and its spines, the rostrum and its ventral and dorsal teeth, the ridges or carinae, the grooves or sulci, telson, appendages and their segments, petasma and appendix masculine in the male, the thelycum in female.

Central Marine Fisheries Research Institute (Govt. of India) has surveyed availability of Penaeid and Non-penaeid prawns in different states of India.

74 species of penaeid prawns are known to exist in India, but only 21 of them significant portion of commercially exploited resources. (Dholakia, A. D. 2004)

The genus Penaeus has a worldwide distribution and the various species belonging to it are found both in tropical and temperate latitudes. Practically all of them are marine although some are know to spend a part of their life in the brackish water and even in freshwater. Of the 28 valid species of the genus, only 8 are represented in Indian waters: they being *Penaeus japonicus, P. latisulcatus, P. canaliculature, P. monodon, P.semisulcatus, P. inducus, P. merguiensis and P. penicillatus.*

Metapenaeus, comprising of 24 species, only 10 species have been recorded to occur in Indian waters. They are *M.dobsoni. M. monoceros, M. affinis, M. brevicornis, M. ensis, M. lysianassa, M. burdenroadi, M. stebbengi, M. kutchensis* and *M. alcocki.*

In the distribution of this species in Indian waters one difference noticed from other species like *M. monoceros* and *M.affinis* is that, it does not occur in the southern area, but contribute a good fishery in the northern region both on the west as well as east coasts. Well represented in estuaries and inshore waters especially in the east coast. In the Gulf of Kutch area the species is mostly distributed in areas with sandy bottom.

The genus Parapenaeopsis comprising of 16 species enjoys a wide distribution. Majority of the species are restricted to tropical and warm temperate shallow seas, but few of them are also recorded from brackish water regions. Most of the species are recorded from Indian region so far. They are *Parapenaeopsis uncta, P. cornntra maxillipedo, P. nana, P. acclivirostris, P. sculptilis, P. hardwickii* and *P. stylifera.* Among them, *P. stylifera, P. sculptilis* and *P. hardwikii* are commercially exploited in India.

Unlike are most abundant from Veraval to Trivandrum coast, but moderately available in the Sind Mekran and Kutch areas. In the southern most part of west coast and in east coast they are found in lesser numbers. It occurs all the year round in the west coast in India, but abounds the shallow inshore waters from January to June and deeper waters in September to October. Their occurrence in the marine regions and the relative abundance during warmer months seems to be due to their inability to tolerate lesser salinity.

Parapenaeopsosis hardwickii ranks third among the commercially exploited species of the genus *Parapenaeopsis* in the Indian region. The general distribution of the species is from the coasts of India through Malayesia to southern China. Although, the species occur on both the coasts of India, it supports a good fishery only in Bombay and in lesser magnitude in Andhra coast.

Marine shrimp farming is the most important aquaculture sector in the world. In South East Asia, Penaeoid shrimp have contributed greatly to the economy of Indonesia, Philippines, Vietnam and Thailand. Thailand in particular managed to maintain its title of "the world's largest cultured shrimp producer" for many years. Over 90 per cent of the shrimp produced in Thailand are cultured. Sea-caught shrimp are mostly white and pink shrimps. The black tiger shrimp, *Penaeus monodon* Fabricius, is the only species in Thailand which has been selected for intensive farming.

The prawn includes two species which are very similar in morphology, namely *Penaeus merguiensis* and *Penaeus indicus*. *P. merguiensis* de Man, 1888 or Banana prawn (Australian and FAO) has the following principal taxonomic features: rostrum with high blade and teeth above and below, hepatic ridge absent and gastro-orbital ridge absent or very feebly defined. Their size for females is up to 240 mm total length and males to 200 mm. This species can be mistaken for *Penaeus indicus*, but generally can be distinguished by the higher rostral crest in adult *P. merguiensis* and the presence of a distinct gastro-orbital ridge in *P. indicus*. *P. merguiensis* is one of the most important commercial species in the Indo-Pacific region, being the basis of extensive fisheries in Australia, New Guinea, Indonesia, the Philippines and to a lesser extent, in Malaysia, India, Pakistan and the Persian Gulf. *P. indicus* forms the basis of major commercial fisheries in East Africa, India, Malaysia, Thailand and Indonesia.

The global annual production of freshwater prawns in 2003 was about 280,000 tonnes, of which China produced some 180,000 tonnes, followed by India and Thailand with some 35,000 tonnes each. Other major producer countries are Taiwan, Bangladesh, and Vietnam. In the United States, there are only a few hundred small farms for *M. rosenbergii* with an overall production of just about 50 tonnes in 2003. (http://en.wikipedia.org/wiki/Freshwater_prawn_farm)

Post larvae and juvenile penaeid prawns to species level-7. Therefore, there is a need to develop an identification technique, which is easy and reliable.

Such a technique would be useful for studying genetic variations of wild populations and in brood stock selection for intensive farming. In this book I have tried to give identification of 95 prawn and shrimp species.

Need for Culture

It can be seen from the Figure 1.5 that total percent share of fish production from India is only 4.36 per cent. It is therefore necessary to increase the production. The production from natural catch is limited to natural stock. Hence to match world demand we have to start culture.

The total export of marine products from the state which was 83,011 tones during 1999–2000, has increased to 1,17,815 tones during 2000–2001 year, earning a foreign exchange worth Rs. 891.42 crores. Frozen shrimp, fish, cuttle fish, lobsters, seer fish

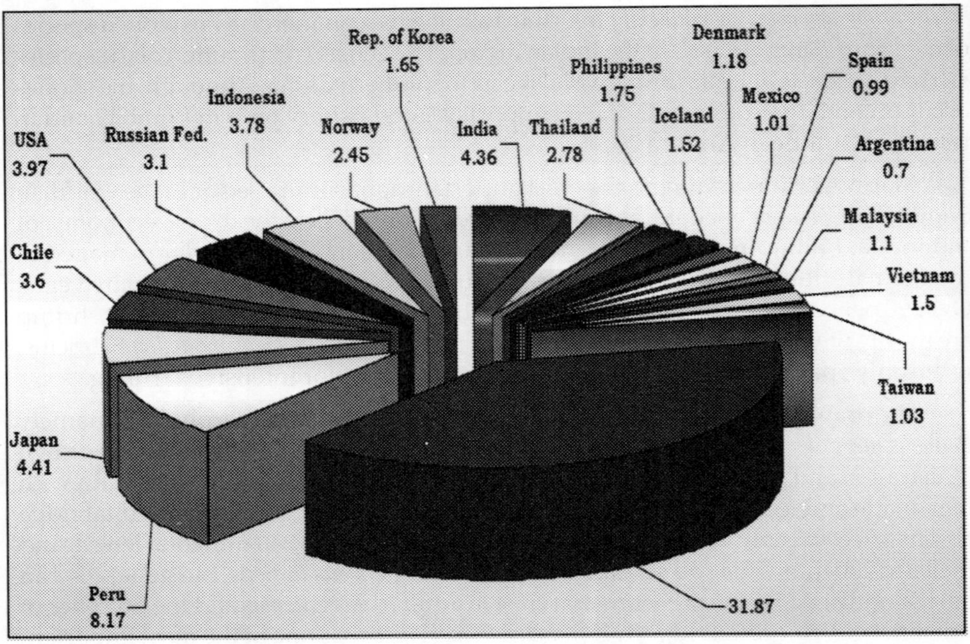

Figure 1.5: Per cent Share of Major Fish Producing Countries in World Fish Production

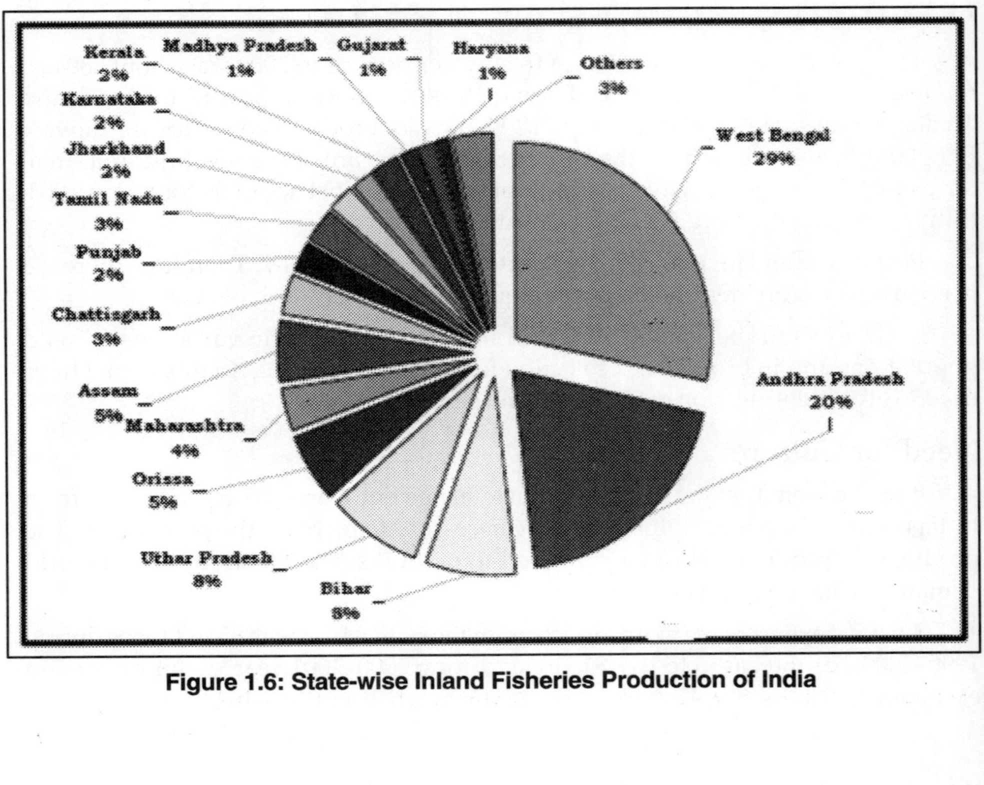

Figure 1.6: State-wise Inland Fisheries Production of India

and pomfrets form the major items of export. The shrimps form one of the important items of export, their export being 10,728 tones during 2000-2001 which was worth Rs. 284.62 crores.

Keeping in view the vast potential of coastal aquaculture in its contribution towards enlarging fish basket of the State, the following areas need to be considered for priority attention.

1. Worldwide Mari-culture production is growing at the rate of 5-7 per cent annually, but in our country it is yet to be taken up on appreciable scale. Mari-culture includes sea farming of shrimps, oysters, mussels, clams, seaweeds and fishes in floating cages/rafts etc. So far, the development has taken place only for shrimp farming. Technologies for cultivation of other species are available and they need to be demonstrated and popularised among the coastal communities.
2. Setting up of diagnostic mobile laboratories for regular monitoring of water and soil qualities of ponds and tanks and identification/treatment of common finfish and shellfish diseases at disease diagnostic centers as service to the farmers/entrepreneurs and for training.
3. Popularization of stabilization ponds for waste water management in shrimp farms.
4. Development of improved traditional shrimp farming.
5. There is lot of scope for finfish culture such as, Jitada, Mullet, Etroplus and Chanos, etc. in the State. Culture of these fin fishes is necessary to popularize among the aquaculturists, which can also provide diversification for uncertain and risky single-crop cultivation such as shrimp farming. There is also scope for crab-culture and crab fattening in coastal areas of the State.
6. Popularization of the freshwater prawn farming, including setting up of small prawn hatcheries. The co-operative societies can be encouraged to establish such Backyard hatcheries.
7. Setting up of medium sized fish feed units (for carp and prawn farming) at various places is essential.
8. The thrust areas understood to have been identified for the cold-water fisheries in the State are establishment of hatcheries and farms for Mahaseer species and development of selected high altitude lakes and hill stream for food and game fisheries development.
9. Utilization of low cost fishes for preparation of value added products and by-products.
10. Ornamental fish breeding and culture for domestic and export market.

Chapter 2
Identification of Mysis Stages

The larval development of 8 commercially important species of penaeid prawns reared in the laboratory. The identification of Mysis stages of these species are described here under. The larval descriptions are very brief; only the changes in structure and setation observed in each substages are given. To understand the exact location and development, drawings are given for each stage. The setules on the plumose setae are not shown to avoid cluttering the figures.

Identification of Mysis Stages

A key for the Identification of Mysis stages of five commercial penaeid prawns of India.

1. a. 5th abdominal segment with a pair of posterolateral spines. Dorsal spines present on 4th, 5th and 6th abdominal segment; sometimes on 3rd segment also .... *Penaeus*

1. b. 5th abdominal segment devoid of posterolateral spines. Dorsal spines present only on 5th and 6th abdominal segment 2

2. a. Telson with 7 + 7 spines. Hepatic spine present in later mysis sub stages .... *Metapenaeus*

2. b.Telson with 8 + 8 spines 6

3. a. Supaorbital spines present .... *Penaeus indicus*

4. a. Median dorsal spine on 5th abdominal segment conspicuous 4

4. b. Median dorsal spine on 5th abdominal segment absent .... *Metapenaeus dobsoni*

5. a. Rostrum extending to 3/4 eye; larvae conspicuously brownish .... *Metapenaeus monoceros*
5. b. Rostrum reaching tip of eye or slightly beyond it; larvae not brownish, ... *Metapenaeus affinis*
6. a. Telson with 8 + 8 spines, Hepatic spine absent in all mysis substages .... *Parapenaeopsis*
6. b. Superaorbital spine absent .... *Parapenaeopsis stylifera*

Mysis Stages of *Penaeus monodon* Fabricius

Mysis–I

MCL; 1.18mm (1.14–1.23 mm); MTL 3.79 mm (3.65–3.96 mm)

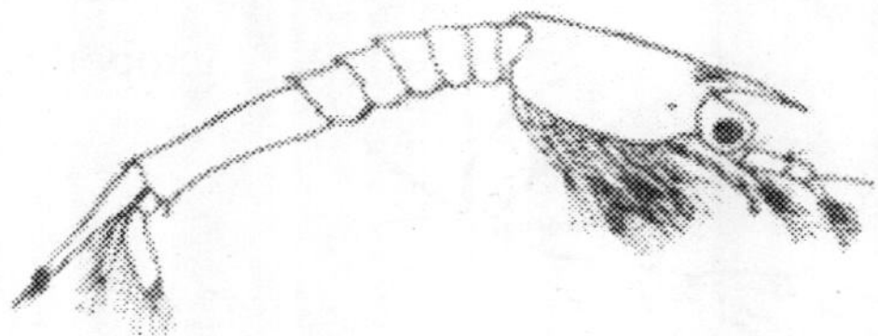

Characteristic	*Figure*	*Details*
Carapace		Rostrum longer than eye, no rostral tooth, superaorbitalo, pterygostomial and hepatic spines present abdominal segments 3–6 with dorsomedian spines, 5th and 6th with prominent posterolateral spines, 6th with prominent posteroventral spine also, curved ventromedian spine present at junction of 6th abdominal segment with telson, minute pleopod buds on abdominal segment.
Antennual–I	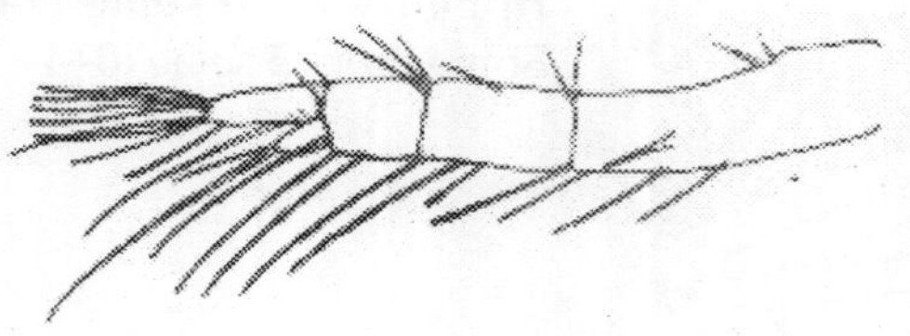	3 segmented, basal segment with anteromedian ventral spine, 2 setae above stylocerite rudiment, outer flagellum with 5 aesthaetes and 3 setae, inner flagellum minute with 2 setae, 1 long and 1 short.

Characteristic	*Figure*	*Details*
Antennual–II	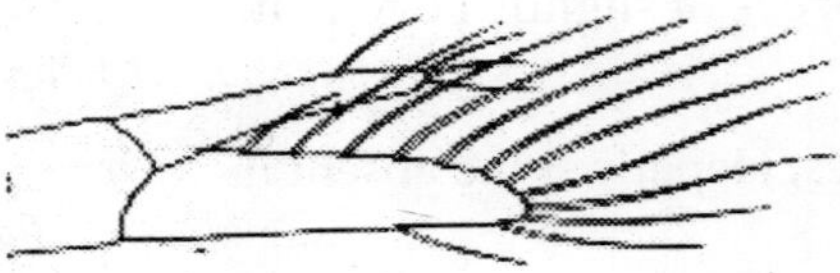	Exopod unsegmented, fringed with 11 setae on inner and distal margin and a plumose outer distolateral seta, endopod tipped with 3 short setae.
Mandible		Right with 3 and left with 7 free standing teeth, palp not seen.
Maxilla-I	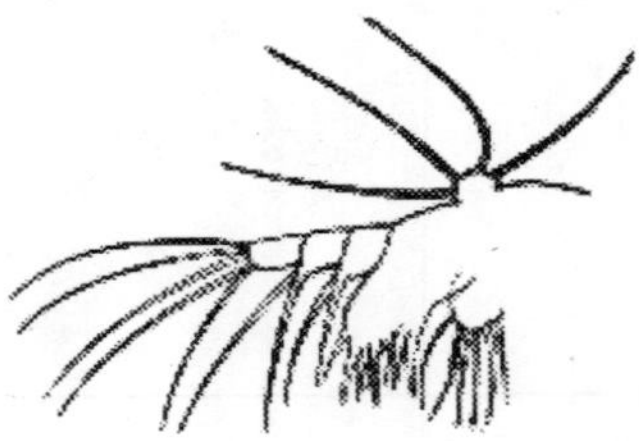	Exopod retained.
Maxillla–II	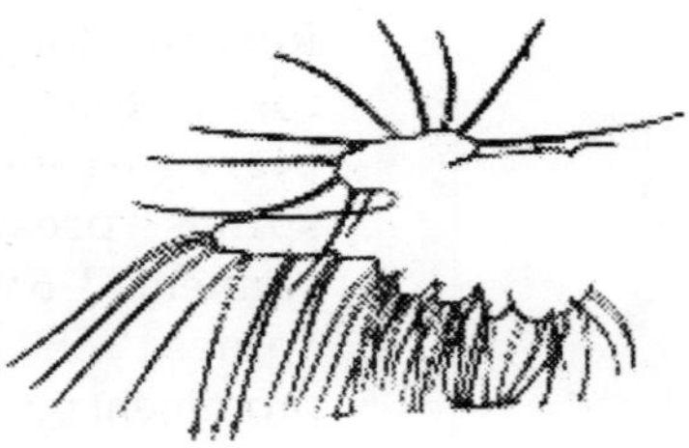	10 setae on exopod.
Maxiliped–I and Maxiliped–II	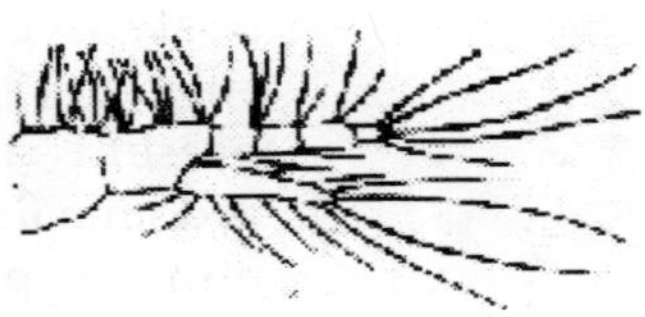	Exopod with 12 setae, outer lateral setae added to 1st and 2nd segment of endopod.
Maxiliped–III	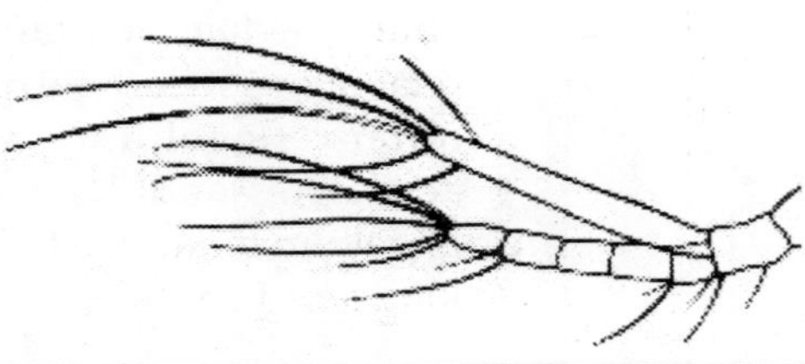	Well developed, with 5 segmented endopod, exopod tipped with 6 setae (1+4+1). Some times 5 setae (0+4+1).

Characteristic	*Figure*	*Details*
Telson	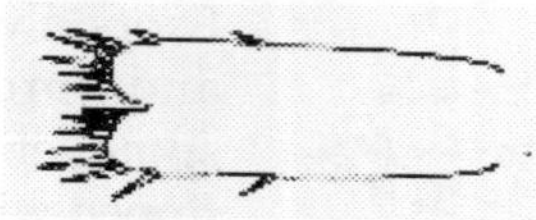	With 8 + 8 short stout setae, cleft deep reaching halfway between level of origin of 2 pairs of outer lateral setae.

Mysis–II (*P. monodon*)

MCL: 1.39 mm (1.34–1.47 mm) MTL: 4.16 mm (3.90–4.37 mm)

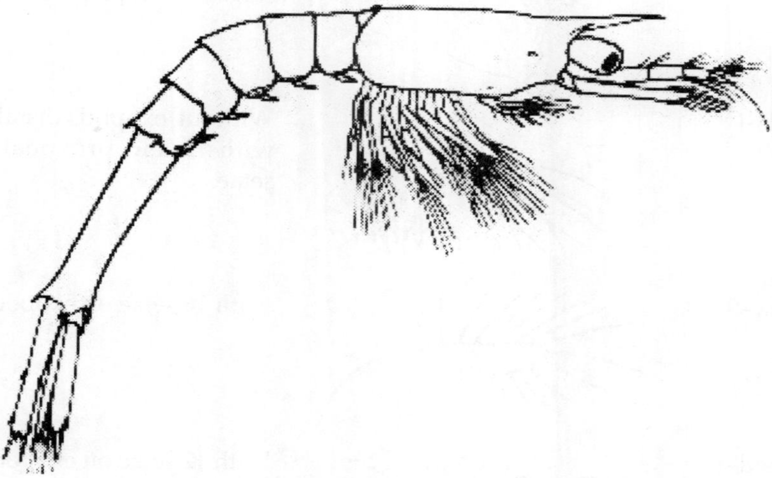

Characteristic	*Figure*	*Details*
Carapace		Rostrum usually without teeth, rarely a minute tooth may be present, no change in spination of carapace and abdomen, pleopods short, unsegmented.
Antennual–I	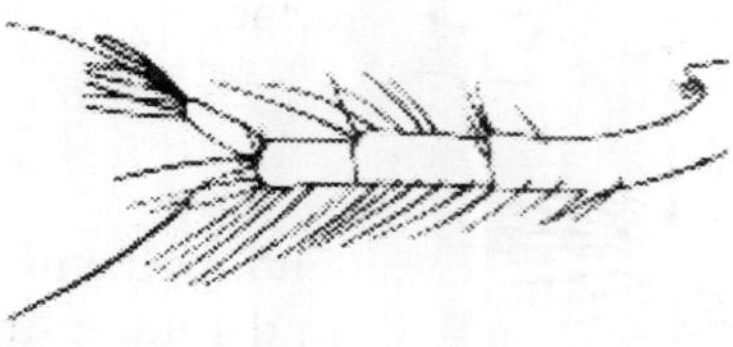	Inner flagellum ¾ length of outer.

Characteristic	*Figure*	*Details*
Antennual–II	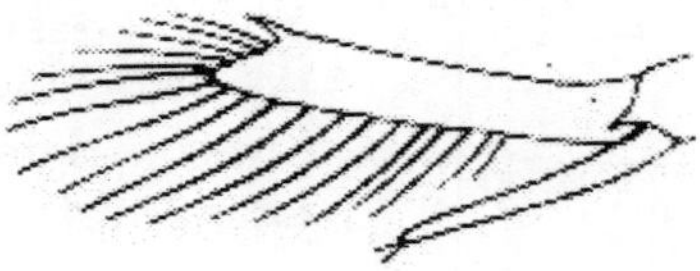	Exopod with 19 plumose setae anda prominent distolateral spine, endopod unsegmented tipped with 2 short setae.
Mandible		With 3 and 7 free standing teeth on right and left mandible respectively, palp developed.
Maxilla-I	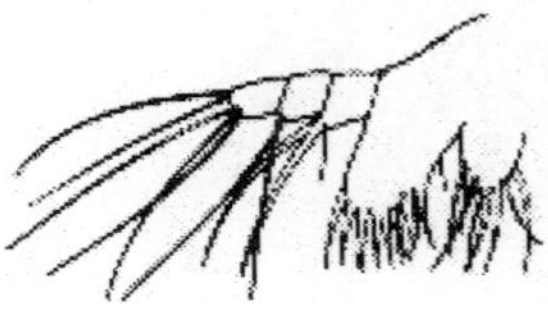	Without exopod, distal endite with 12 and proximal with 8 setae.
Maxillla–II	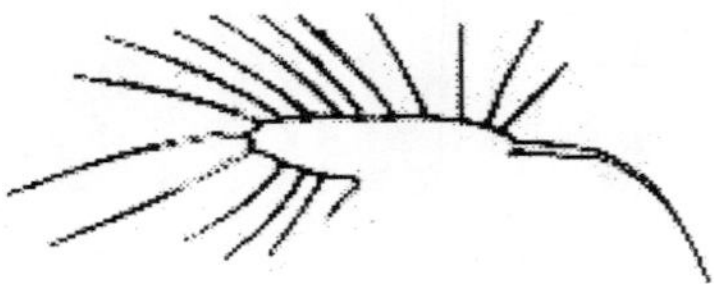	With 16 setae on exopod.
Maxiliped–I	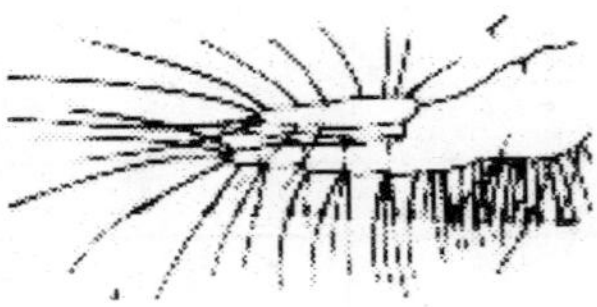	With 12 setae on exopod, rarely with 13 setae.
Maxiliped–II	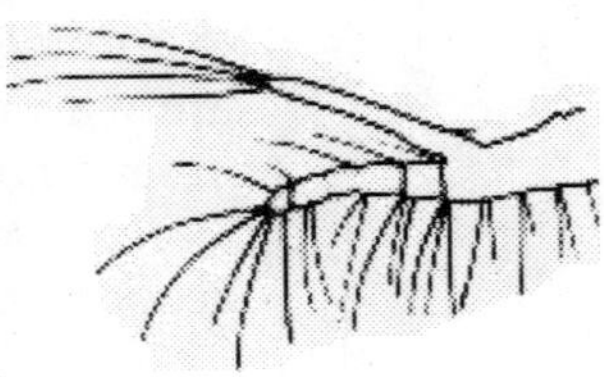	With endopod with penultimate segment indistinctly divided in to 2, distal segment with 6 setae, exopod with 4 terminal setae.
Maxiliped–III	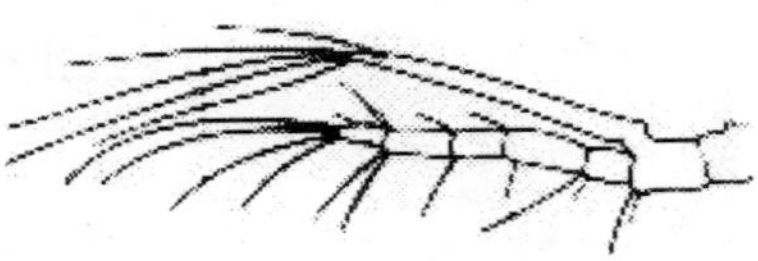	Exopod with 0+1+1 or 1+4+1 setae.
Uropod		Endopod with 20 plumose setae and 1 non-plumoseseta shorter than prominent diostolateral spine.

Characteristic	*Figure*	*Details*
Telson	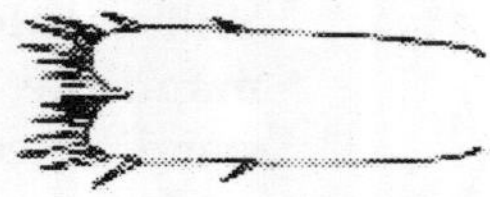	With cleft reaching to level of origin of penultimate pair of outer setae.

Mysis–III (*P. monodon*)

MCL: 1.40 mm (1.32–.51 mm) MTL: 4.2 mm (4.00–4.40 mm)

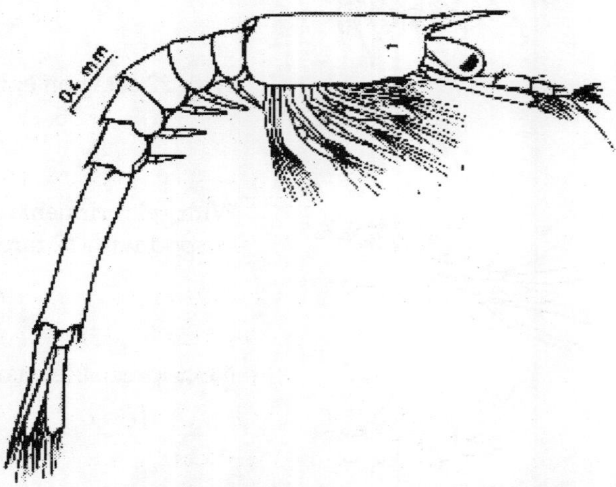

Characteristic	*Figure*	*Details*
Carapace		Rostrum with 1 tooth in some specimens, no changes in spination of carapace and abdomen, pleopods long, 2 segmented without setae.
Antennual–I	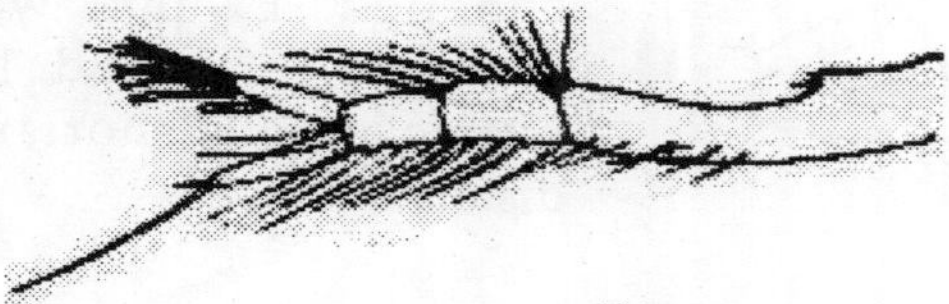	With flagella unsegmented, more or less equal in length, outer with 7 aesthaetes and 2 setae.
Antennual–II		With 22–23 plumose setae and a prominent distolateral spine, endopod 2 segmented.

Characteristic	*Figure*	*Details*
Mandible		With palp longer, right and left, mandible with 3 and 7 free standing teeth respectively.
Maxilla-I	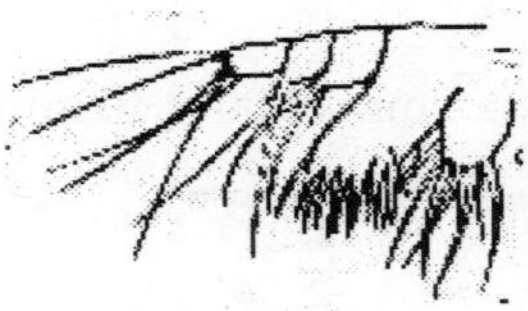	With 13 setae on distal endite.
Maxillla–II		With 22 setae on exopod.
Maxiliped–I	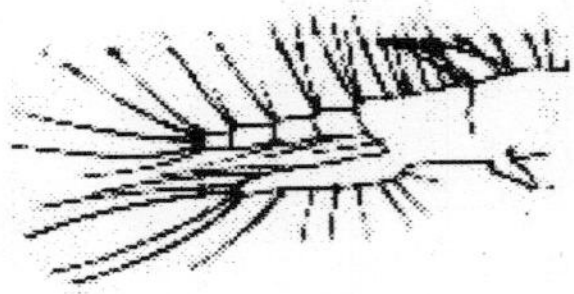	With gill rudiment on protopod, exopod with 12 rarely 13 setae.
Maxiliped–II	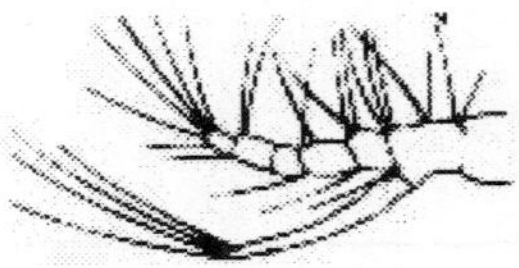	No appreciable change.
Maxiliped–III	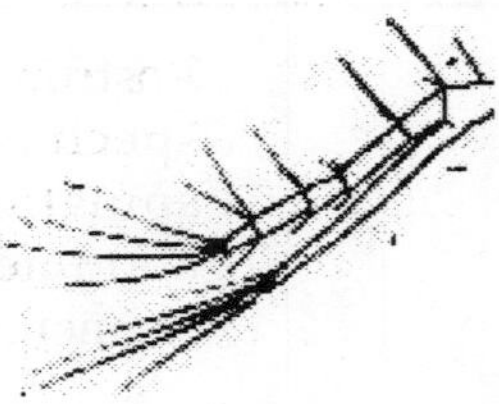	Endopod longer than exopod which has 1+4+1 or 0+4+1 setae.
Uropod	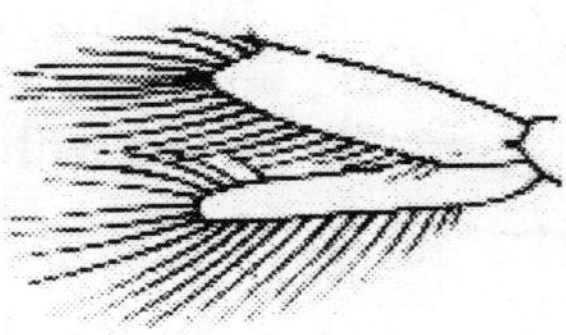	With 22 plumose setae on endopod. Exopod with 20 plumose setae and 1 non-plumose seta shorter than distolateral spine.
Telson		Cleft in telson reaching level of origin of 3rd pair of outer lateral setae on telson.

Source: Silas *et al.*, 1978; 6-8.

Mysis Stages of *Penaeus indicus*

Mysis–I

MTL: 3.36 mm (3.07–3.65mm); MCL 1.17 mm (1.12–1.26 mm)

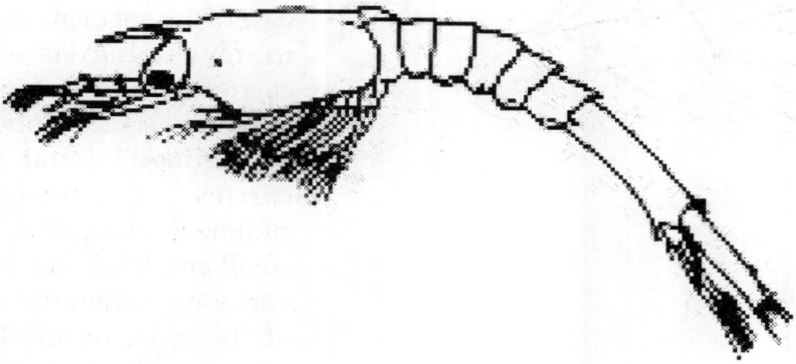

Characteristic	*Figure*	*Details*
Carapace		Rostrum long and curved extending beyond eye, devoid of rostral spines, supraarobital prominent, a small spine present at anteroventral angle of carapace. Hepatic spine well developed. Carapace covers thoracic region completely and thoracic appendages are well developed; posterolateral spines persist on 5th and 6th abdominal segments. Dorsal spines present on posterior margin of 4th, 5th and 6th abdominal segments, in some specimens on 3rd segment also, in rare cases even the 1st and 2nd abdominal segments possess a dorsal spine; minute pleopods buds seen on 1st five abdominal segments; 6th abdominal segment develops a ventromedian curved spine at junction with telson, ventrolateral spines on posterior end retained.

Characteristic	*Figure*	*Details*
Antennual–I	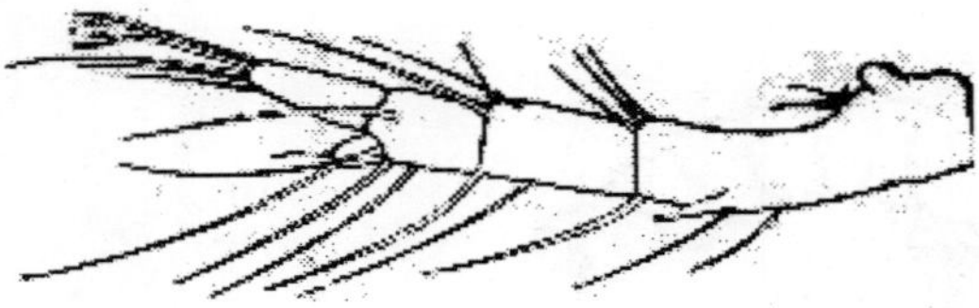	With 3 segmented peduncle 1st segment longest with a ventromedian serrated spine, base of this segment swollen due to developing statocyst and carries 2 shortplumose setae, numerous setae occurring along appendage, distal segment carries 2 unsegmented rudiments of flagellae, inner one small and knob like carrying 1 very long and another short seta at its apex, outer flagellum carries on distal margin 3 setae and 4 aesthetes. Size is 0.1 mm.
Antennual–II	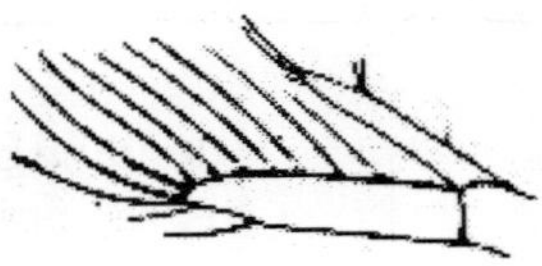	With endopos unsegmented carrying 3 terminal setae, one proximal setae on inner margin and 2 small setae near a very small know- like projection on inner side distally, exopod unsegmented, leaf like, with a distolateral setae on outer margin and 11 setae on distal and inner lateral margin.
Mandible		Asymmetrical with 7 free standing teeth in left mandible and 3 in right mandible, molar part shows a number of hard ridge bearing small teeth.
Maxilla-I		Proximal segment of protopod with 8 setae.

Characteristic	*Figure*	*Details*
Maxillla–II	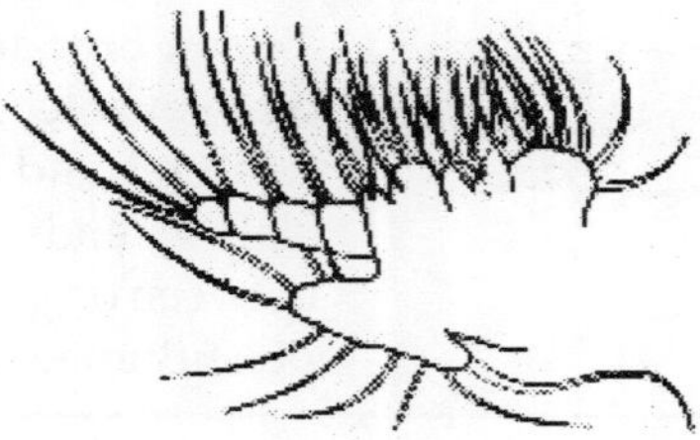	With exopod enlarged to form scaphognathite carrying 10 pulmose setae, proximal one being long and thick.
Maxiliped–I	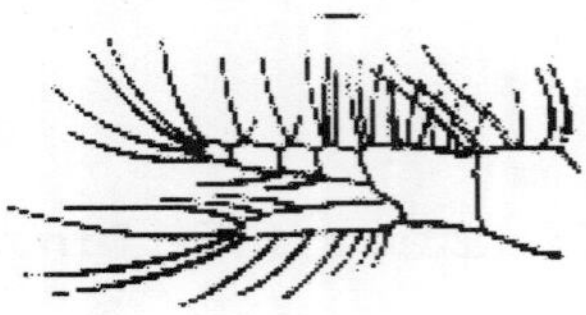	With some setae on inner side of protopod longer and stouter, setae on coax reduced to 5, exopod with 12 plumose setae, one seta each exopod with 12 plumose setae, one seta each added to outer margin of 1st and 2nd segments of endopod.
Maxiliped–II	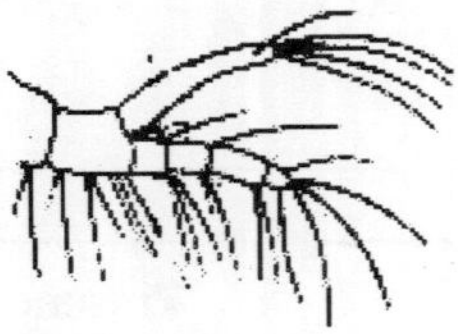	With 7 setae on basis of protopod, exopod as long as endopod carrying only 6 setae, 4 apical and 2 subapical, endopod 4 segmented, first 2 segments carry 1 setae on the outer side, terminal segment with 5 setae. Size is 0.3 mm.
Maxiliped–III		Well developed, protopod with 3 setae on basis, coax without setae, endopod 5 segmented terminal segment with 1 short and 5 long setae. 1st 2nd and 4th segments each with 2 setae 3rd.
Uropod	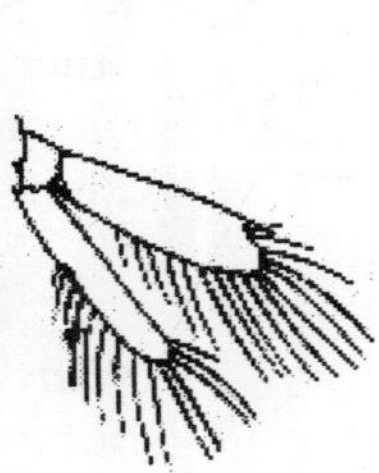	Uropods well developed protopod with a large posteroventral spine, exopod with a prominent posterolateral spine followed by a shortnonpulmose seta and about 15 plumose setae along distal and inner margin, endopod with 14 plumose setae along inner and distolateral margin.

Characteristic	*Figure*	*Details*
Telson		Telson broader distally with a median notch, each lobe bearing 2 lateral and 6 terminal setae, claft extends to level half way between origin of outermost and peneultimate pair of setae.

Source: Muthu, M. S. *et al.* (2) 1978; 17–19.

Mysis–II (*Penaeus indicus*)

MTL: 3.5 mm (3.39–3.58 mm); MCL 1,20 mm (1.15–1.26 mm).

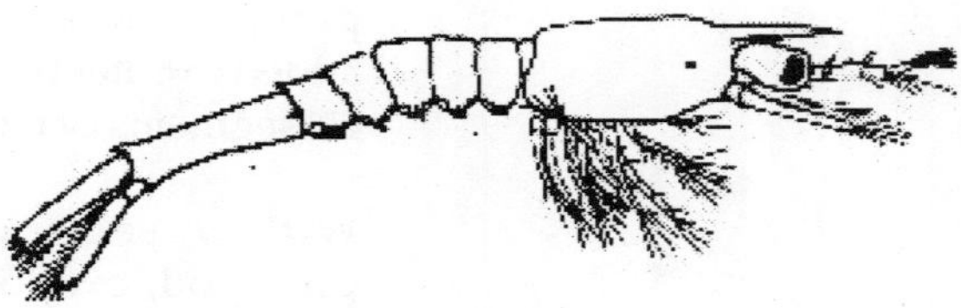

Characteristic	*Figure*	*Details*
Carapace		Presence of pine on scaphocerite and appearance unsegmented pleopods buds distinguish this substage from mysis I. No change in spination of carapace and abdomen.
Antennual–I	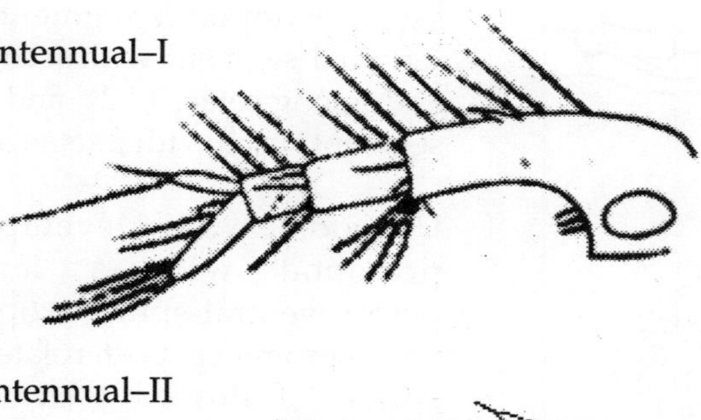	With increased number of setae on peduncle, inner flagellum has increased in length and outer flagellum which is longer than inner with 6 aesthetes and 1 or 2 setae at distal end.
Antennual–II		With a small ventral spine on outer distal end of 2nd segment of protopod, endopod nearly half length of exopod bearing a short apical seta, exopod with 19 long plumose setae along inner and distal margin and 1 spine at distal lateral angle.

Characteristic	*Figure*	*Details*
Mandible		With small unsegmented palp, 8 free standing teeth on left and 3 on right.
Maxilla-I		Without exopod, size of endopod reduced.
Maxillla–II		With 14 to 15 plumose setae on exopod.
Maxiliped–I	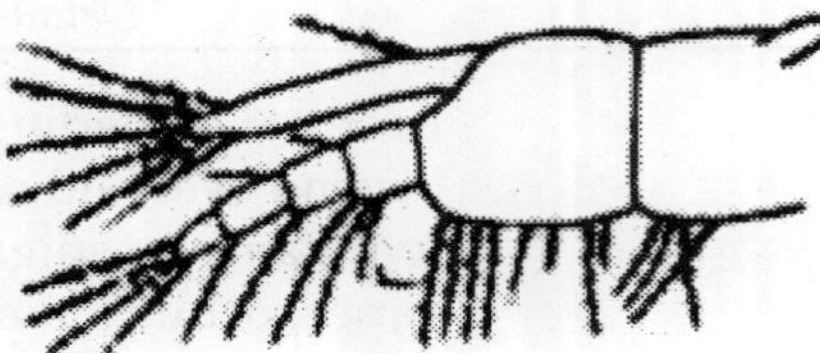	With 12 setae on exopod.
Maxiliped–II	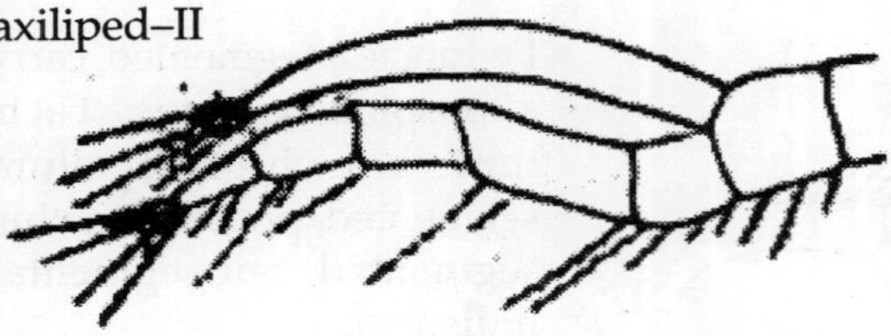	With 5 segmented endopod, with newly added segment in middle without setae, terminal segment with 6 setae.

Characteristic	*Figure*	*Details*
Maxiliped–III		With endopod longer than exopod, 3rd segment with 2 setae a seta added to outer distal margin of 4th segment.
Uropod	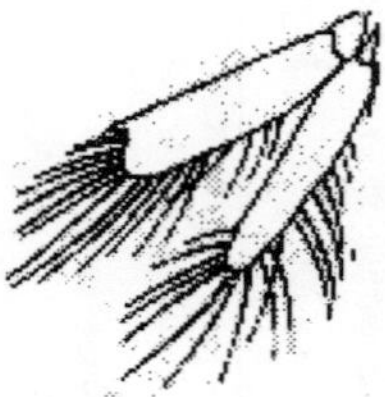	Exopod and endopod of uropod with 18 setae.
Telson		Cleft on telson extends to level of origin of penultimate pair of lateral telesonic setae.

Source: Muthu, M. S. *et al.* (2) 1978; 17–19.

Mysis–III (*Penaeus indicus*)

MTL: 3.90 mm (3.43–4.17 mm): MCL: 1.26 mm (1.12–1.37 mm)

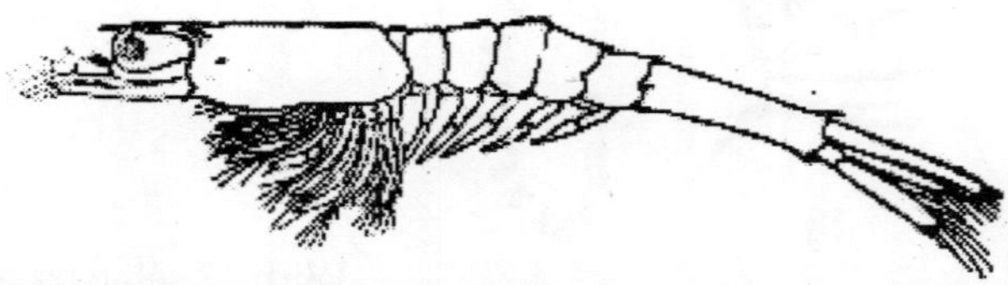

Characteristic	*Figure*	*Details*
Carapace		Rostrum extending to slightly beyond eyes and with a dorsal tooth and an spigastric spine; superaorbital and anetennal spines small; hepatic spine well developed.
Antennual–I		Peduncle 3 segmented, carrying a conspicuous statocyst at base; inner branch of flagellum 3-segmented; pouter flagellum 2-segmented, but segmentation indistinct.

Characteristic	*Figure*	*Details*
Antennual–II	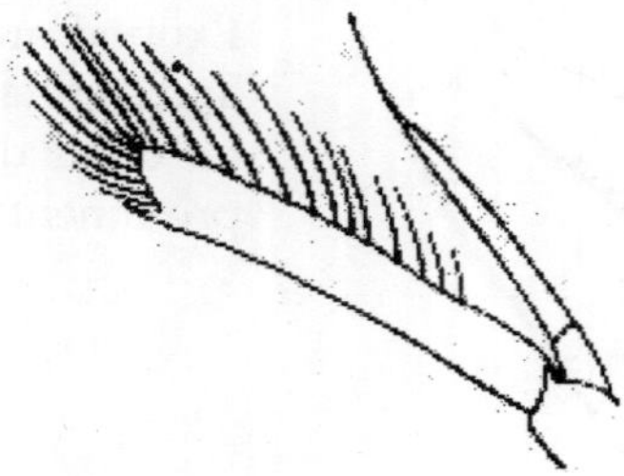	Flagellum segmented, 4–5 segments at tip indistinct; a pair of subequal apical setae on flagellum present; scale with 26 or 27 pulmose setae, anterolateral spine well developed, but not reaching tip.
Mandible		Same as in Mysis II palp not segmented.
Maxilla-I		Same as in previous substage.
Maxilla–II	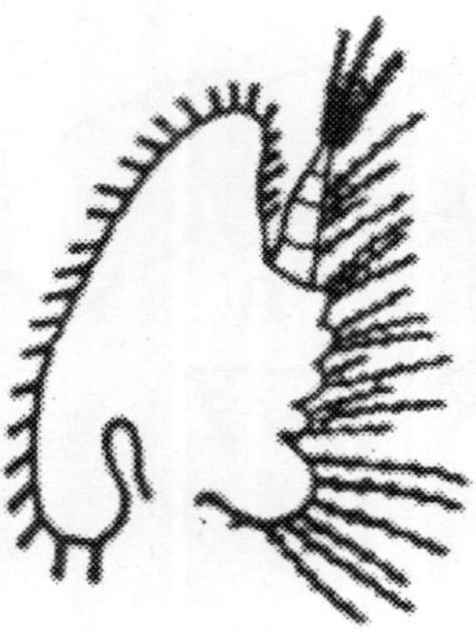	Protopod with 5 endites; endopod 4 segmented and with 3 setae at tip; scaphoganthite with 26 or 27 setae.
Maxiliped–I and Maxiliped–II	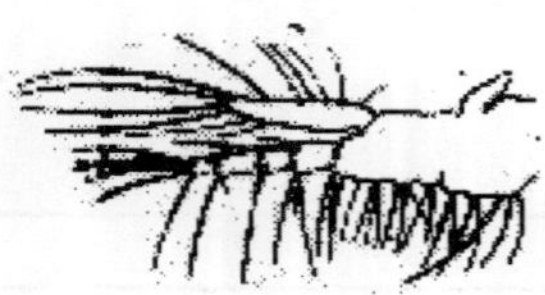	Same as in Mysis II.
Maxiliped–III	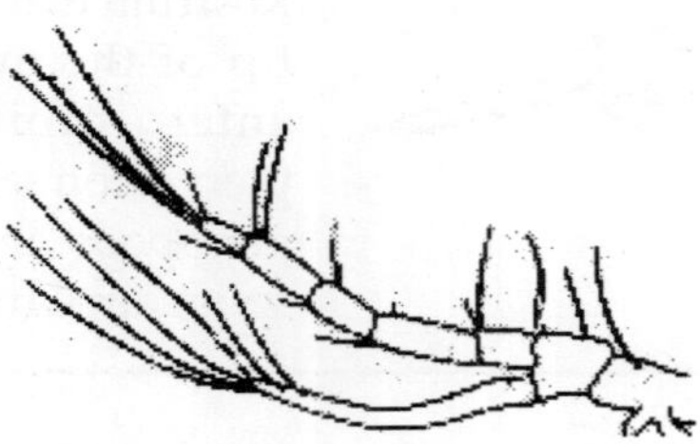	Endopod 5 segmented, exopod terminally 3 segmented, segments indistinct.

Characteristic	*Figure*	*Details*
Uropod		Exopod and endopod with 28 and 24 setae respectively; spine on outer distal end of exopod prominent.
Telson		Changes only a little truncate with 8 + 8 spines.

Source: Dholakia, A. D. 1994, Muthu, M.S. *et al.*, 1978, Ved Vyas Rao, 1973.

Mysis Stages of *Penaeus merguiensis* (de Man)

Mysis–I

MTL 3.075 mm (2.99–3.1 mm).

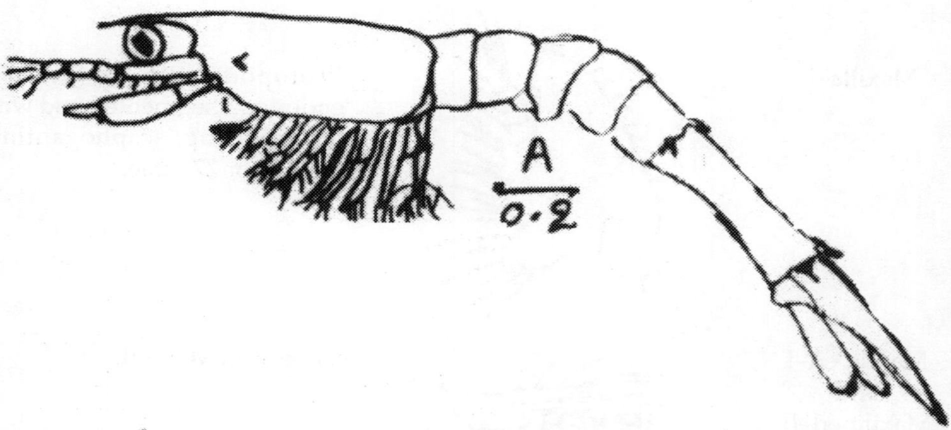

Characteristic	*Figure*	*Details*
Carapace		Rostrum extending beyond the tip of the eye and have one antero orbital spine and a pteryostomial spine, biramous periopods and a small hepatic spine appears.

Characteristic	*Figure*	*Details*
Antennual–I	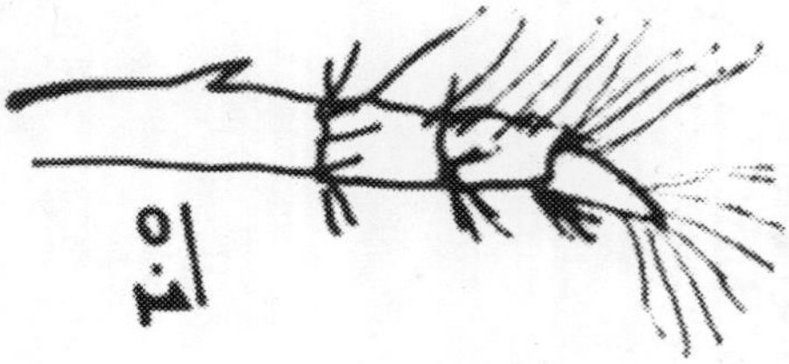	3 segmented with the basal segment being longest.
Antennual–II	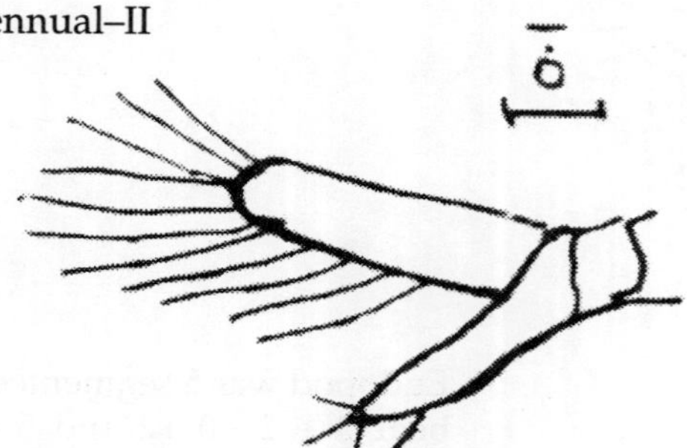	Distal segment gives 2 unsegmented branches of which outer one longer than inner. Shows both endopod and exopod bear 11 setae along the inner margin.
Mandible	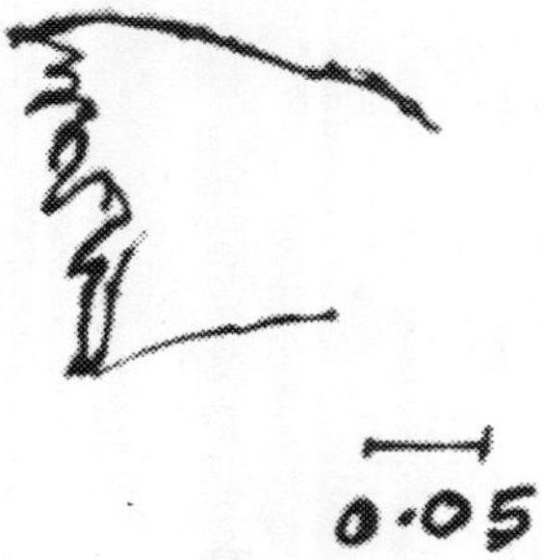	Longer with teeth and molar processors.
Maxilla-I	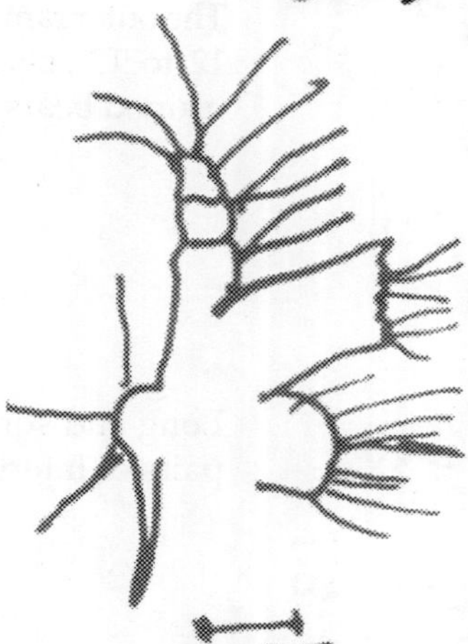	Exopod bears 9 setae.

Characteristic	*Figure*	*Details*
Maxilla II	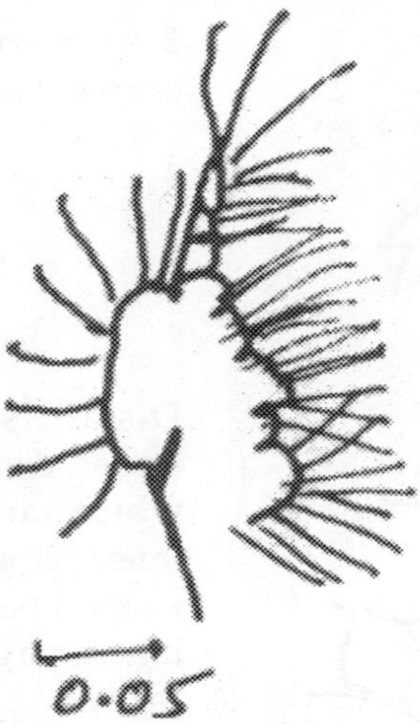	It bears 9 setae.
Maxiliped–III	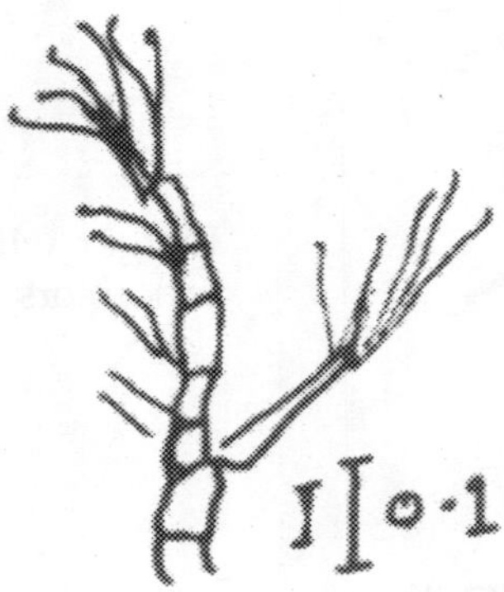	Endopod was 5 segmented and bear 2 + 2 +0 +2 and 5 setae respectively exopod bear 5 setae.
Uropod	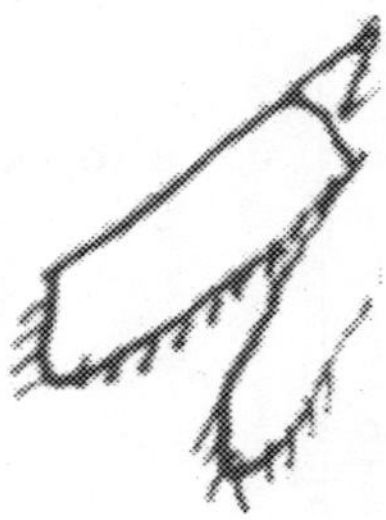	The outer ramus of uropod bears 12 to 13 setae while the inner uropod bears 8 to 9 setae.
Telson		Long and square and bear two pairs of internal spines.

Source: Dholakia, A. D. 1994.

Mysis–II

MTL:3.63 mm (3.61–3.64 mm).

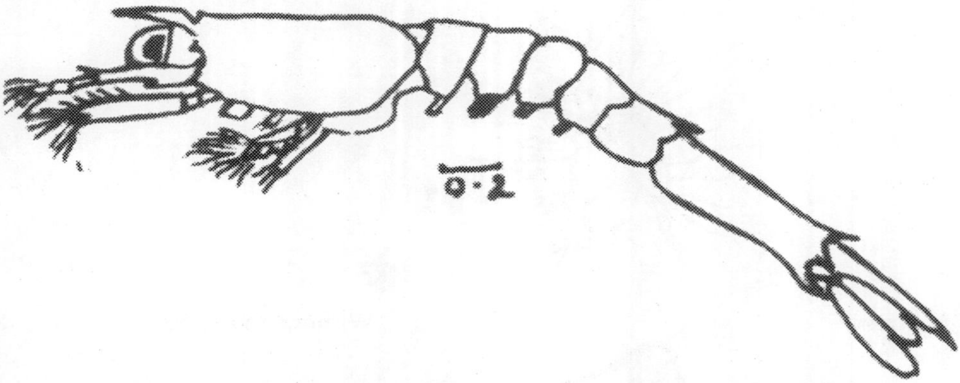

Characteristic	*Figure*	*Details*
Carapace		Development of supraorbital spine and the buds of the plecopods were the major development.
Antennual–I	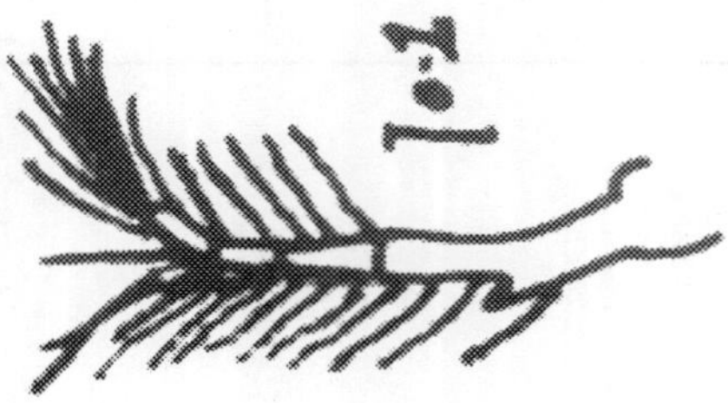	24 plumose setae.
Antennual–II	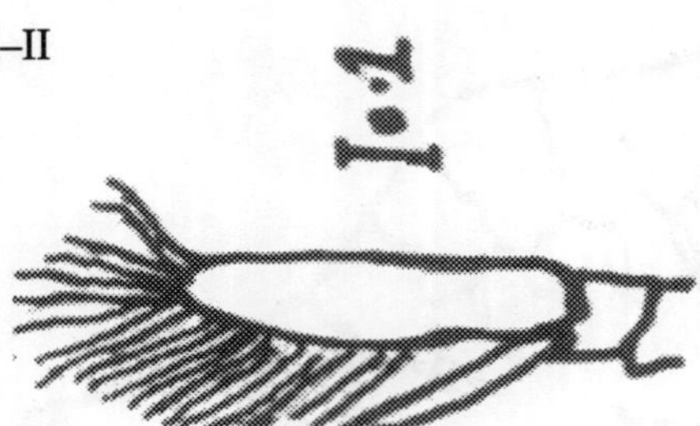	Exopod bears 21 setae. The endopod shows three small hairs at the tip but no setae. Statocyst appears at the base.
Mandible	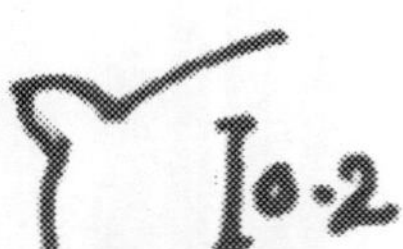	A small bud like palp seen.

Characteristic	*Figure*	*Details*
Maxillla–II		Exopod bears 16 setae.
Uropod		Without any setae.
Telson	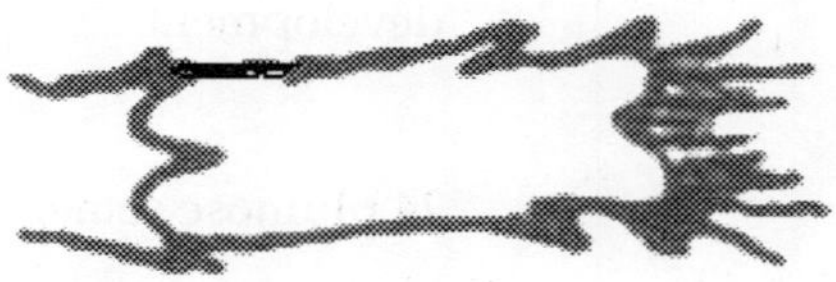	Two lobed telson with 6 + 6 spine formula.

Source: Dholakia, A. D. 1994.

Mysis–III

MTL 4.13 mm (4.11–4.14 mm).

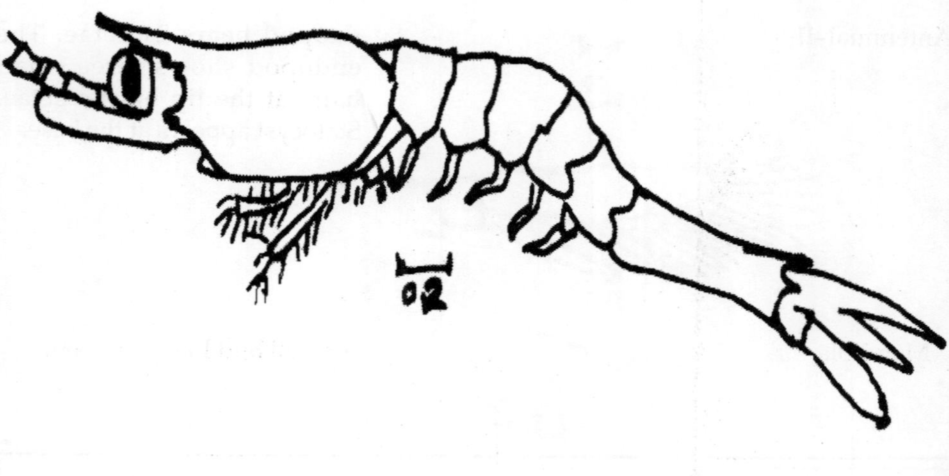

Characteristic	*Figure*	*Details*
Carapace		Two segmented pleopods developed, distal segment of the protopod developed antero-median spine.
Antennual–I	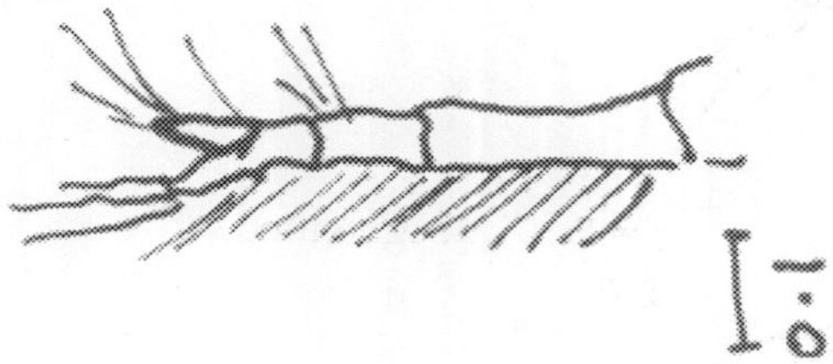	3 segmented.
Antennual–II	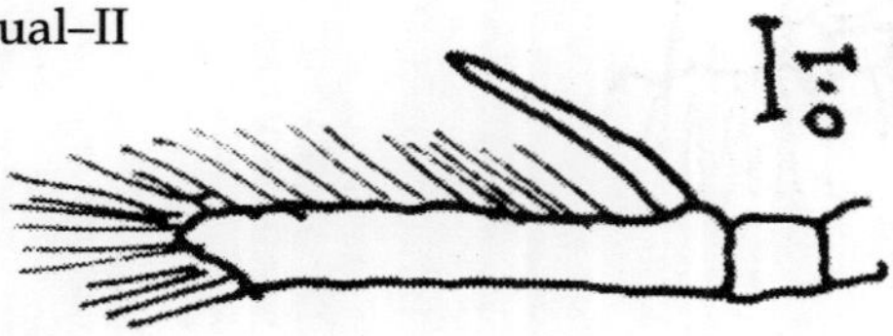	2 segmented, exopod bears 17 setae, endopod 2 segmented.
Mandible		Palp 2 segmented but no setae.
Maxilla-I	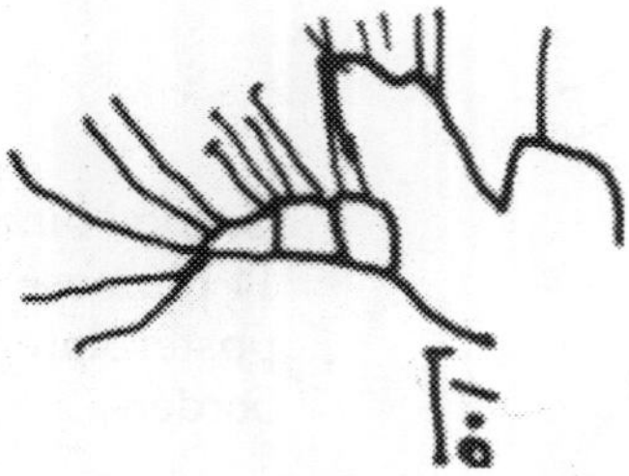	3 segmented.
Maxillla–II	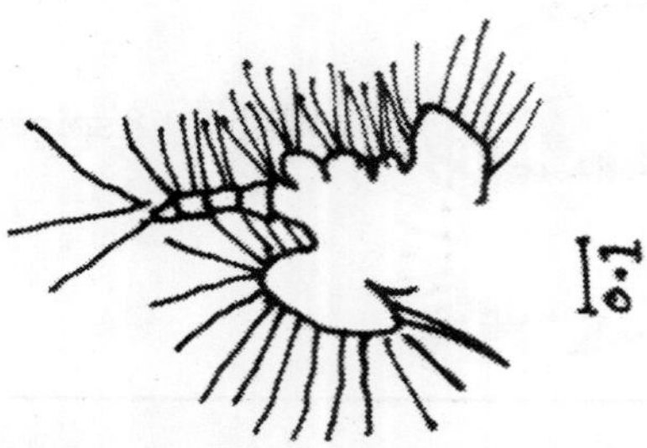	4 segmented, bears numerical setae.

Characteristic	*Figure*	*Details*
Maxiliped–I	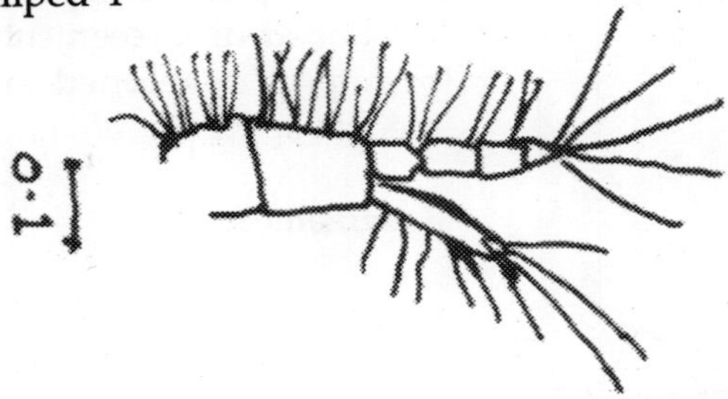	Endopod shows 4 setae at the end, exopod unsegmented.
Maxiliped–II	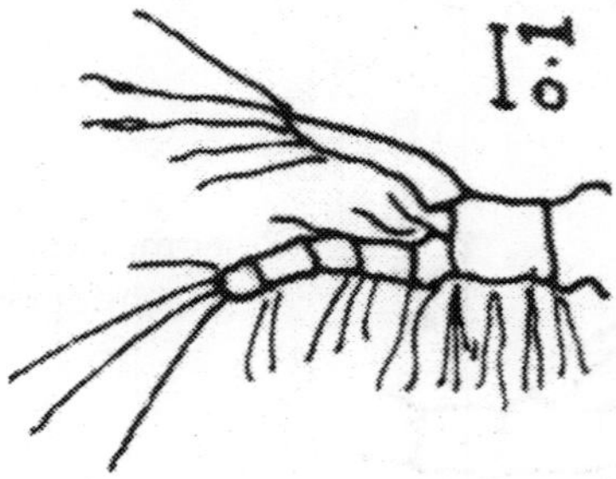	5 segmented.
Maxiliped–III	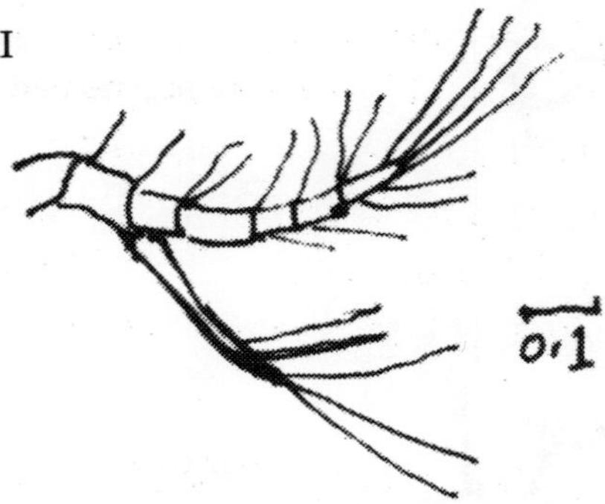	Endopod 4 segmented, terminal segment with 5 long setae.
Uropod		Uropod biramous exopod with 14 plumose setae, a prominent posterolateral spine on outer border.
Telson	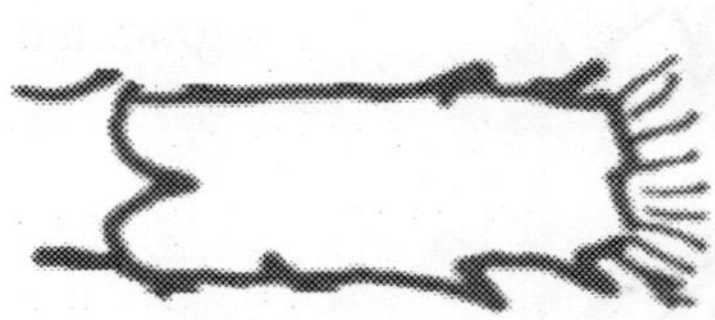	With 8 + 8 spine formula.

Source: Dholakia, A. D. 1994.

Mysis Stages of *Penaeus semisulcatus* De Haan

Mysis–I

MCL: 3.15 mm (2.99–3.29 mm); MCL: 1.10 mm (1,00–1.15 mm)

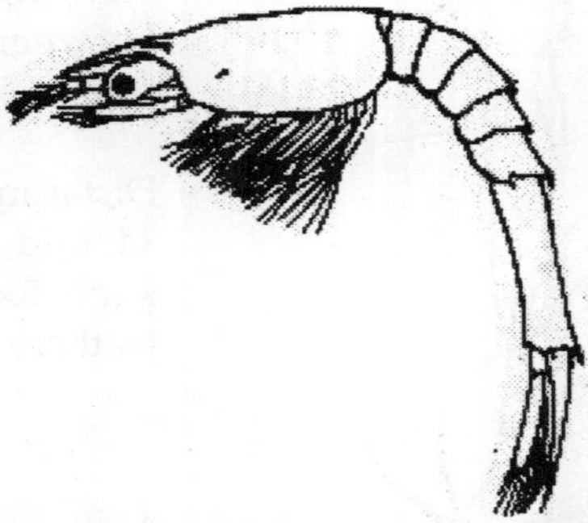

Characteristic	*Figure*	*Details*
Carapace		Rostrum long, extending beyond eye and devoid of spines; carapace with hepatic superobital and ptergostomial spines 4th, 5th and 6th abdominal segments with dorsal spine on posterior margin, 5th and 6th with a pair of lateral spines and 6th in addition bears a ventromedian curved spine at junction with telson.
Antennual–I	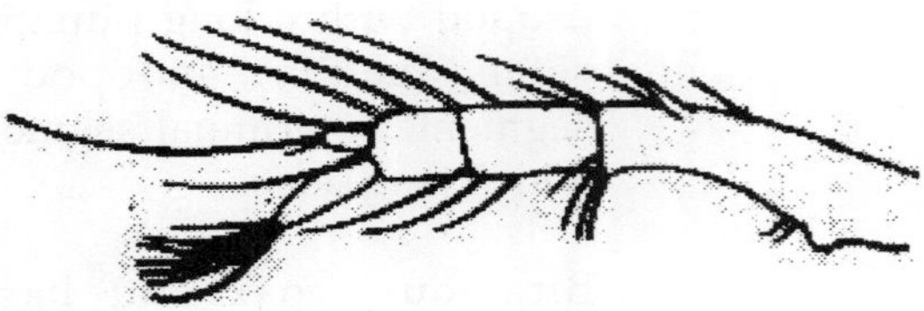	3 segmented, basal segment with anteromedian spine and 2 setae above stylocerite rudiment, distal segment with 2 flageller reduments, outer rudiments with 6 aesthaetes and 1 seta, inner only half size of outer, bearing 1 long and 1 short setae terminally.
Antennual–II		Exopod unsegmented, leaf like with a distolateral seta on outer margin and 11 setae along inner and distal margin, endopod unsegmented carrying 3 terminal setae and 2 distolateral setae placed on a minute projection.

Characteristic	*Figure*	*Details*
Mandible	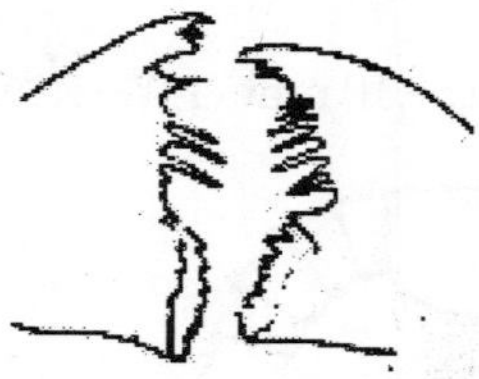	Asymmetrical, right and left mandible with 3 and 7 free standing teeth respectively between incisor and molar processes.
Maxilla-I	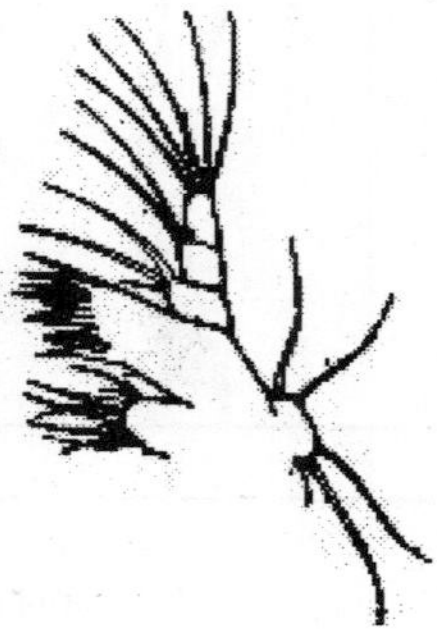	Distal and proximal endite with 11 and 8 seae respectively, knob–like exopod with 4 long feathery setae.
Maxillla–II		Exopod expanded to form scaphognathite, with 10 plumose setae.
Maxiliped–I	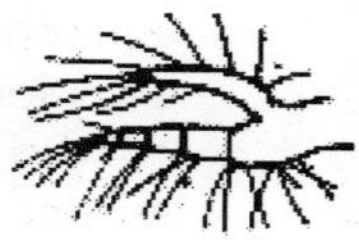	Exopod with 11 plumose setae, 1 seta added to 1st segment of endopod along outer margin.
Maxiliped–II		Basis with 7 setae on inner side, exopod with 6 long plumose setae distally, endopod 4 segmented, terminal segment with 5 long setae.
Maxiliped–III	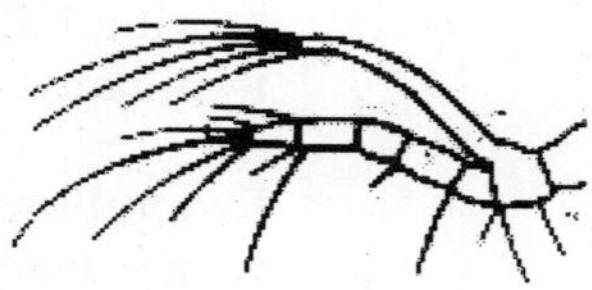	Biramous, coax and basis bearing 1 and 3 setae on inner side respectively, exopodas long as endopod bearing 4 apical and 2 subapical plumose setae, endopod 5 segmented, terminal segment with 5 setae, 1st 2nd and 4th bearing 1, 1, 3 setae respectively.

Characteristic	*Figure*	*Details*
Uropod		Uropod biramous exopod with 14 plumose setae, a prominent posterolateral spine on outer border and short non–plumose seta between the post erolateral fixed spine and the plumose setae. Endopod with 13 plumose setae along inner and distal margin.
Telson		Broader distally with mediam notch, each lobe with a 1 lateral and 7 terminal setae.

Source: K. Devrajan, *et al.*, 1978.

Mysis–II

MCL: 3.50 mm (3.48–3.52 mm); MCL: 1.33 mm (1.10–1.16 mm).

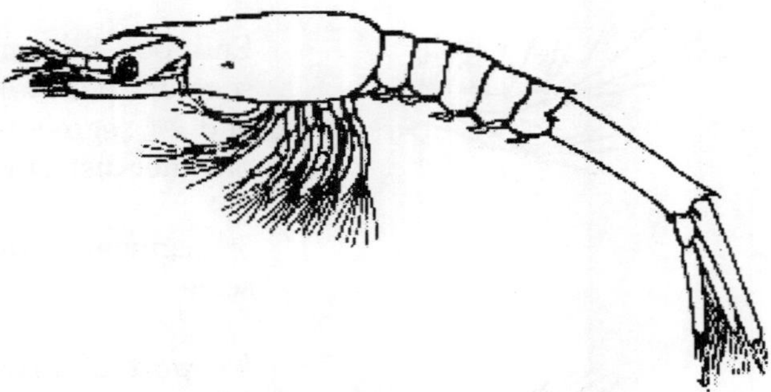

Characteristic	*Figure*	*Details*
Carapace		Rostrum extending to tip of eye or slightly beyond, devoid of any spine, carapace with spination as in the previous substage.
Antennual–I	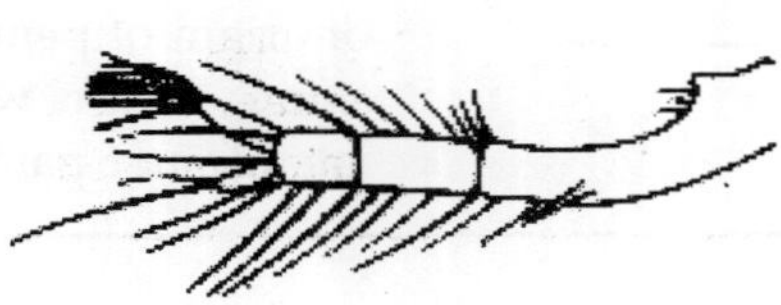	Number of setae on antennual segments increased, basal segment with swollen base showing the developing statocyst and carries 3 plumose short setae.

Characteristic	*Figure*	*Details*
Antennual–II		Exopod with 18 plumose setae along inner and distal margin, a spine replaces the seta on distolateral outer margin, endopod unsegmented and devoid of setae.
Mandible		Unsegmented small palp developed.
Maxilla-I		Except for absence of exopod no other change in appendage.
Maxiliped–I		Except for absence of exopod no other change in appendage.
Maxiliped–II	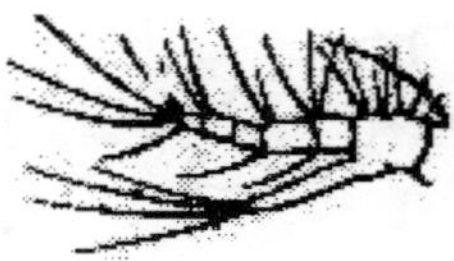	Endopod 5 segmented, distal segment with 6 long setae, 1st, and 2nd segments 1 setae each on outer distal margin.
Maxiliped–III		3rd segment of endopod with 2 setae.
Uropod		Exopod of uropod with 16 plumose setae and 1 non–plumose seta in addition to posterolateral fixed spine, endopod with 14 plumose setae.
Telson	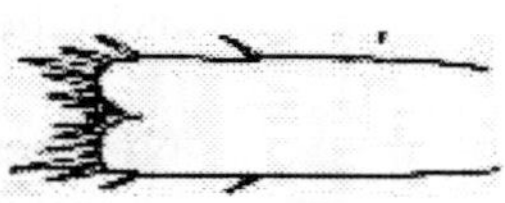	Telson almost rectangular, middle claft extending to level of origin of penultimate lateral setae, telson with 2 pairs of lateral and 6 pairs of distal setae.

Source: K. Devrajan, *et al.*, 1978.

Mysis–III

MCL: 4.43 mm (4.38–4.48 mm); MCL: 1.50 mm (1.48–1.51 mm).

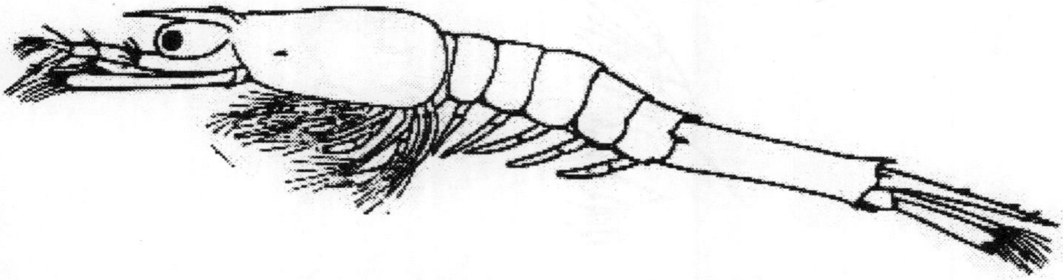

Characteristic	*Figure*	*Details*
Carapace		Rostrum extends beyond eye in some specimens a rudimentary rostral tooth may be present, supraorbital, hepatic and ptrerygostomial spines present; pleopods 2 segmented exopod and endopod of uropod with 19 and 23 setae respectively.
Antennual–I		Statocyst seen at base of 1st segment, flagella of equal size, outer flagellum 2 segmented with 7 aesthaetes in 3 groups of 3 +2+2 and 2 terminal seta, inner flagellum indistintctly 2 segmented with 4 apical setae.
Antennual–II		Exopod with 22 plumose setae and 1 outer lateral spine, endopod 2 segmented bearing 4 apical setae.
Mandible	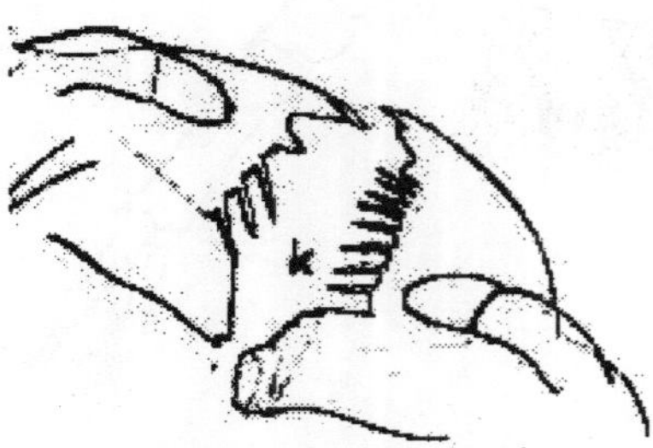	Palp 2 segmented but devoid of setae.

Characteristic	*Figure*	*Details*
Maxilla-I	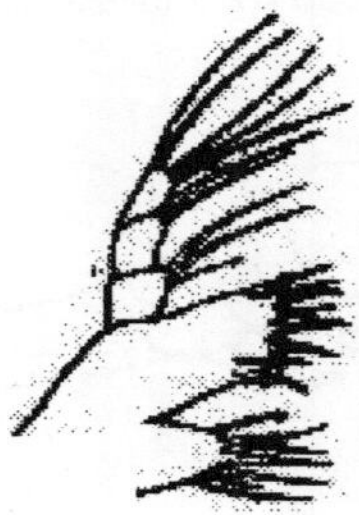	No change from previous stages.
Maxiliped–II	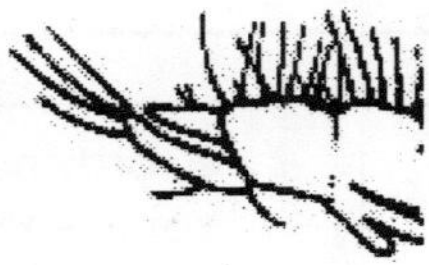	Rudiment of gill developed, exopod bears 4 long plumose setae apically.
Maxiliped–III	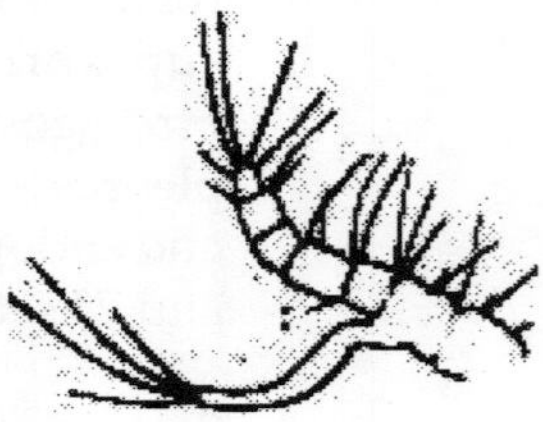	Rudimentary gills present on coax, endopod longer than exopod and 5 segmented, distal segment with 4 setae.
Telson		Median cleft of telson reduced.

Source: K. Devrajan, *et al.*, 1978.

Mysis Stages of *Metapenaeus affinis* (H. Milne–Edwards)

Mysis–I

TL–2.13–2.40 mm; CL–0.55–0.68 mm.

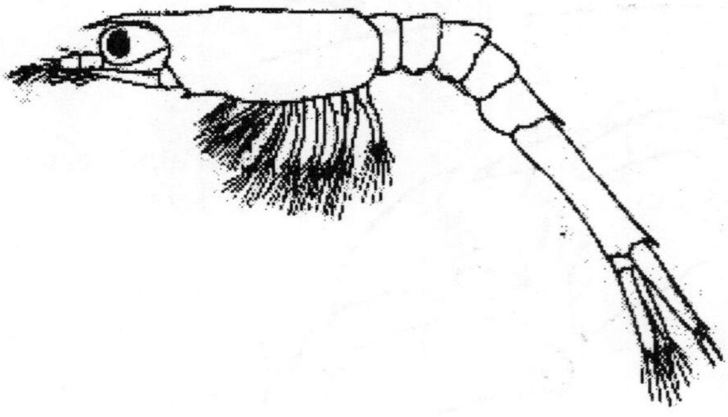

Characteristic	*Figure*	*Details*
Carapace		Rostrum long and pointed, extends to tip of eye or slightly beyond it reaching distal end of first antennular segment; carapace with a small antennual and conspicuous prerygostomial spine.
Antennual–I	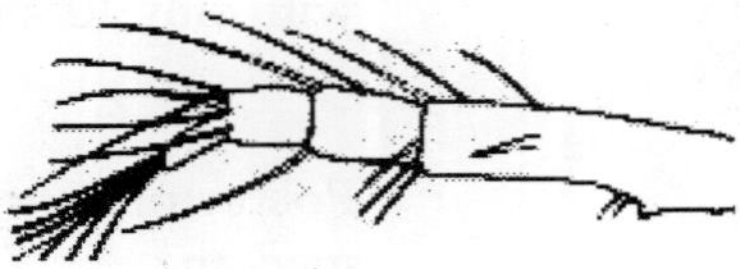	First segment of peduncle longer than 2nd and 3rd which are equal in length; flagella unsegmented, inner very much smaller; rudiment of stylocerite present.
Antennual–II		Scale which 8–9 setae and without anterolateral spine; flagella unsegmented, with a single terminal setae and minute one at subterminal outer margin.
Mandible		It is with a conspicuous ventral spine and 6 or 7 serrated teeth; no palp.
Maxilla-I	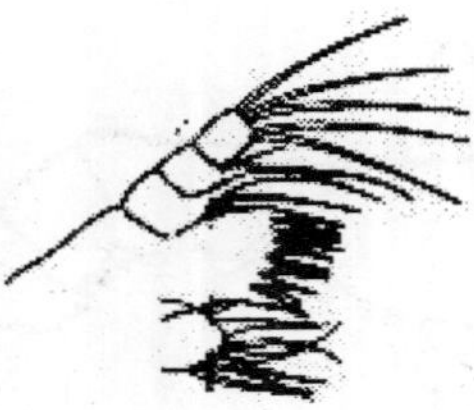	2 endites with stout setae; endopod 3 segmented, first 2 segments with 2 setae each and 3rd with 3 terminal setae; knob-like, and bears 4 plumpose setae.
Maxillla–II		5 endites proximately, lower most largest; endopod 4 segmented; exopod with 5 setae along its border.
Maxiliped–I	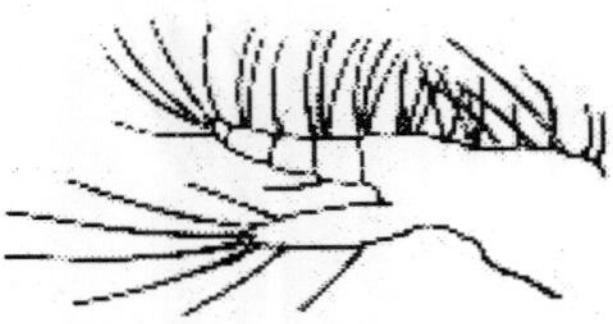	Protopod 2 segmented, carrying long setae on the inner margin; endopod 5 segmented and exopod unsegmented.

Characteristic	*Figure*	*Details*
Maxiliped–II and Maxiliped–III	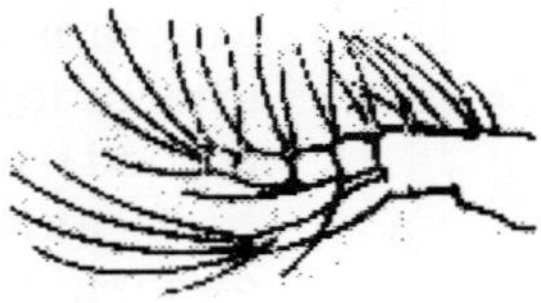	Portopod 2 segmented; endopod.
Uropod		Well developed, endopod and exopod with 14 or 15 setae respectively. While endopod with only 10 setae.
Telson	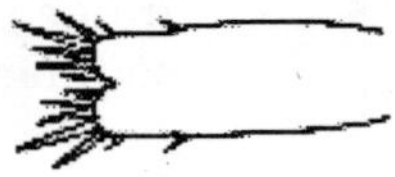	Posterior margin with deep median cleft; 7 spines on each lobe, 4th from outer lateral longest.
Colouration		Live larvae slightly brownish; eye stalk a little yellowish; ventral side of abdomen provided with branched brownish chromatophores; dorsal portion of telson with brownish pigments.

Source: Vedvyas Rao 1973 with modification.

Mysis–II

TL–2.38–2.90 mm; CL–0.55–0.78 mm.

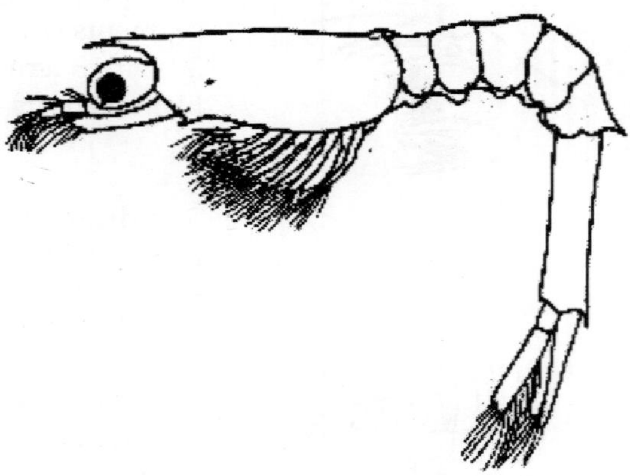

Characteristic	*Figure*	*Details*
Carapace		C-Rostrum develops a small dorsal tooth; antennal, pterygostomial and a minute hepatic spine present.
Antennual–I	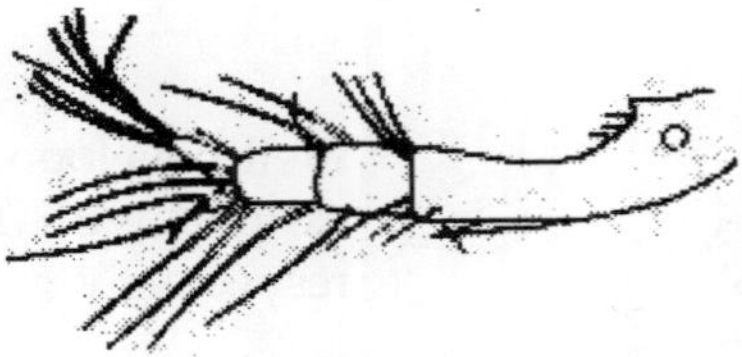	Peduncle 3-segmented with a developing statocyst and stlocerite at base of first segment; flagella unsegmented, inner slightly shorter than the outer.
Antennual–II		Scale with an anterolateral spine on outer margin and 15 setae along inner margin and tip; flagellum unsegmented and about half length of scale. One distolateral spine.
Mandible	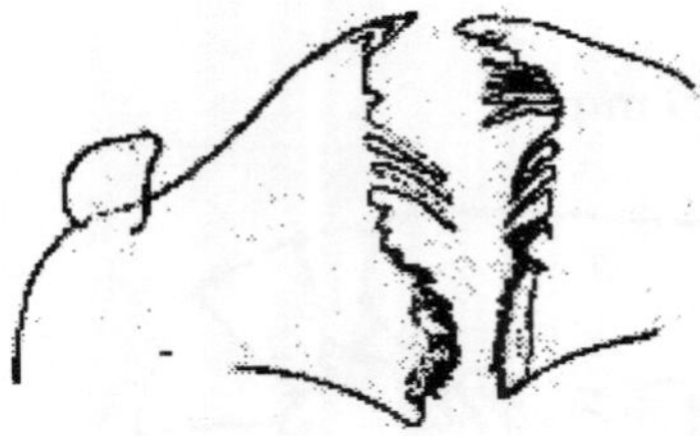	Spine on ventral side strong bud like palp present. Right and left mandible with 3 and 8 free standing teeth respectively.
Maxilla-I		Similar to that of previous sub-stage, except for absence of exopod.
Maxillla–II	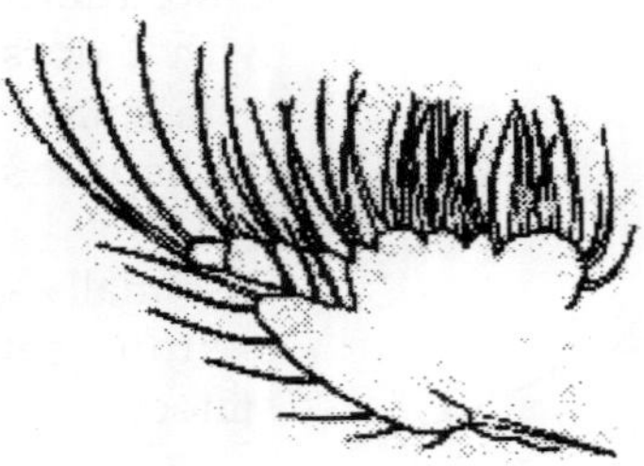	Setae scaphognathite increase to 9. Exopod with 11 setae.
Maxiliped–I		It develops small gill at basal region of protopod.

Characteristic	*Figure*	*Details*
Maxiliped–II and Maxiliped–III	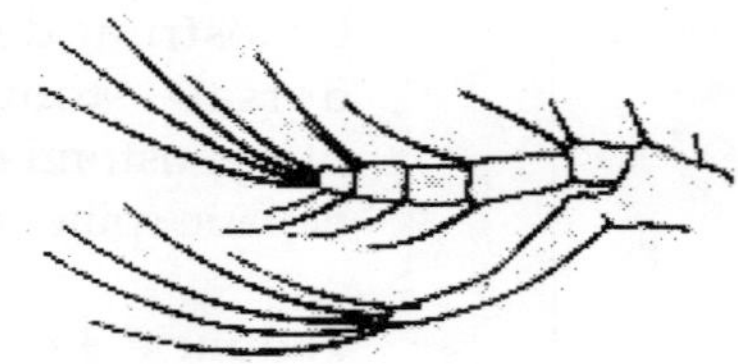	Characteristics similar to those of previous substage; pre-endopod of 1–3 legs 3 segmented being chelate.
Uropod		Well developed, endopod and exopod with 12 and 15 setae respectively.
Telson		Spination same as in previous substage; median cleft of posterior margin shallow.

Source: Vedvyas Rao 1973 with modification and Muthu M. S. *et al.*, 1978.

Mysis–III

TL–2.88–3.28 mm; CL–0.75–0.79 mm.

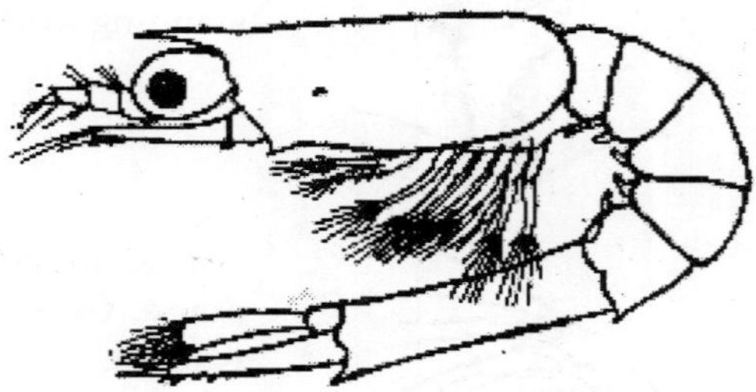

Characteristic	*Figure*	*Details*
Carapace		C-Rostrum reaches tip of eye and with 2 dorsal teeth; carapace with spines as in mysis II, but the hepatic spine larger.
Antennual–I	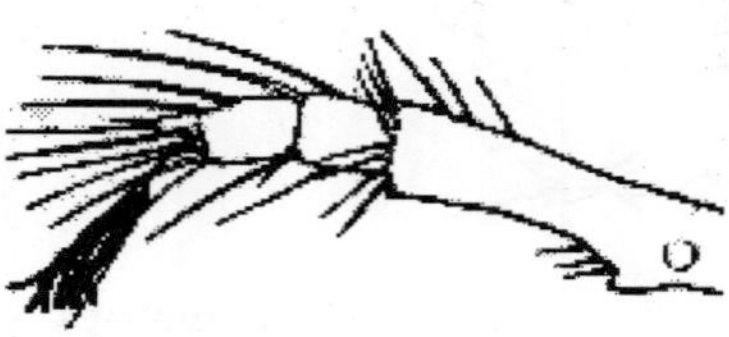	Essentially same as in mysis–II; 3 stout setae present over statocyst.

Characteristic	*Figure*	*Details*
Antennual–II		Flagellum 2 segmented, basal segment short, tip carries 2 short setae; scale with 16 or 17 setae and outer anterolateral spine almost reaches tip of scale.
Mandible		Only 3 serrated teeth; palp well developed.
Maxilla-I	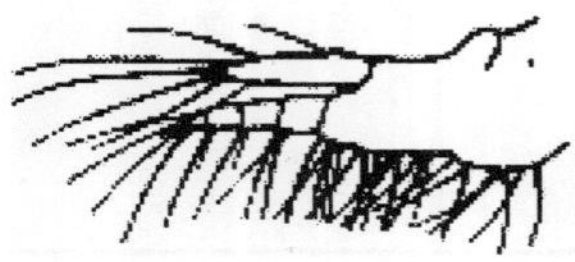	Morphological structure and pereopods essentially similar to those of preceding substage, except for increased number of setae in sca.
Maxillla–II	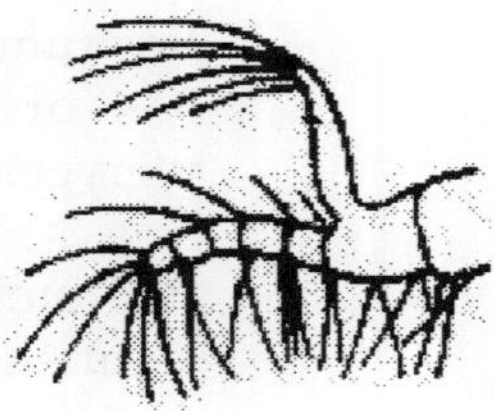	Morphological structure and pereopods essentially similar to those of preceding substage.
Maxiliped–I		Morphological structure and pereopods essentially similar to those of preceding substage.
Maxiliped–II and Maxiliped–III		Morphological structure and pereopods essentially similar to those of preceding substage, except for increased number of setae in scaphoganathite of Maxiliped–II and size at par.
Uropod		Exopod develops a small spine at distal outer margin and beset with 16 setae; endopod with 15 or 16 setae.
Telson	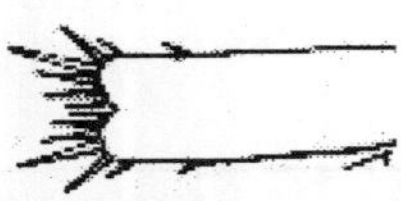	7 + 7 spines; median cleft of posterior margin disappears.

Source: Vedvyas Rao 1973 with modification.

Mysis Stage of *Metapenaeus dobsoni*

Mysis–I

MCL: 0.63 mm (0.60–0.66 mm); MTL: 2.04 mm (1.97–2.14 mm).

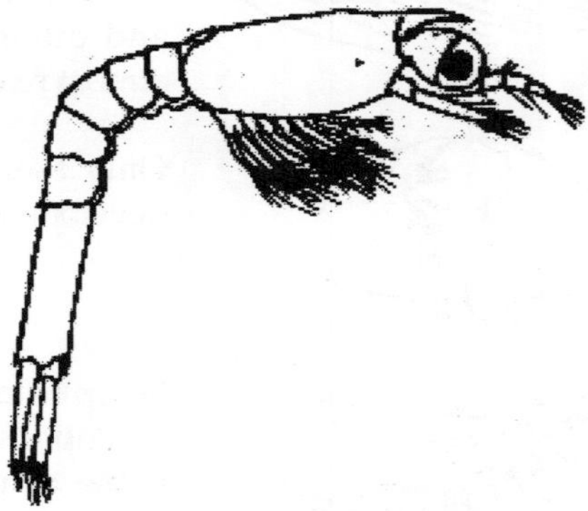

Characteristic	*Figure*	*Details*
Carapace		Rostrum without tooth, reaches 3/4 of eye; antennual and pterygostomial spine present.
Antennual-I	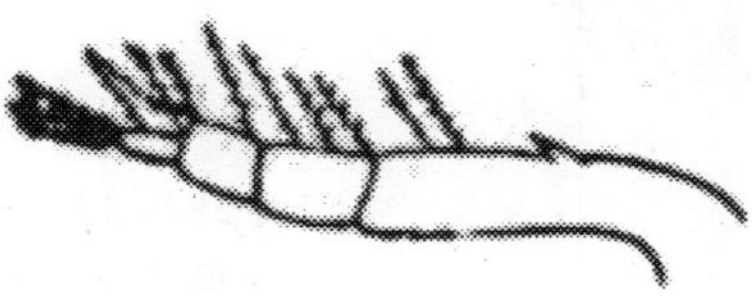	Peduncle 3 segmented, first segment longest, with a prominent spine on inner side; 2nd segment slightly longer than 3rd which carries 2 unsegmented sub-equal flagella, inner flagellum being short and knob-like.
Antennual–II	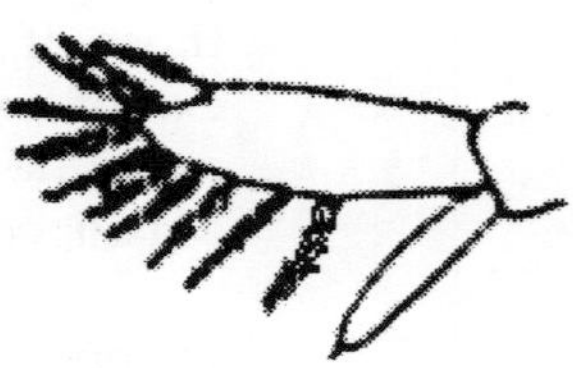	Scale with setae, 8 along inner margin and extremity and 3 along outer distal margin; flagellum un-jointed, half as long as scale, tip carrying a single short setae.
Mandible	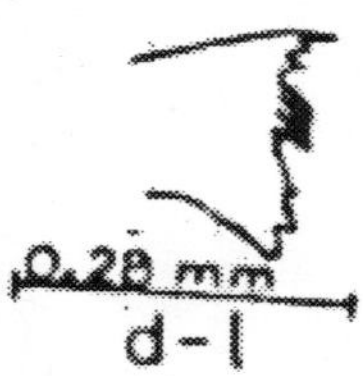	Cutting edge with 3 long, slender and serrated tooth incisor region with a conspicuous tooth; molar region with 7 or 9 small and short teeth.

Characteristic	*Figure*	*Details*
Maxilla–I		2 endites, each with a number of short setae; endopod 3 segmented and exopod knob–like with 4 plumose setae.
Maxilla–II	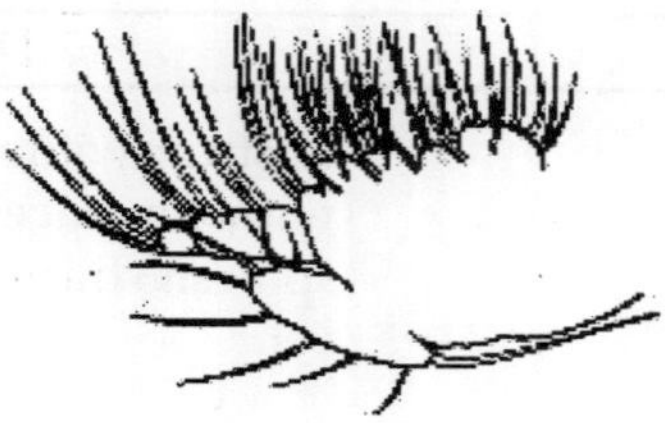	Protopod with 5 endites, proximal endite with 5 setae and rest with 2 or 3 setae; endopod 4 segmented, distal segment carries 3 terminal setae exopod well developed with 5 or 6 plumose setae.
Maxiliped-I	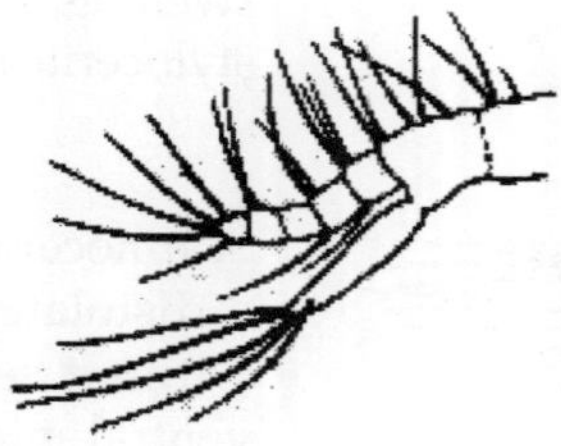	Protopod possesses many setae at inner margin; endopod 4 segmented, distal segment carrying 4 apical setae; exopod about 3/4 of endopod and with 4 terminal setae.
Maxiliped-II		Almost similar to that of Maxiliped-I.
Maxiliped-III	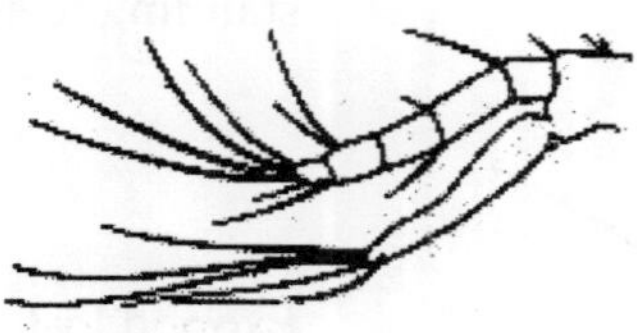	Consists of a protopod of 2 segments, 5 segmented endopod and a small unsegmented exopod; lateral and terminal setae present on endopod; per-well developed, 1–3 chelate each composed of 2 segmented protopod and endopod and an unjointed exopods of 4–5 legs unjointed.
Telson		Posterior margin deeply cleft medially, each lobe with 7 spines, 4th spine from outer side being longest.

Source: Vedvyas Rao 1973 with modification.

Mysis–II

(MCL: 0.64 0.63–0.67mm); MTL: 2.16 mm (2.08–2.31 mm).

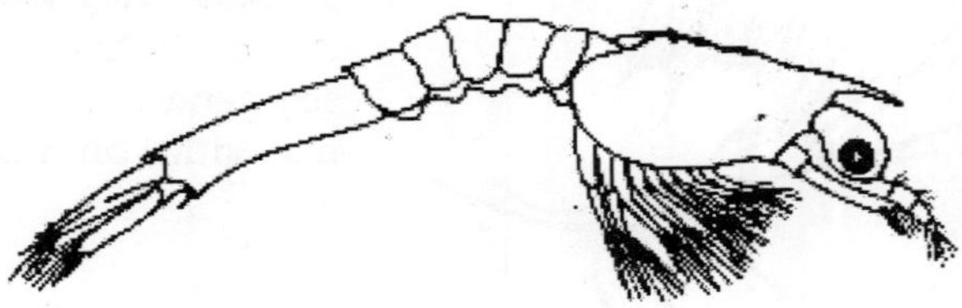

Characteristic	*Figure*	*Details*
Carapace		Rostrum without teeth; carapace with no trace of supraorbital spinesno trace of pleopods buds yet.
Antennual-I	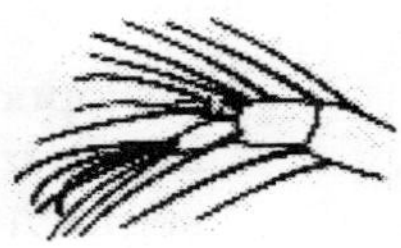	Small otolith visible in baal swelling. 3 short setae just above stylocerite rudiment.
Antennual–II		Scaphocerite with 13 setae and a distolateral spine, endopod without setae, protopod with ventral spine on distal segment.
Mandible		Right with 3 and left with 8 free standing teeth, palp larger.
Maxilla–I		Exopod lost.
Maxilla–II	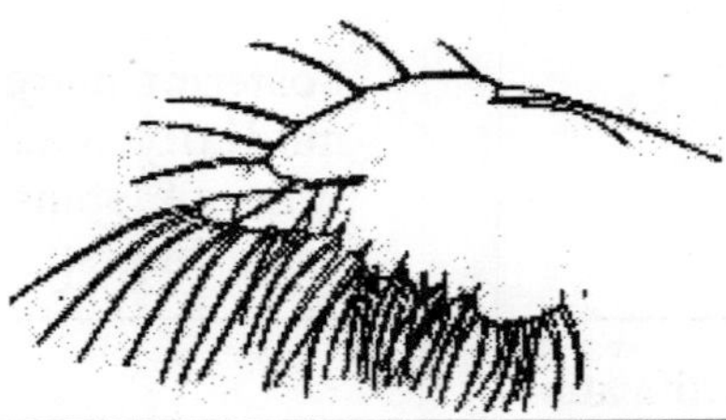	Exopod with 10 setae, more setae added to endites.

Characteristic	*Figure*	*Details*
Maxiliped-II	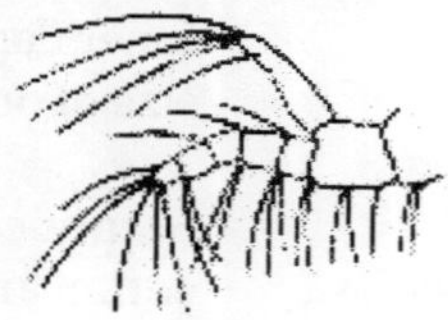	Endopod 5 segmented, one more seta added to distal segment and 1 outer lateral seta to 3rd segment.
Uropod	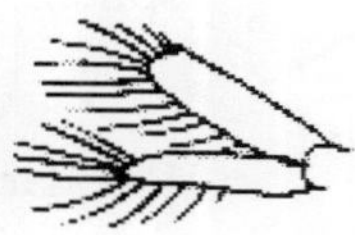	Exopod with 12 plumose setae and 1 short non-plumose setae outer distal angle, at the base of this seta outer margin of exopod produced into a minute tooth which becomes a well defined fixed spine in later sub stages, a short spine on ventral aspect of protopod.
Telson		Cleft of telson reaching only level of origin of penultimate pair of lateral setae.

Source; Muthu, M.S. *et al.* (3), 1978.

Mysis–III

MCL: 0.71 mm (0.67–74 mm); MTL: 2.35 mm (2,17–2.67 mm).

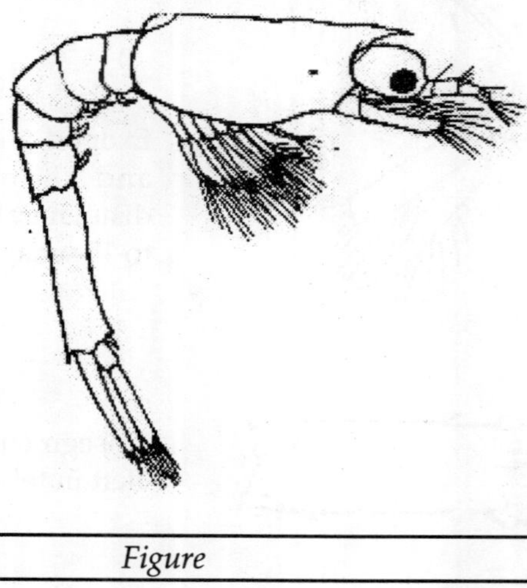

Characteristic	*Figure*	*Details*
Carapace		One rostral tooth present no other change in the spination of carapace and abdomen.

Characteristic	*Figure*	*Details*
Antennual-I		Inner flagellum of this stage half size of outer.
Antennual–II		Scaphocerite with 15 plumose setae and one anterolateral spine.
Mandible		Palp larger.
Maxilla–II		Exopod with 13 plumose setae.
Maxiliped-I	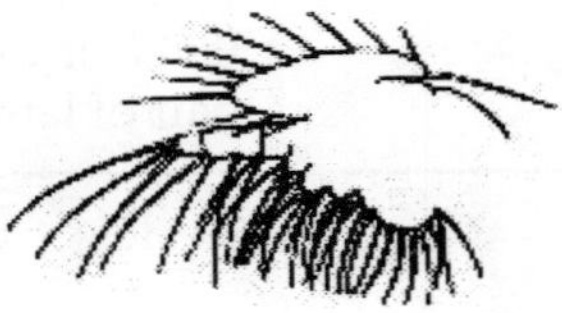	One outer seta added to basal segment of endopod.
Maxiliped-II	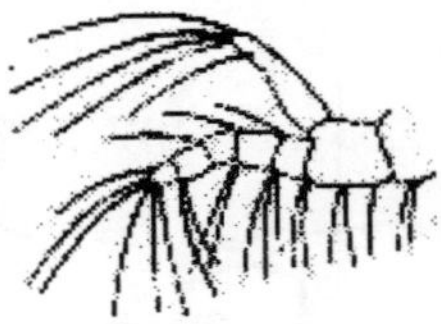	With one short seta added to junction of exopod and endopod.
Uropod	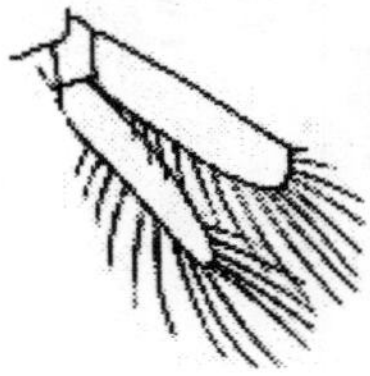	Exopod with 14 plumose setae and 1 non–plumose seta distolaterally, endopod with 14 to 15 setae.
Telson	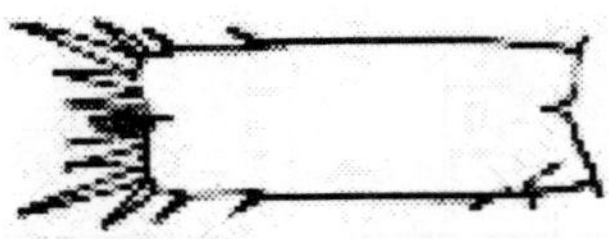	Unsegmented buds developed cleft in telson shallow.

Source: Muthu, M.S. *et al*. (3), 1978.

Mysis Stage of *Metapenaeus monoceros* (Fabricius)

Mysis–I

MCL:2.4 mm (2.37–2.45 mm); MTL: 0.86 mm (0.84–0.87 mm).

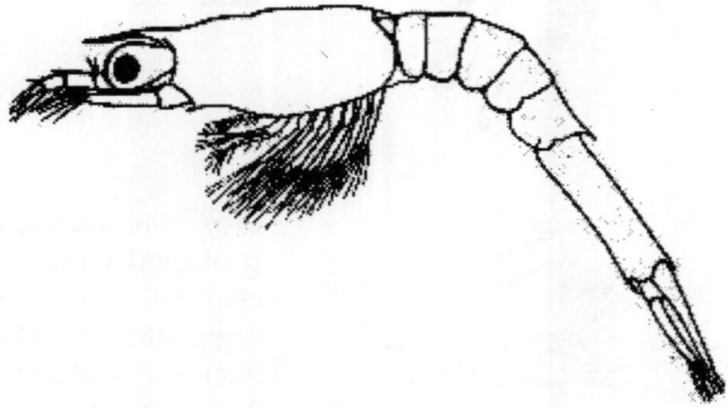

Characteristic	*Figure*	*Details*
Carapace		Rostrum extends beyond eye, devoid of teeth, carapace with pterygostomial and antennal spines, 5th and 6th abdominal segments with dorsal median spine 6th abdominal segment with a prominent ventral median spine on posterior end.
Antennual-I		3 segmented, proximal segment with a prominent ventral spine, plumose setae present on junction of segments and on inner side, distal segment bearing 2 flageller rudiments inner one small, bud like bearing 2 setae; outer unsegmented bearing 6 aesthaetes and 1 seta.
Antennual–II	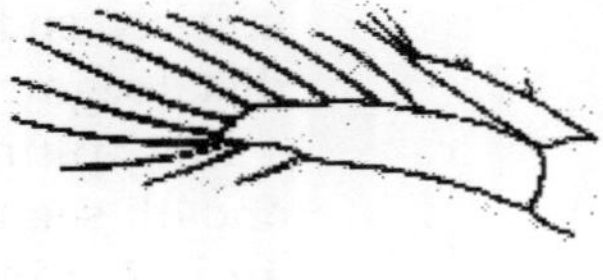	Exopod unsegmented, scale like bearing 10 setae along inner and distal margin and 1 seta on distolateral angle, endopod half length of exopod, bearing 3 short setae apically and 1 + 2 setae on inner margin.

Characteristic	*Figure*	*Details*
Mandible		With 8 free standing teeth on left mandible and 3 on right mandible.
Maxilla–I	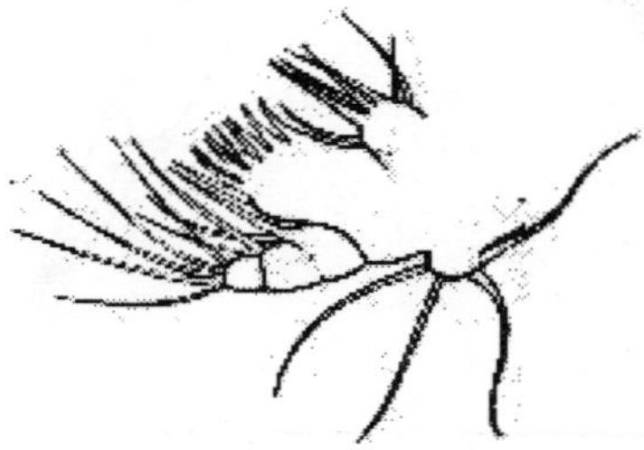	Proximal and distal enditute of protopod with 7 and 10 setae respectively, endopod 3 segmented, distal segment with 5 setae, 1st and 2nd segments with 2 to 3 and 2 setae respectively, exopod with 4 feathery setae.
Maxilla–II	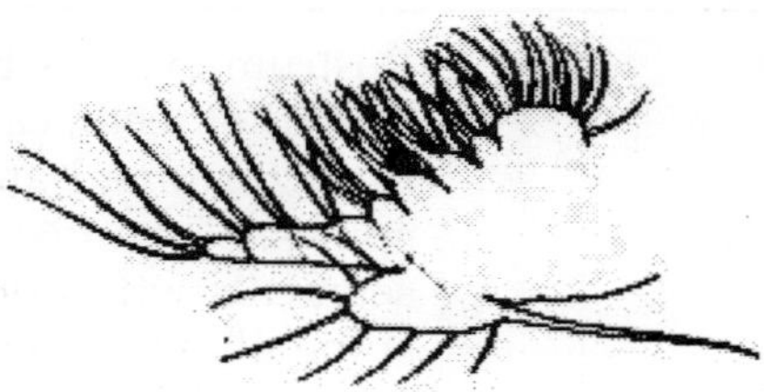	With 9 plumose setae.
Maxiliped-I	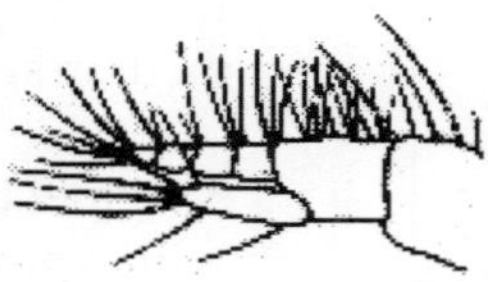	Exopod with 7 plumose setae.
Maxiliped-II	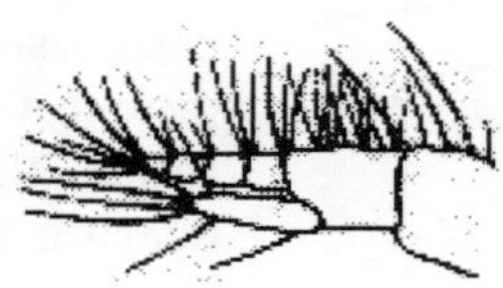	Exopod with 4 apical and 2 subapical plumose setae, endopod 4 segmented, 1st and 2nd segments carrying 1 seta on outer margin.
Maxiliped-III	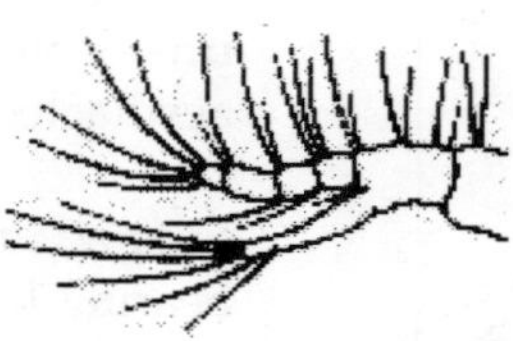	Biramous, fully developed, protopod 2 segmented; exopod with 4 long plumose setae apically, endopod 5 segmented, distal segment with 1 short and 4 long plumose setae, 1st 2nd and4th segments carrying 2, 1 and 2 inner lateral setae respectively outer distal margin of 2nd, 3rd and 4th segments with 1 seta each.

Characteristic	*Figure*	*Details*
Uropod	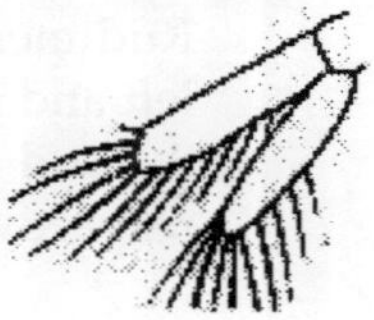	Biramous, exopod with 1 short nonplumose outer seta and 11 long plumose setae; endopod with 10 plumose setae.
Telson		Broader distally with deep median cleft carrying 7 + 7 setae.

Source: Mohmed, K. H. *et al.*, 1978.

Mysis–II

MCL: 0.87 mm (0.84–0.90 mm); MTL: 2.50 mm (2.38–2.55 mm).

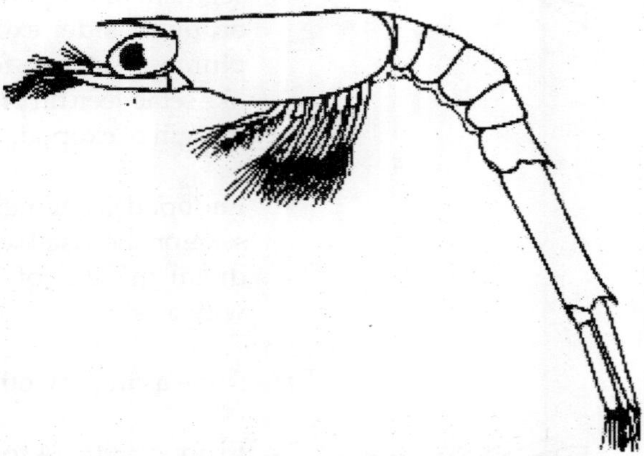

Characteristic	*Figure*	*Details*
Carapace		Rostrum with dorsal tooth, pleopods buds still absent.
Antennual-I	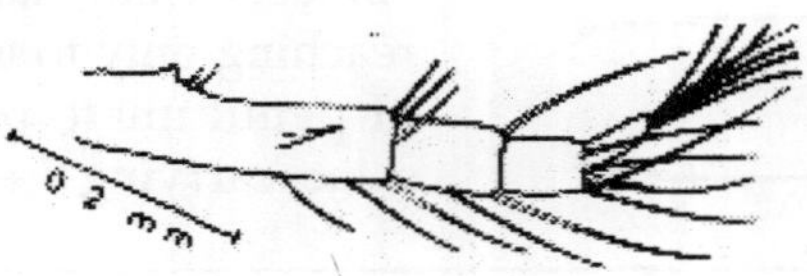	Number of setae on segments increased, inner flagellum reaching 2/3rd of outer flagellum.
Antennual–II		Exopod with 14 plumose setae and 1 distolateral spine.

Characteristic	*Figure*	*Details*
Mandible		Rudimentary palp developed, left and right mandible with 7–8 and 3 standing teeth respectively.
Maxilla–II		Exopod with 11 plumose setae.
Maxiliped-I	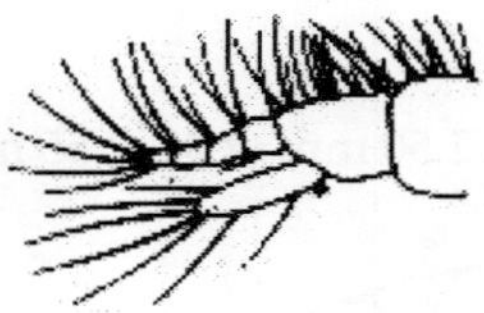	Proximal and distal segments of protopod with 8 and 14 setae respectively on inner side, 1st segment of exopod with 1 setae on outer side, exopod with 7 plumose setae, a few fine hair like setae seen on proximal outer margin of exopod.
Maxiliped-II		Endopod 5 segmented, with 6 setae on the distal segment, outer distal margin of 3rd segment with a seta.
Maxiliped-III		Same as in previous stage.
Uropod	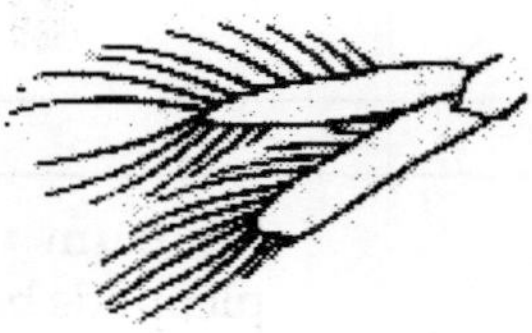	Exopod with 14 to 15 plumose setae, 1 nonplumose seta and 1 short spine distolaterally.
Telson	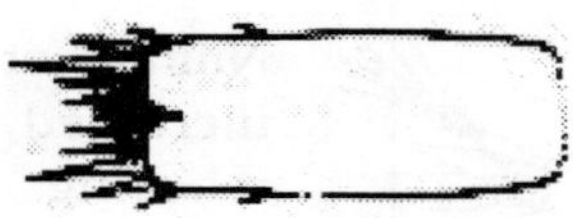	Almost rectangular, median cleft reaching only to level of origin of penultimate pair of lateral setae, carrying 7 + 7 setae.

Source: Mohmed, K. H. *et. al.*, 1978.

Mysis–III

MCL: 0,95 mm (0.92–0.98 mm); MTL: 2.7 mm (2.58–2.81 mm).

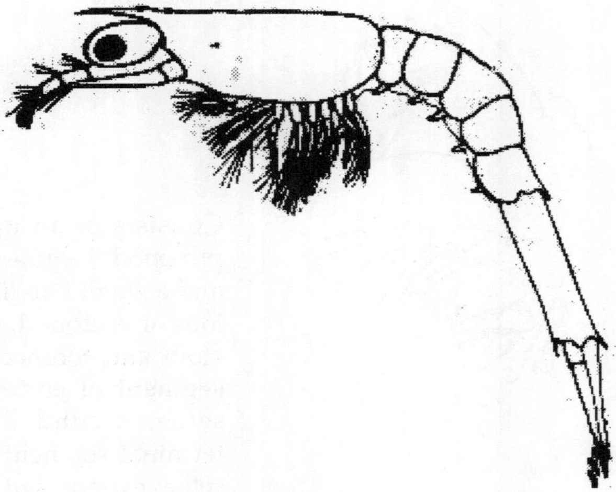

Characteristic	*Figure*	*Details*
Carapace		Rostrum with 2 dorsal teeth in some specimens a small hepatic spine seen,pleopods buds small and unsegmented.
Antennual-I	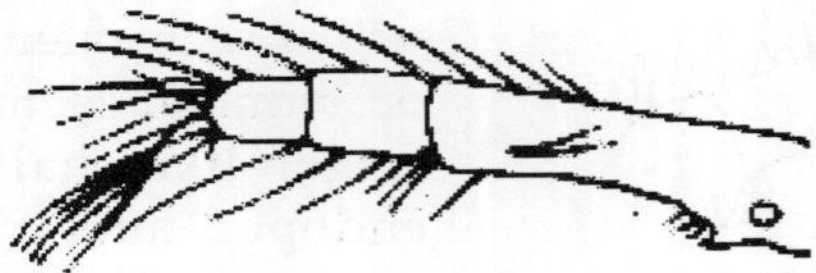	Peduncle 3 segmented, proximal segment longest and terminal segment shortest; flagella unjointed, inner branch small and outer longer carrying 5 aesthetes; a conspicuous inner marginal spine on proximal segment.
Antennual–II	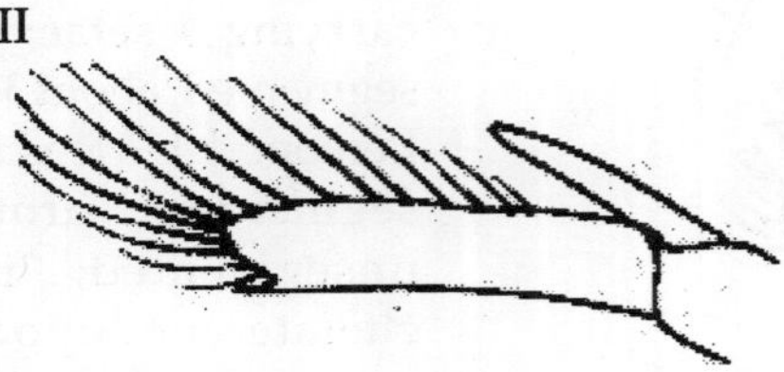	Endopod unsegmented not reaching half of exopod, exopod with 16 plumose and setae and 1 distolateral spine.

Characteristic	*Figure*	*Details*
Mandible		Developed into a finger shaped projection.
Maxilla–I	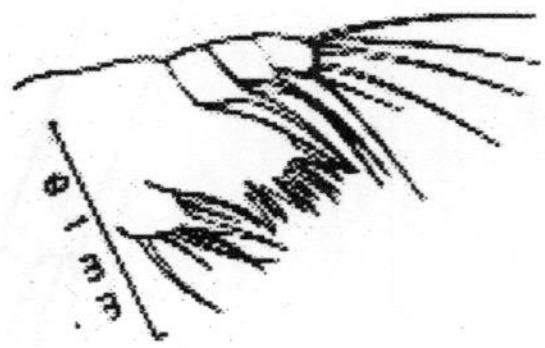	Consists of an unsegmented protopod, 3 segmented endopod and a small kno-like exopod; 2 lobs of protopod, carry several stout and toothed spines first segment of endopod bears 3 setae, second 2 setae and terminal segment with 4 long setae; exopod with 4 pulmose setae.
Maxilla–II		Exopod with 13 plumose setae.
Maxiliped-I	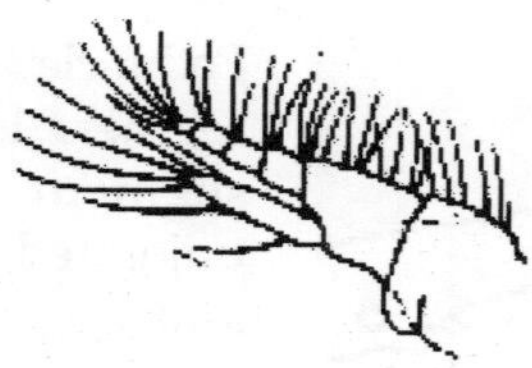	Protoped 2 segmented; endopod with 4 segmented with 5 setae on terminal segment; exopod finger–like, half length of endopod and with 4 apical setae.
Maxiliped-II Maxiliped-III	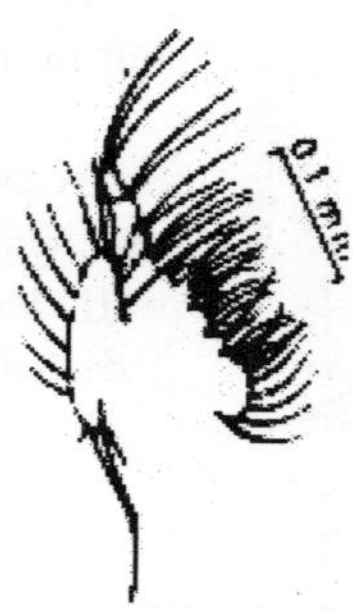	Protopod 2 segmented; endopod 5 segmented distal segment carrying 4 setae; exopod unsegmented, about 3/4 endopod; Pereopod 1-3 legs formed by 2 segmented protopod, an unsegmented, indistinctly chelate endopod and an unsegmented exopod; Pereopod 4 and 5 with protopod of 2 segments enopod indistinctly segmented at tip and an unsegmented exopod.

Characteristic	*Figure*	*Details*
Uropod		Well developed, endopod with 11–12 setae and without spine exopod with 12 setae and outer distal spine.
Telson		Telson posteriorly lobed with 7 spines on each lobe.

Source: Mohmed, K. H. *et al.*, 1978 and Vedvyas Rao 1973.

Mysis Stage of *Metapenaeus brevicornis* (H. Milne Edwards)

Mysis–I

MCL: 0.69 mm (0.67–0.70 mm): MTL: 1.97 mm (1.96–1.00 mm).

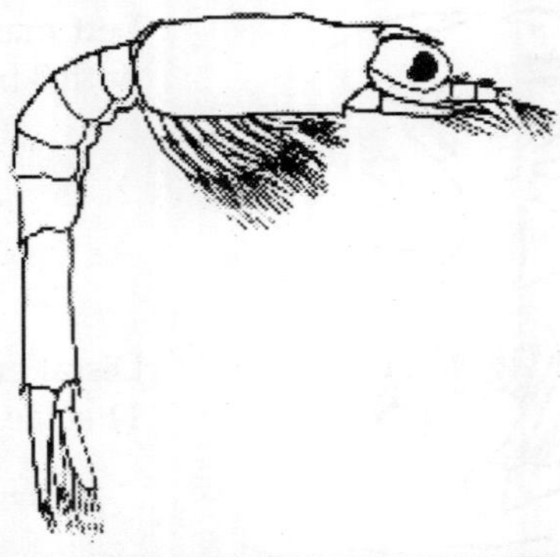

Characteristic	*Figure*	*Details*
Carapace		With rostrum just falling short of anterior end of eye, rostrum devoid of teeth, antennal and pterygostomial spines well developed, small but distinct supraorbital spines present, 5th and 6th abdominal segments with dorsal spines, no lateral spine on 5th and 6th abdominal segments a prominent ventro-median spine on posterior end of 6th abdominal segment at junction with telson.

Characteristic	*Figure*	*Details*
Antennual-I	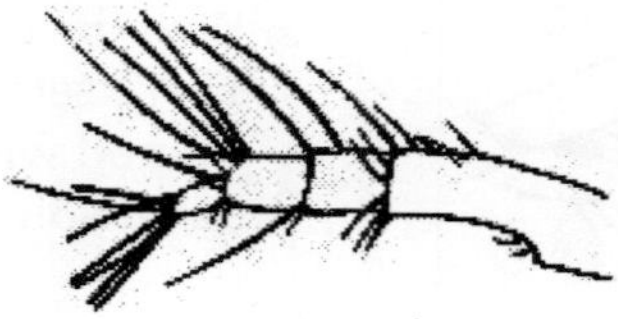	3 segmented, basal segment with prominent ventral spine, slight basal swelling with 2 setae just above stylocerite rudiment, outer flagellum with 6 aesthaetes and 1 setae, inner flagellum short bearing 1 short and 1 long seta spically.
Antennual–II		Exopod and endopod unsegmented, exopod scale-like bearing 10 setae along inner and distal margin and 1 seta at distolateral angle, endopod half length of exopod bearing short setae apically and inner side.
Mandible		Left mandible with 8 and right with 3 free standing teeth.
Maxilla–I	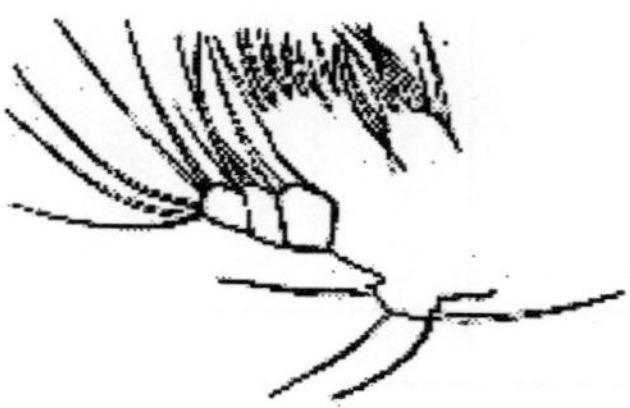	Distal endite of protopod with 11 setae.
Maxilla–II		Exopod with 9 plumose setae.
Maxiliped-I	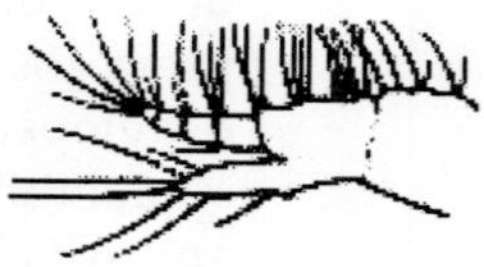	Exopod with 7 plumose setae, below the proximal outer lateral seta a few small hair–like setae seen, 1 outer lateral seta added to 1st segment of endopod.
Maxiliped-II	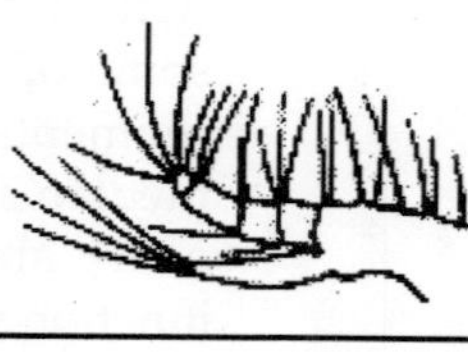	1st and 2nd segments of endopod acquired 1 distal outer seta each.

Characteristic	*Figure*	*Details*
Maxiliped-III	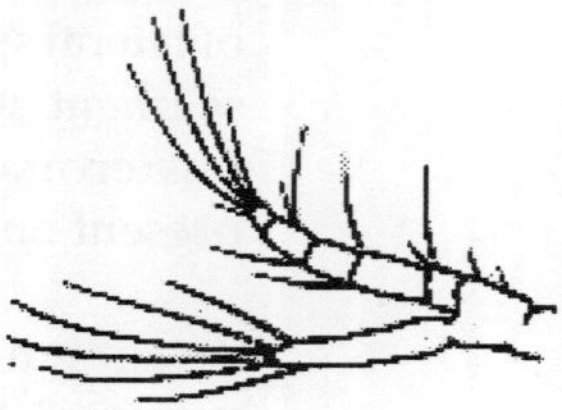	Fully developed 5 segmented endopod, distal segment with 5 setae; 1st, 2nd and 4th segments carrying 2, 1 and 2 setae respectively, along the distal inner margin, 2nd and 3rd segments each carrying one seta on distal outer margin.
Uropod		With 12 plumose setae and 1 nonplumose seta, endopod with 10 plumose seta.
Telson		Telson with 7 + and setae, deep cleft reaching level of origin of outermost pair of setae.

Source: G. Sudhakar Rao, 1978.

Mysis Stage of *Parapenaeopsis stylefera* (H. Milne Edwards)

Mysis–I

MCL: 2.00 mm (1,8–2.1 mm); MTL: 0.64 mm (0.58–0.66 mm)

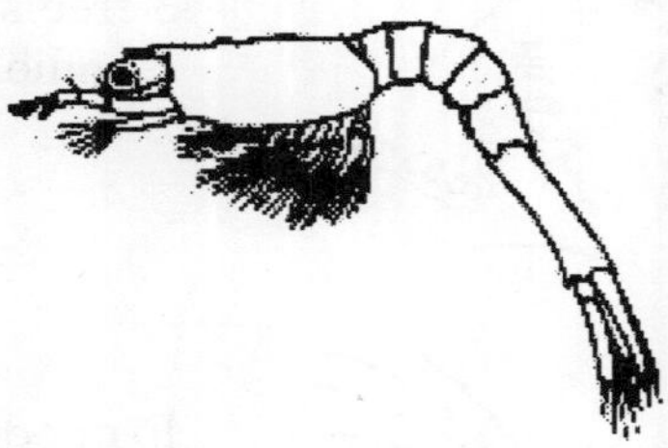

Characteristic	*Figure*	*Details*
Carapace		With rostrum shorter than eye, a minute supraorbital spine, a prominent pterygostomial spine and a smaller antennal spine present; dorsal organ prominent in lateral view, dorsomedian spine present only on 5th and 6th abdominal segments; a vesting

Characteristic	*Figure*	*Details*
		of lateral spine on 5th abdominal segment still present; a ventral posteromedian curved spine present on 6th segment.
Antennual-I	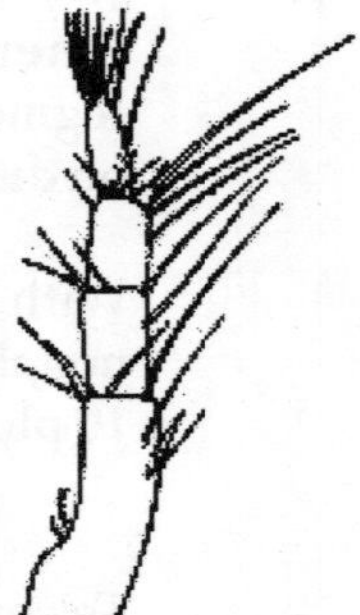	3 segmented, distal segment carries 2 flagellar rudiments, larger one tipped with 6 aesthaetes and 1 small seta and smaller bud-like one tipped with 2 setae, 1 long and 1 short, the 1st segment has a basal swelling wherein statocyst is developed in later substages, 2 short setae present above this swelling distally, on the ventromedian aspect a prominent spine present.
Antennual–II		Scaphocerite with 10 setae on inner and distal margin and 1 distolateral setae, endopod very short with 3 terminal setae and 2 + 1 lateral setae.
Mandible	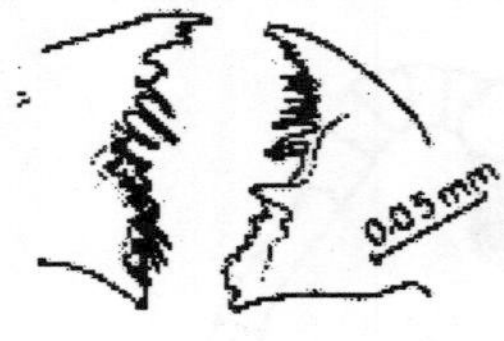	3 free standing teeth in right mandible and 7 in left.
Maxilla–I	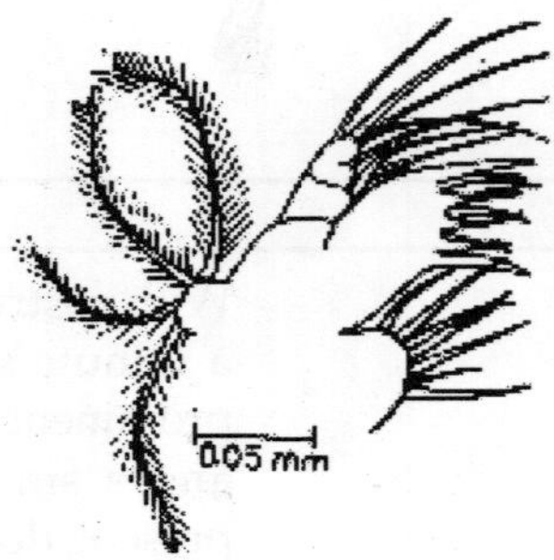	Exopod with 4 feathery setae, still present, distal endite with 9 setae.

Characteristic	*Figure*	*Details*
Maxilla–II	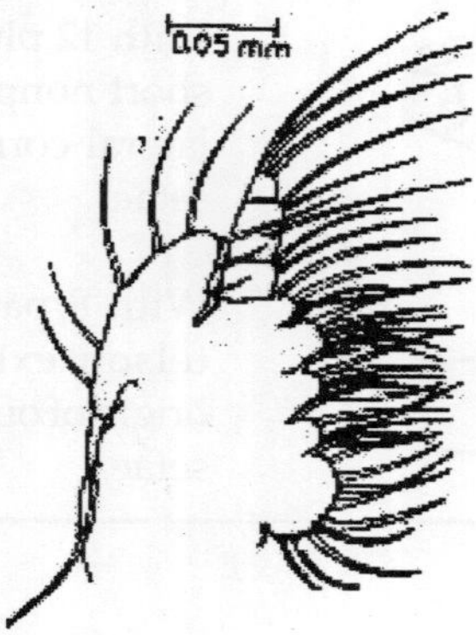	Exopod with 10 setae, distal endite with 3 and 3 middle endites with 5 to 7 setae.
Maxiliped-I	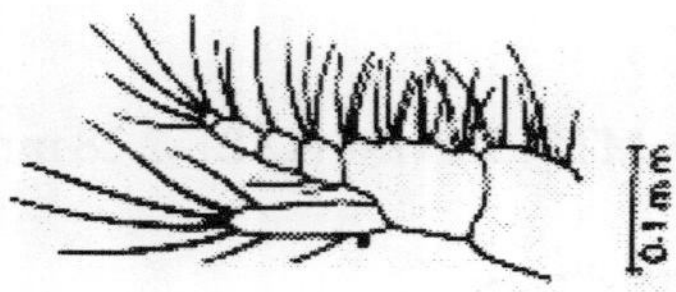	Protopod broader, exopod with 7 long setae and a few hair-like setae below the proximal outer setae, 1 outer lateral seta added to 1st segment of endopod, 1 outer seta at junction with protopod, terminal setae of endopod shorter than length of endopod.
Maxiliped-II	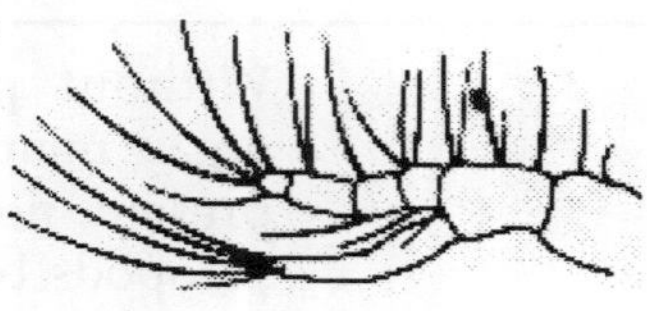	Exopod with 6 setae, 4 terminal and 2 subterminal setae on 1st and 2nd segments and also at junction with exopod, 1 more inner seta added to basal segment of endopod.
Maxiliped-III	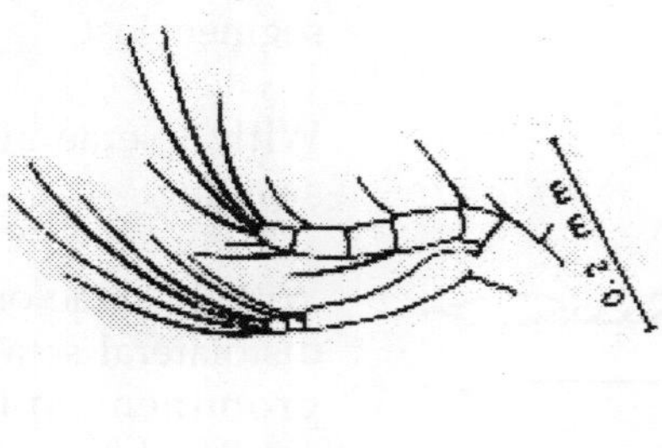	Well developed, endopod 5 segmented distal segment with 5 setae, 4th segment with 1 outer and 1 inner lateral setae, 3rd segmented distal without any seta, 2nd with 1 outer and inner seta and the 1st with one lateral seta, exopod as long as endopod, with 4 long terminal setae and 2 pairs of subterminal setae, 2 indistinct segmentations seen at distal end.

Characteristic	*Figure*	*Details*
Uropod	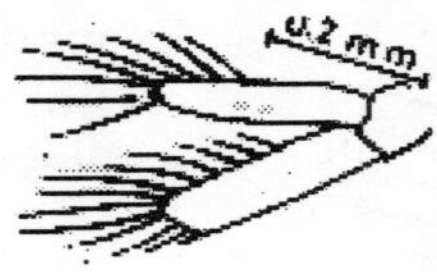	With 12 plumose setae and one short nonplumose seta at outer lateral corner, endopod with 8 setae.
Telson		With 8 pairs of setae; cleft on telson extending to level of origin of outermost pair of lateral setae.

Source: Muthu.M. S. *et al.*, (4), 1978.

Mysis–II

MCL: 0.69 mm (0.67–0.71 mm); MTL: 2.25 mm (2.21–2.28 mm).

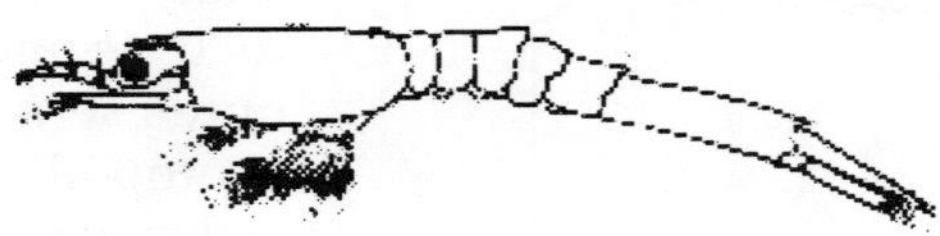

Characteristic	*Figure*	*Details*
Carapace		Without prominent dorsal organ, minute supraorbital still present, no hepatic spine, no pleopods buds on abdominal segments, vestiges of the pair of lateral spines on 5th abdominal segment lost.
Antennual–I		With 3 setae above statocyst swelling.
Antennual–II		With 14 setae on scaphocerite, distolateral seta replaced by a prominent spine, endopod small with a short basal segment.
Mandible	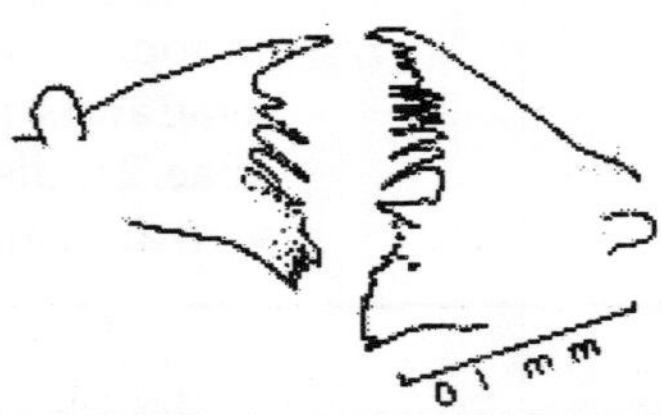	Small palp seen.

Characteristic	*Figure*	*Details*
Maxilla–I	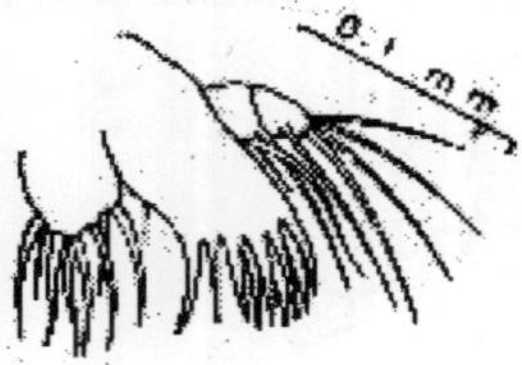	Exopod lost.
Maxilla–II		Exopod with 13 setae
Maxiliped-I		Terminal setae on endopod slightly shorter than endopod, 1 outer lateral seta added to 2nd segment.
Maxiliped-II	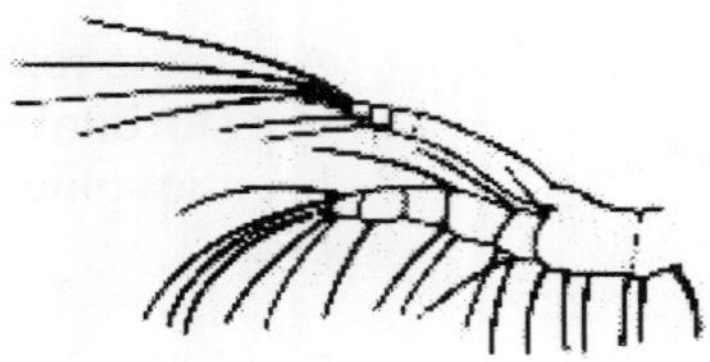	Endopod 5 segmented, penultimate segment of mysis 1 divided into 2.
Maxiliped-III	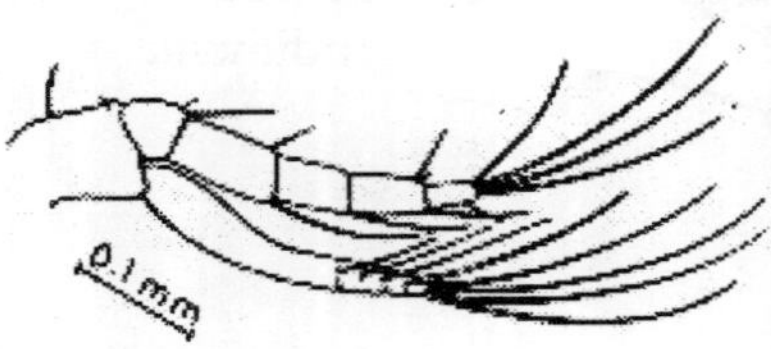	1 more sea added to inner margin of basal segment of endopod, exopod with 4 terminal and 3 pairs of subterminal setae.
Uropod	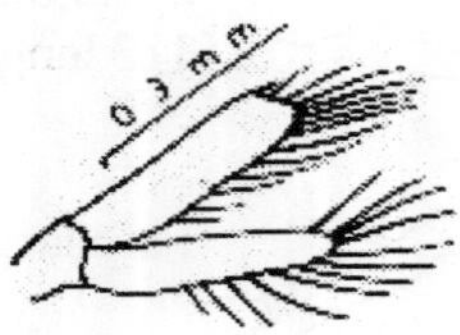	Exopod with 12 to 13 plumose setae and 1 nonplumose, short distolateral seta. Outer margin of exopod not produced into a fixed spine, endopod with 11 to 12 setae.
Telson	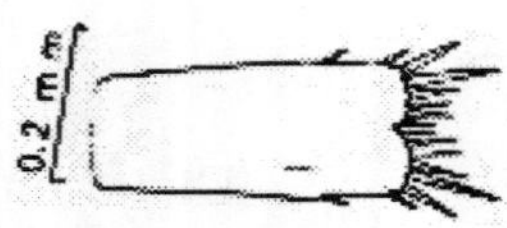	With cleft extending only up to level of origin of penultimate pair of lateral setae.

Source: Muthu.M. S. *et al.*, (4), 1978.

Mysis–III

MCL: 0.75 mm 90.74–0.76 mm). MTL: 2.39 mm (2.28–2.41 mm).

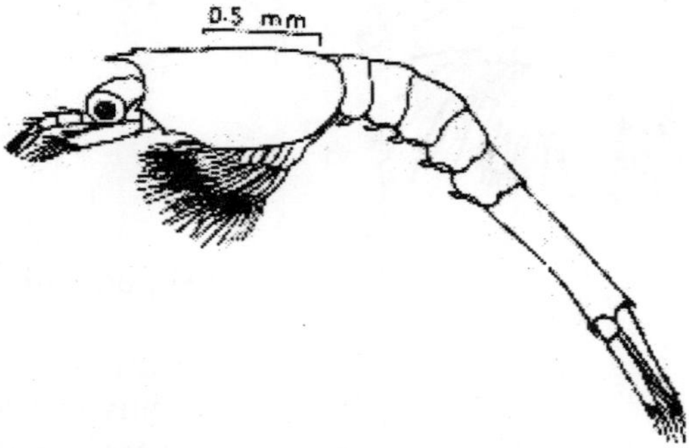

Characteristic	*Figure*	*Details*
Carapace		A minute rostral tooth present supraorbital spine absent, no hepatic spine, still no pleopods buds.
Antennual-I	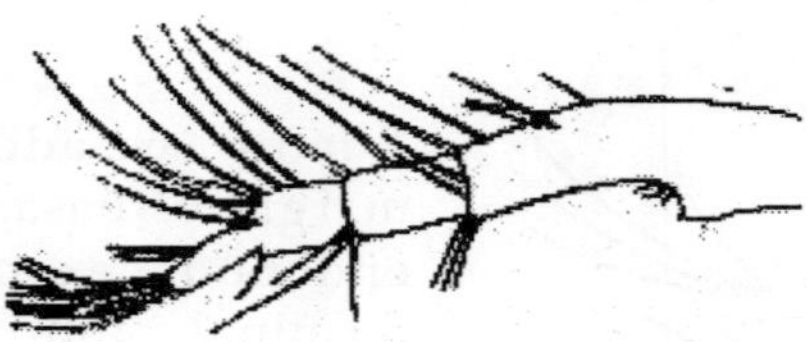	2 subterminal aesthaethaetes added to outer flagellar rudiment.
Antennual–II		Scaphocerite with 16 setae and a distolateral spine, endopod 1/3 length of scale.
Mandible	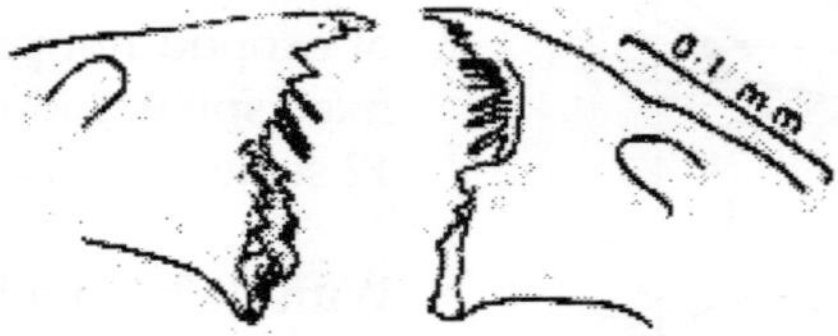	Palp larger.
Maxilla–I		Distal endite with 10 setae, proximal endite with 8 setae.
Maxilla–II		Exopod with 15 setae.

Characteristic	*Figure*	*Details*
Maxiliped-I	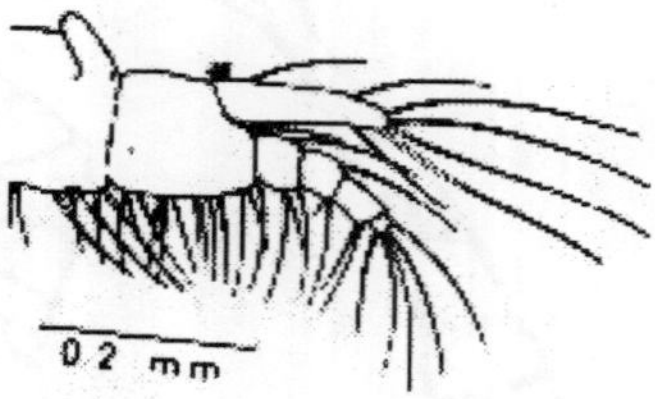	Small gill rudiment present on outer distal corner of basal segment of protopod, inner lateral sea added to 2nd segment of exopod.
Maxiliped-II	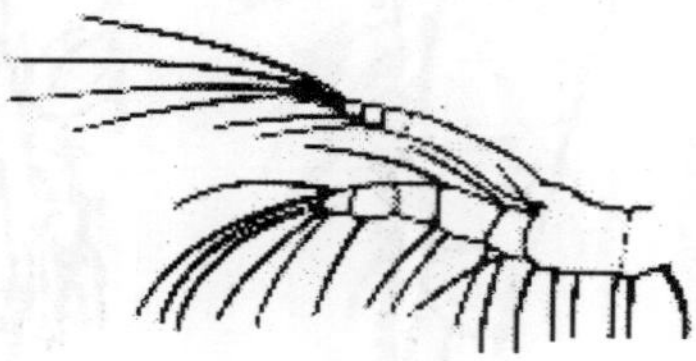	One more inner lateral sea added to 2nd segment and another terminal seta to distal segment, 1 seta added to junction of exopod and endopod, exopod with 8 setae, 4 terminal and 2 pairs of subterminal.
Maxiliped-III	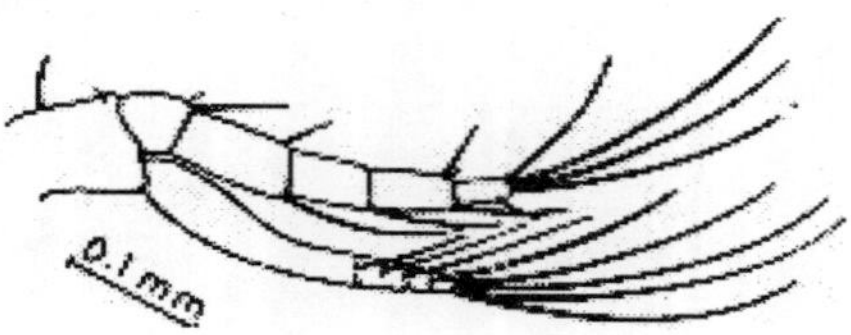	Exopod with 12 setae, 4 terminal and 4 pairs of subterminal, 1 outer lateral seta added to 3rd segment of endopod.
Uropod	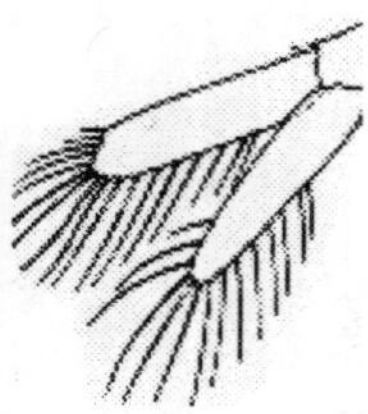	Exopod with 16 plumose setae and 1 nonplumose disto-lateral seta, endopod with 14 setae.
Telson	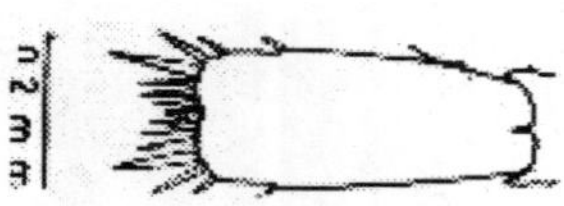	With shallow cleft.

Source: Muthu, M. S. *et al.*, (4), 1978.

A Guide to Identification of the Larval Stages of *M. rosenbergii*

Larvae Stage	Age (Days)	Characteristics		Pictures
I	1	Sessile eyes	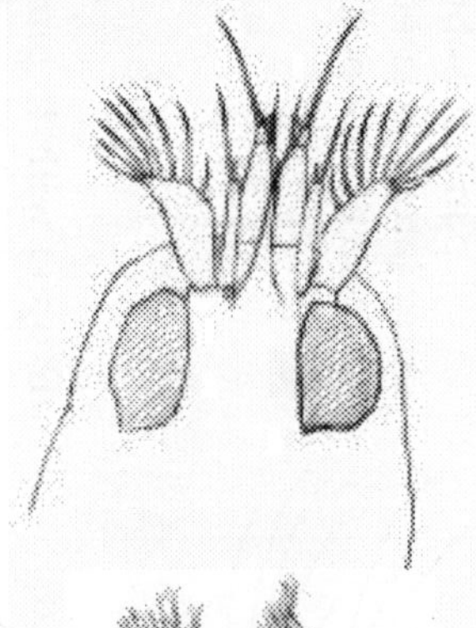	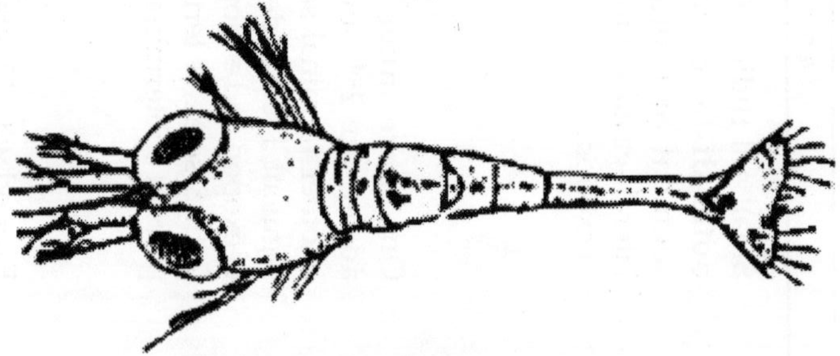
II	2	Stalked eyes	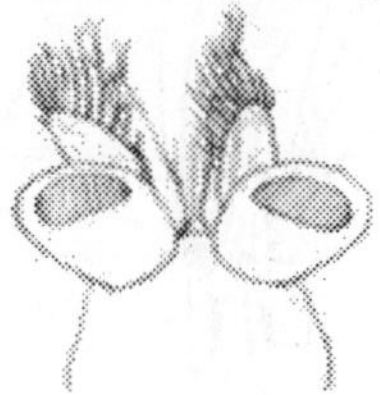	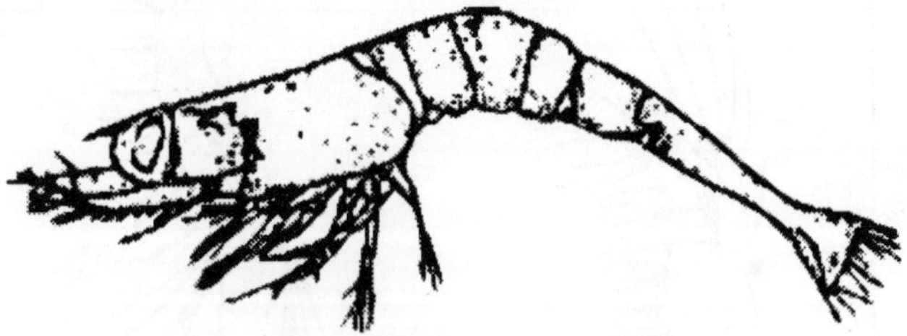
III	3	Uropods appear	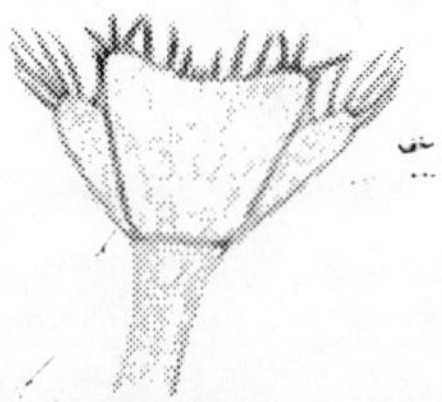	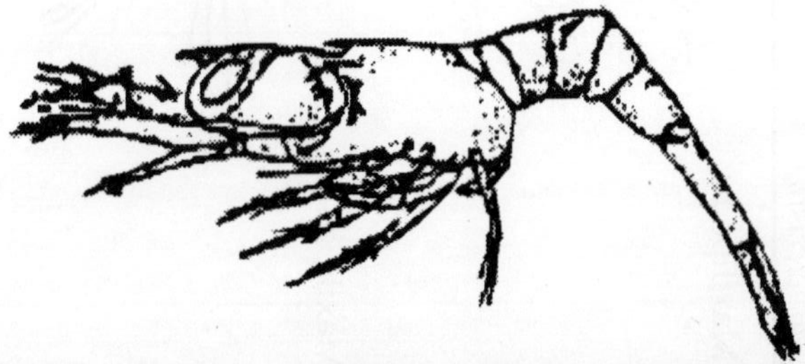

Larvae Stage	*Age (Days)*	*Characteristics*		*Pictures*
IV	4–6	Two dorsal teeth on rostrum	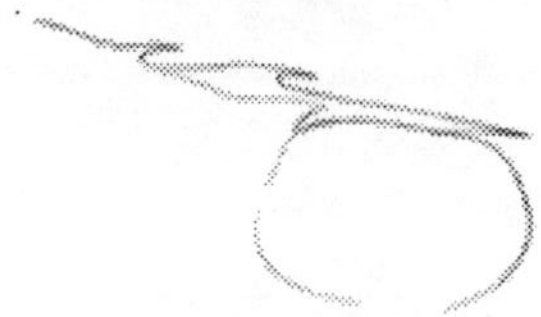	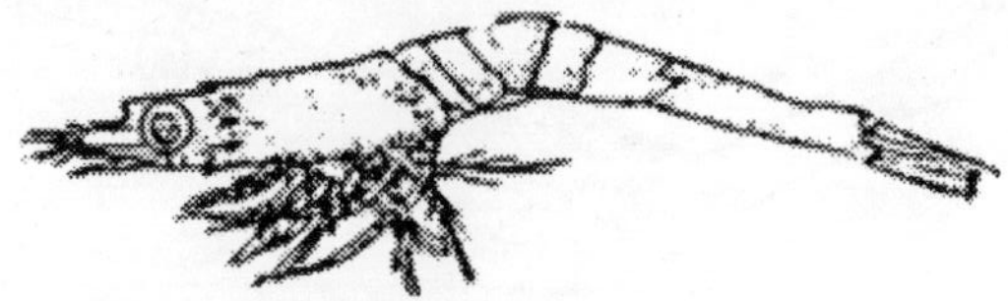
V	5–8	Telson narrower and elongated	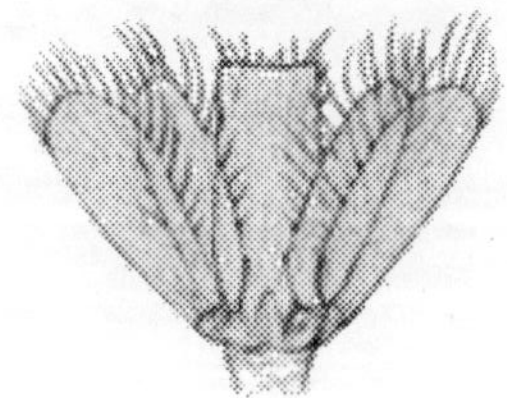	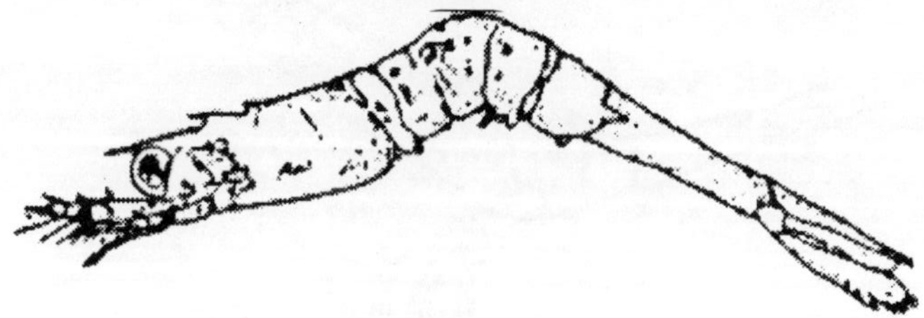
VI	7–10	Pleopod buds appear	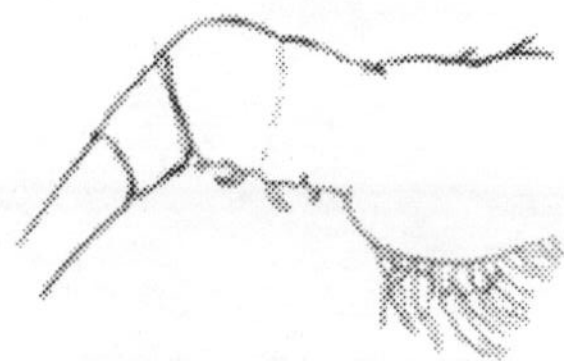	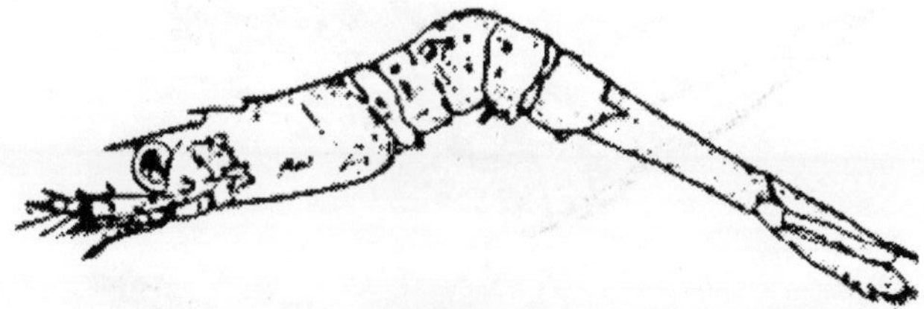
VII	11-17	Pleopods biramous and bare	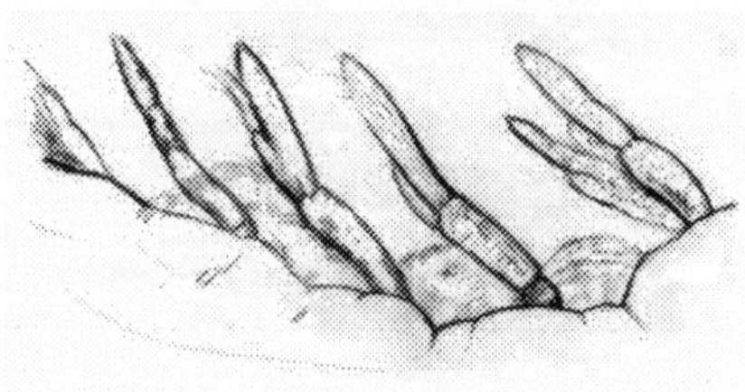	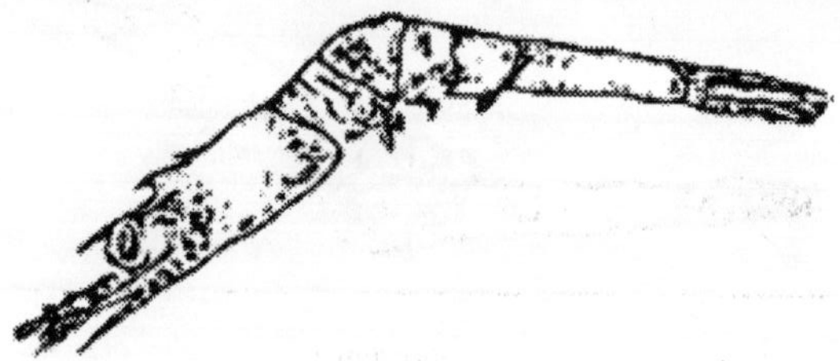

Larvae Stage	Age (Days)	Characteristics		Pictures
VIII	14–19	Pleopods with setae	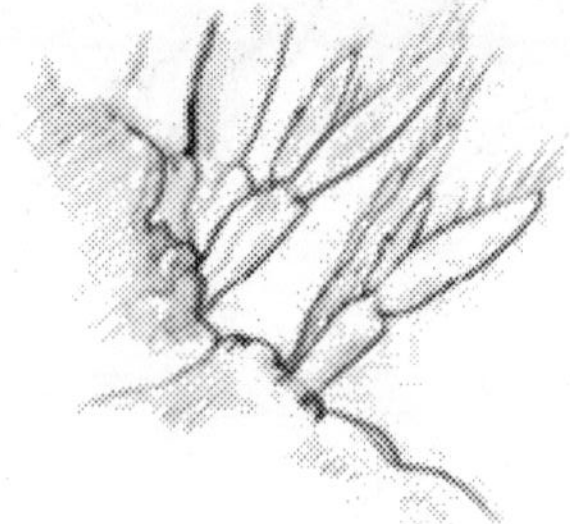	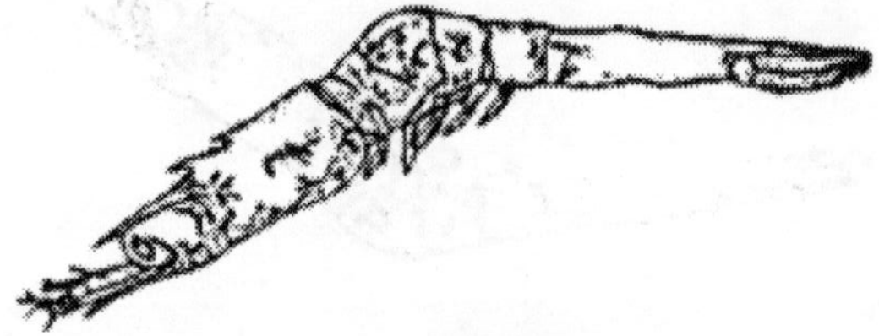
IX	15-22	Endopods of pleopods with appendices internae		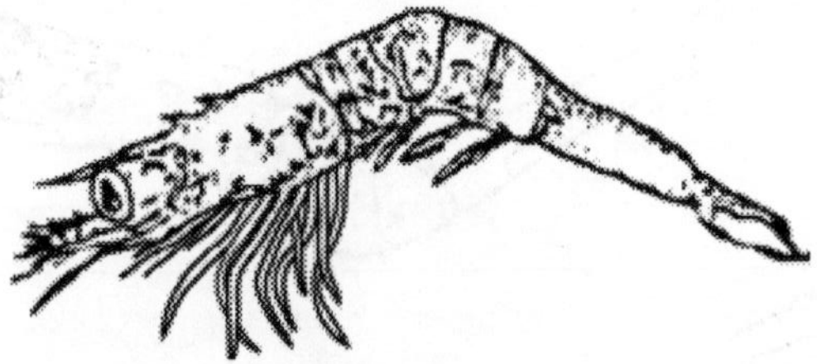
X	17-24	3–4 dorsal, teeth on rostrum	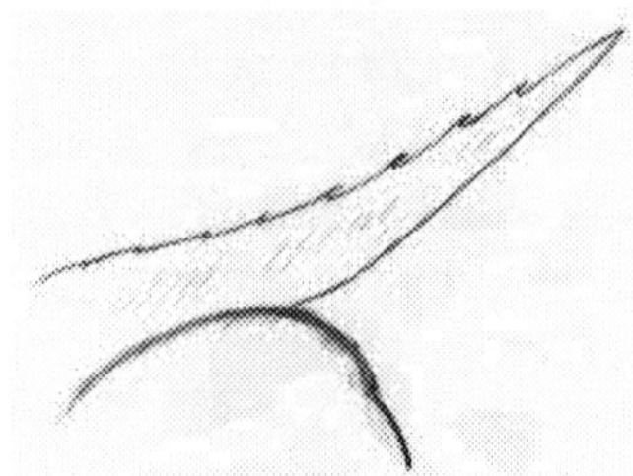	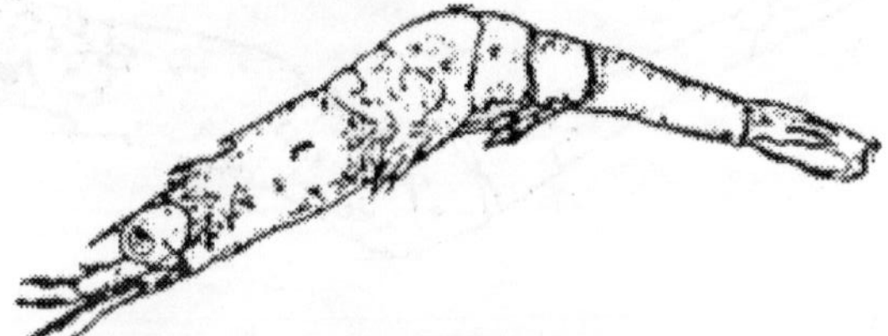
XI	19–26	Teeth on half of upper dorsal margin	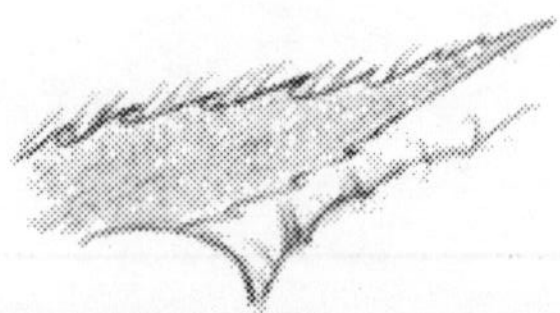	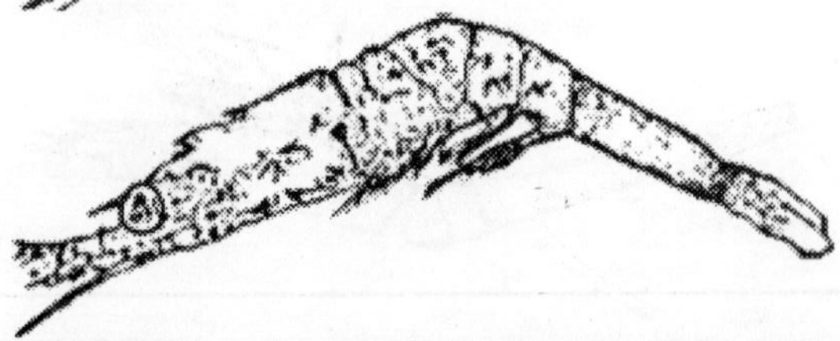

Source: Satya Nandlal and Timothy Pickering.

Macrobrachium rosenbergii go through eleven distinct larval stages before metamorphosing to become postlarvae. Photographs of some of them is shown as under.

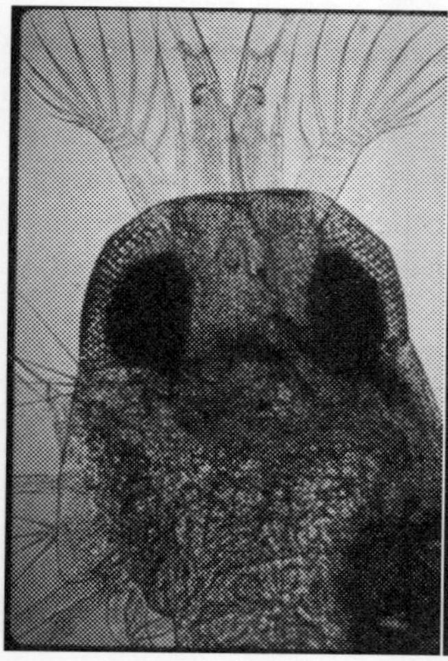

Figure 2.1

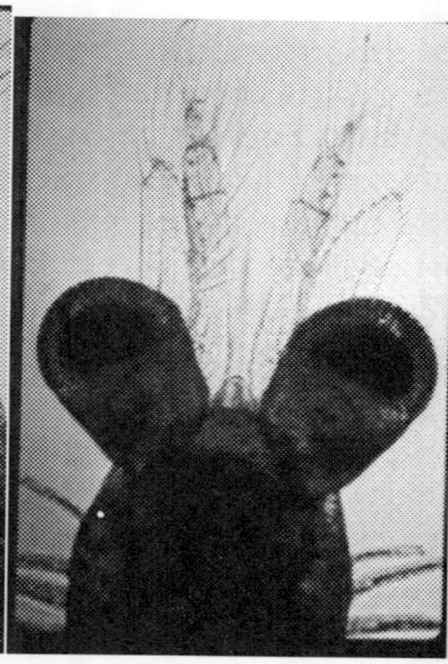

Figure 2.2

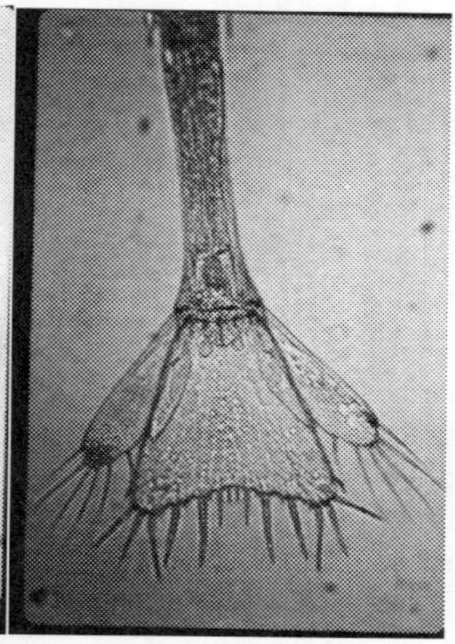

Figure 2.3

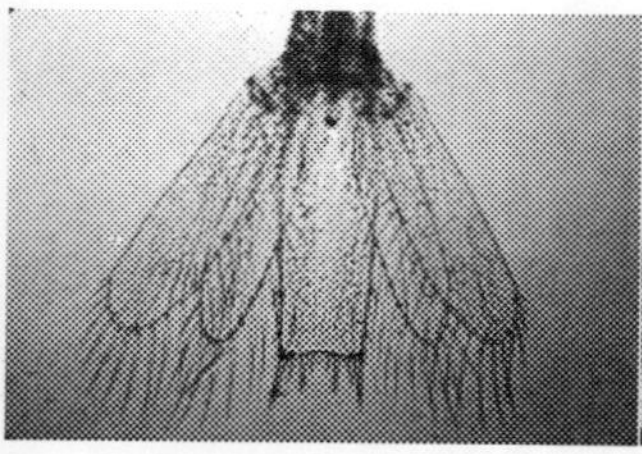

Figure 2.4

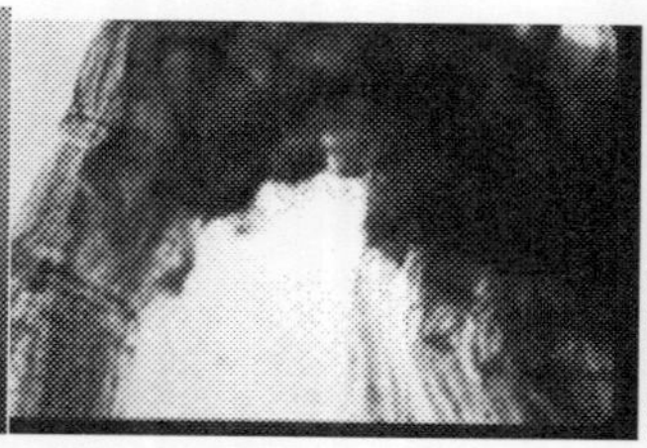

Figure 2.5

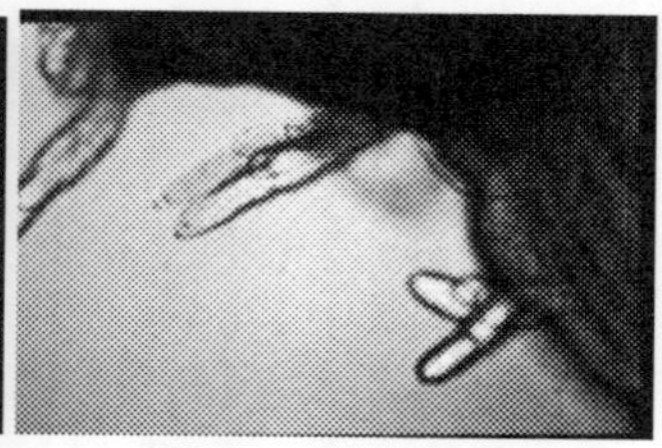

Figure 2.6

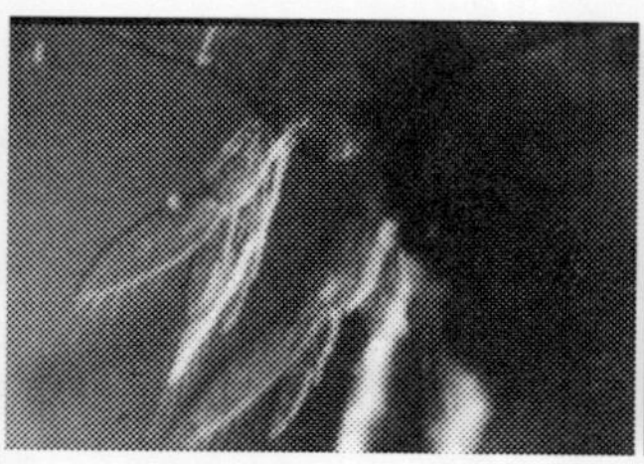

Figure 2.7

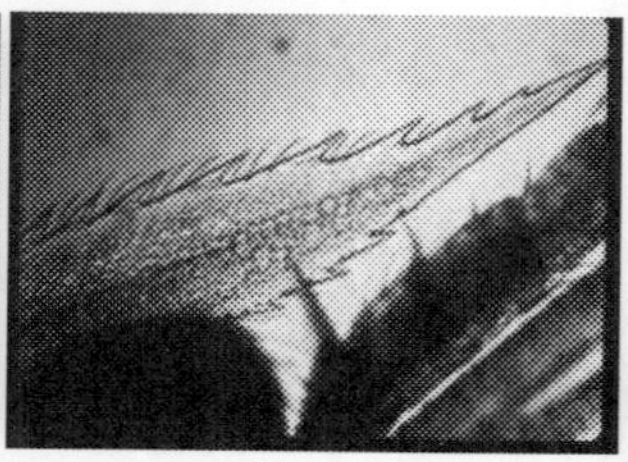

Figure 2.8

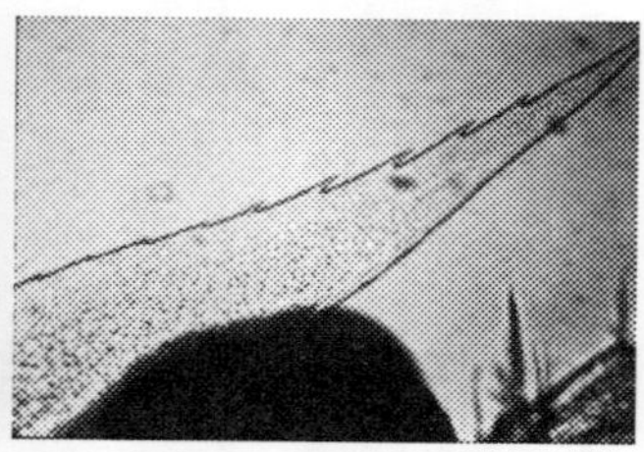

Figure 2.9

Source: Takuji Fujimura.

Freshwater Larval Development

Though the prawn inhabits rivers and estuaries, the adults normally prefer freshwater environments. Thus in the wild, mating and incubation take place in freshwater. The berried females may migrate from freshwater to brackish water regions, where the eggs hatch; or, alternatively, the hatched larvae flow down the river to the coastal zone. The larvae complete their development in the estuarine environment. The larvae swim in the water column of estuaries and coastal lagoons as part of the zooplankton community.

The newly hatched larvae require brackish water within 1–2 days, or they will die. They swim actively, upside-down and tail-first with their eyes looking upwards towards the surface. They are attracted to light, and will aggregate together in dense groups at the brightest spots of a tank if the water is still. The larvae eat continuously. In nature their diet is primarily zooplankton and larvae of other aquatic invertebrates. As the larvae moult, they not only increase in size but also increase in complexity, with new body features appearing at each stage.

There are 11 distinct larval stages and it takes about 22–35 days for a larva to complete these 11 stages, to become a post larva (PL). The change from the larva form to the PL form is called metamorphosis. In captivity, all the larvae develop at the same pace up to stage IV (synchronous larval development), then the timing of moulting and appearance of the developmental stages differs between individuals until the PL stage.

Chapter 3
Identification of Post Larvae Stages

Many research workers have tried to identify post larvae of shrimp and prawn. Some have given steps to find out up to species level. Some workers have given identification up to class. Similarly some workers have classified how to identify post larvae from the same group. Author have tried to collect all different identification and given identification of post larvae of each species along with drawings of each characteristic.

We know that for prawn culture pure seed is necessary. It is likely that at the time of supply we may get mix species of post larvae (PL). It is therefore comparison of penaeus, metapenaeus and parapenaeopsis PL is also given in this chapter. So as to differentiate the mix species supply if any.

Early Post Larvae (Up to PL-2)

1a. 5th abdominal segment with posterolateral spine .... *Penaeus*

1b. 5th abdominal segment without posterolateral spine 2

2a. Telson with 7 pairs of spine .... *Metapenaeus*

2b. Telson with 8 pairs of spines on either side of a mediam spine .... *Parapenaeopsis*

Source: Muthu, Pillai and George (1978)

Key to the Identification of Early Post Larvae Stages Found in the Brackish Waters

1a. Telson bears 8 pairs of spines on side and distal margin 2

1b. Telson bears 7 pairs of spines on sides and distal margin 6

2a. Telson with chromatophores in distal half only .... *Penaeus indicus*

2b. Telson with chromatophores from base to distal end 3

3a. Outer remi of uropods without chromatophores 4

3b. Outer and inner remi of uropods with chromatophores 5

4a. Inner rami of uropods with a single chromatophore near base on medial aspect (usually hidden by telson) .... *Penaeus japonicus*

4b. Inner rami of uropods with 3–4 chromatophores in the middle region on medial aspect .... *Penaeus merguiensis*

5a. Distal half of inner and outer rami of uropods with numerous chromatophores .... *Penaeus semisulcatus*

5b. Inner and outer rami of uropods with a row of chromatophores along the medial margin .... *Penaeus monodon*

6a. Chromatophores present on outer and inner rami of uropods 7

6b. Chromatophores present on inner rami of uropods, outer rami colourless .... *Metapenaeus monoceros*

7a. A single chromatophore at tip of each uropod ramus .... *Metapenaeus affinis.*

7b. A prominent chromatophore in the middle of each uropod ramus .... *Metapenaeus dobsoni*

Source: Muthu M. S. 1978.

Identification of Juveniles of *Metapenaeus*

1a. Rostrum long, without basal crest, extending middle of eye in younger individuals and beyond eye in older forms; abdominal carination commences from 3rd segment; ground colour of body bush brown to brownish; chromatophores closely spaced. Forming short band on the lateral region of carapace 2

1b. Rostrum short, with basal crest, extending to base of eye in younger individuals and tip of eye in older forms; abdominal carination commences from 4th segment; ground colour of body cream white; chromatophores widely spaced, not forming bands .... *Metapenaeus dobsoni*

2a. Rostrum reaching middle of eye or little beyond it; the short band on carapace less conspicuous; 'M'–like patch on mid–dorsal region of 4th abdominal tergum absent .... *Metapenaeus affinis*

2b. Rostrum reaching tip of eye or beyond it; the short band on carapace conspicuous; a distinct 'M' like patch on mid–dorsal region of 4th abdominal tergum present .... *Metapenaeus monoceros*

2c. Rostrum straight with small crest; mid–dorsal carination of the abdominal segment commences from 4th segment; ischial spine present on 1st periopod. Body partly covered with harsh and very short tergum .... *Metapenaeus kutchensis*

Source: Dholakia, 1986.

Key for the Identification of First Post Larvae of Five Commercial Prawns of India

1a. Telson with 7 + 7 spines 2

1b. Telson with more than 7 + 7 spine 4

2a. Rostrum with 4 spines (2 large and 2 small); long setae on distal lateral aspect of 6th abdominal segment 3

2b. Rostrum with 2 spines; no setae on distal lateral aspect of 6th abdominal segment .... *Metapenaeus dobsoni*

3a. Epigasrtic spine on carapace absent; larvae brownish .... *Metapenaeus monoceros*

3b. Epigastric spine on carapace present; larvae not brownish .... *Metapenaeus affinis*

4a. Telson truncate, with 8 + 8 spines, median posterior spines equal; median dorsal spine present on 5th and 6th abdominal segment. .... *Penaeus indicus*

4b. Telson tapering between postero lateral spines, with 8 + 1 + 8 spines, median posterior spine very much larger than adjacent spine; median dorsal spine on abdominal segment absent. ...... *Parapenaeopsis stylifera.*

Identification on Base of Carapace Length

Juveniles Upto 5 mm of Carapace Length of *Metapenaeus affinis*

Details of Characters

1. Rostrum very short and pointed.
2. One–chromatophores present on each inner and outer uropod ramus at distal end.
3. On 6th abdominal segment one anterolateral chromatophore is present and on ventral side two chromatophores are present.
4. Chromatophores present on telson from base to distal end.

5. Telson bears 7 pairs of spines on side and distal margin.
6. Colour of chromatophores in live specimen is brown.
7. 1 to 2 spinules on dorsal carina on 6th abdominal segment may be present near posterior end.
8. Antennal spine on carapace present.
9. 'M' like patch on mid–dorsal region of 4th abdominal tergum absent.

Source: Dholakia, A. D. 2004.

Juveniles Up to 5 mm of Carapace Length of *Metapenaeus dobsoni*

Details of Characters

1. Rostrum long with adrostral crest reaching as far as epigastric tooth.
2. Two prominent chromatophores present in the middle on telson
3. One prominent chromatophore present in the middle of each inner and outer uropod ramus.
4. Tip of the rostrum is actually blunt.
5. Body colour is pale yellow with red brownish or greenish specks.
6. Swims with jerky movement.
7. No setae on distal lateral aspect of 6th abdominal segment.
8. Median dorsal spine on 5th abdominal segment absent.
9. 2 or 3 spinules may be present near posterior end on dorsal carina of 6th abdominal segment.
10. Telson bears 7 pairs of spine on side and distal margin.
11. Abdominal carination commences from 4th segment.
12. Ground colour of body cream white.
13. Chromatophores on body widely spread, not forming bands.

Source: Dholakia, A. D., 2004.

Juveniles Upto 5 mm of Carapace Length of *Metapenaeus monoceros*

Details of Characters

1. Rostrum extending to 3/4 eye.
2. 2–3 chromatophores on inner uropod ramus present and absent non outer uropod ramus.
3. Chromatophores on 6th abdominal segment present and 4th and 5th on ventral side only chromatophores are absent.
4. Tip of the rostrum is pointed and serrated on its upper margin.
5. Rostrum with 4 spine (2 large and 2 small)
6. Epigastric spine on carapace absent in post larvae.
7. Long setae on distal lateral aspect of 6th abdominal segment.

8. Swims in a straight line direction without any jerky movements.
9. 'M'–like patch on mid–dorsal region of 4th abdominal tergum present.
10. The short band on carapace conspicuous.

Species-wise Identification of First Post Larvae

Post Larvae of *Penaeus monodon* Fabricius

MCL: 1.46 mm (1.40–1.54 mm), MTL: 4.56 mm (4.45–4.70).

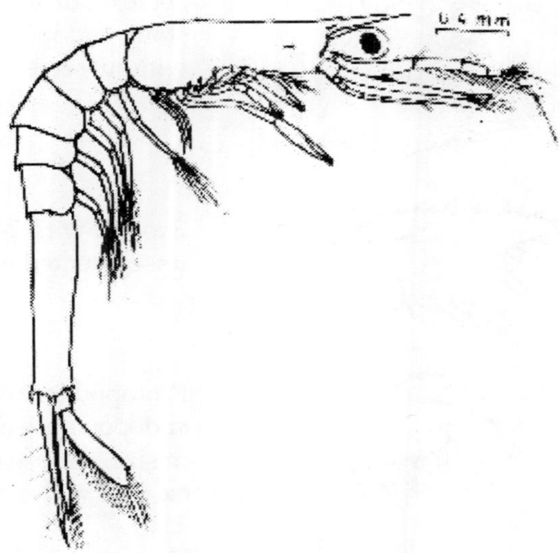

Characteristic	*Figure*	*Details*
Carapace		Rostrum with 1 distinct tooth, supraorbital spine reduced in size, no change in spination of carapace and abdomen, pleopods setose exopods on Mxp2.
Antennual–I	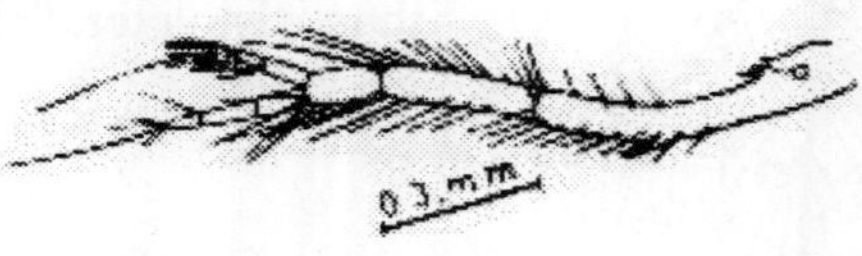	3 segmented inner flagellum twice as long as segmented outer flagellum.
Antennual–II	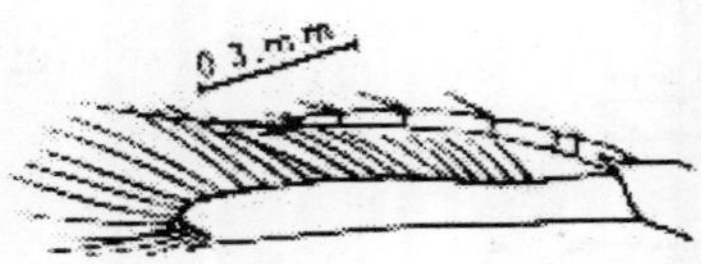	Endopod 6 segmented, exopod with 25 plumose seta and prominent distolateral spine.

Characteristic	*Figure*	*Details*
Mandible	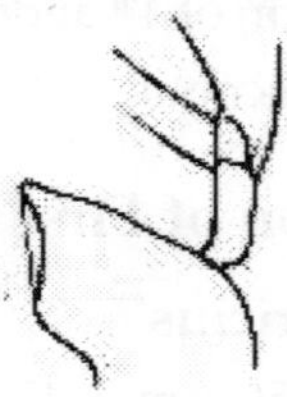	Cutting edge developed, paip 2 segmented, short distal segment with 2 and longer proximal segment with 4 setae.
Maxilla-I	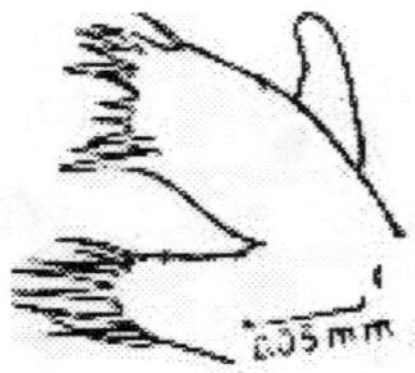	Setae on distal endite short and stout, endopod reduced in size, without setae.
Maxillla–II	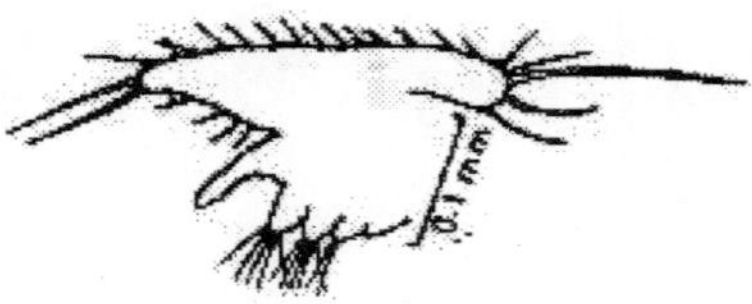	Exopod with 24 setae endopod unsegmented reduced in size.
Maxiliped–I	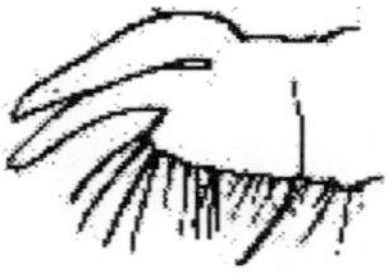	Protopod enlarged exopod and endopod unsegmented, reduced in size, endopod without setae, exopod with single outer seta.
Maxiliped–II		Recurved endopod with stout setae on distal segment, exopod shrunken without setae.
Maxiliped–III	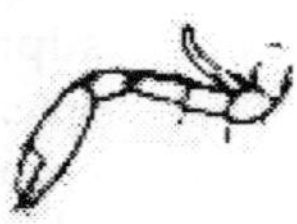	Shrunken without setae.
Uropod	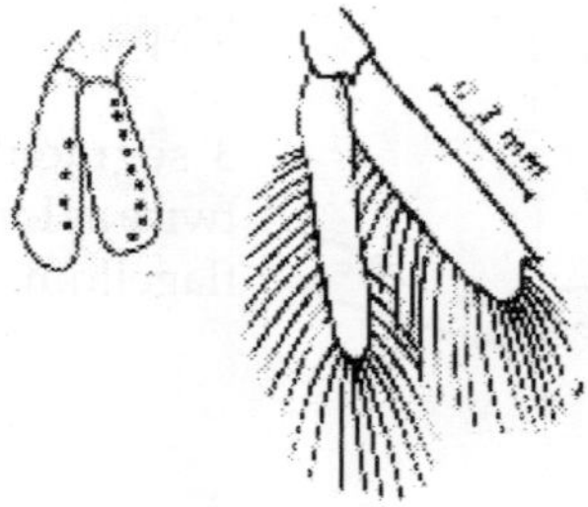	Endopodwith 25 plumose setae, exopod with 22 plumose setae and 1 non plumose seta shorter than distolateral spine.
Telson	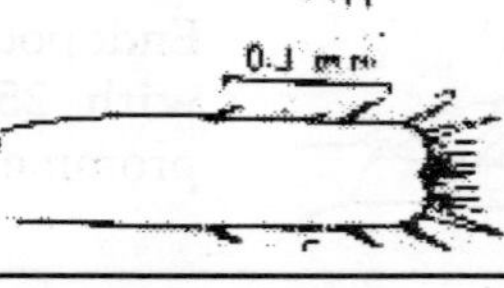	Sallow cleft in telson still present.

Source: Silas E. G. *et al.*, 1978, Dholakia, A. D. 2004.

Penaeus indicus H. Milne Edwards

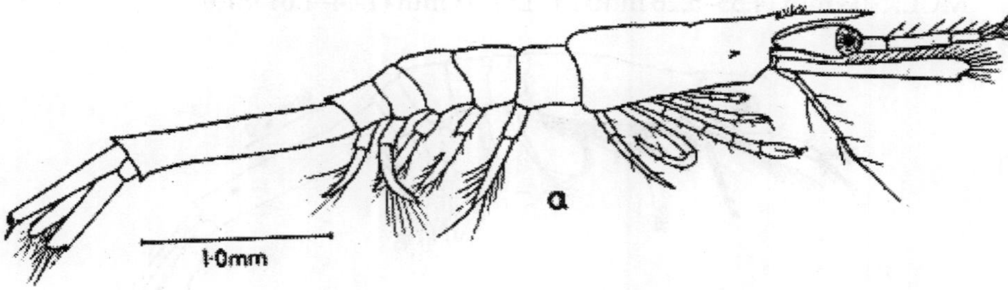

Figure 3.1: First PL of *Penaeus indicus*

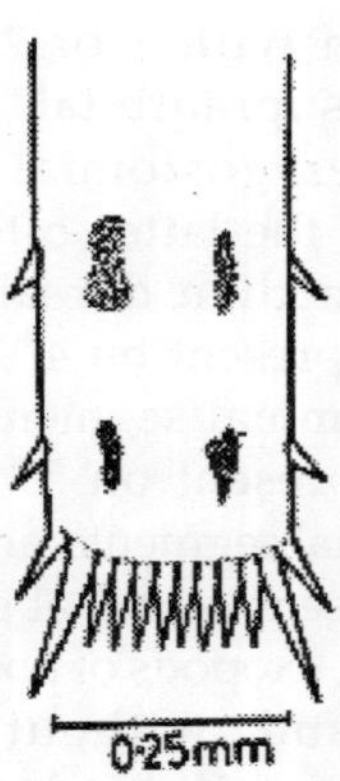

Figure 3.2: Telson

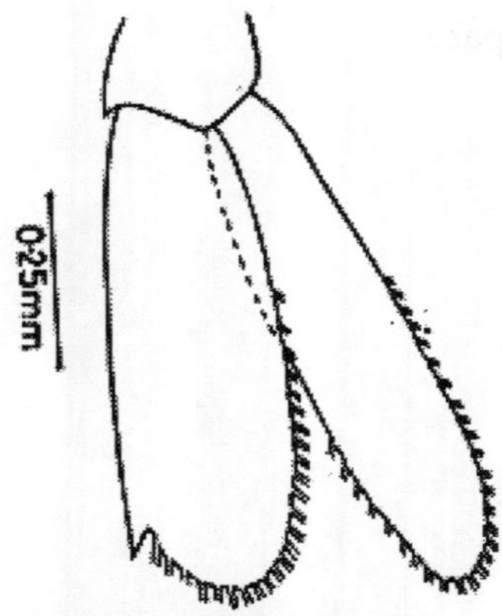

Figure 3.3: Uropod

Although different stages of the post larvae were encountered in all the collections, the earliest stage was obtained only from the offshore collections. They are relatively few in number and their sizes range from 5.22 to 5.95 mm in total length and 1.28 to 1.35 mm in carapace length.

Carapace

Rostrum bears a single tooth and slightly exceeds the eye. A small supraorbital spine present, but the anterolateral angle of the carapace is devoid of any spine. Hepatic spine well developed.

Antennule

Peduncle 3-segmented, with well developed statocyst at the base. The proximal segment longest; the inner branch of the distal segment 3-jointed, while the outer is only 2-jointed.

Post Larvae of *Penaeus indicus* H. milne Edwards

MCL: 5.03 mm (4.55–5.26 mm) MTL: 1.53 mm (1.44–1.61 mm).

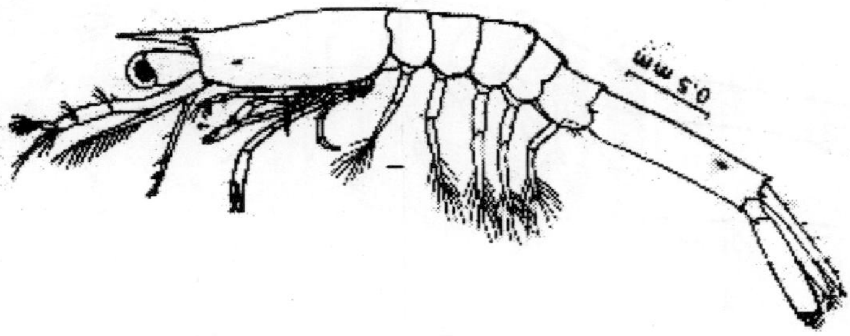

Characteristic	*Figure*	*Details*
Carapace		Rostrum with 1 or 2 dorsal spines, supraorbital, hepatic and pterygostomial spines present, the latter often very small, median dorsal spines usually present on 4th, 5th and 6th abdominal segment lateral spines present on 5th and 6th abdominal segments, anal spine still present on 6th abdominal segment, exopods of post larvae small and without setae, pleopods well developed and setose.
Antennual–I	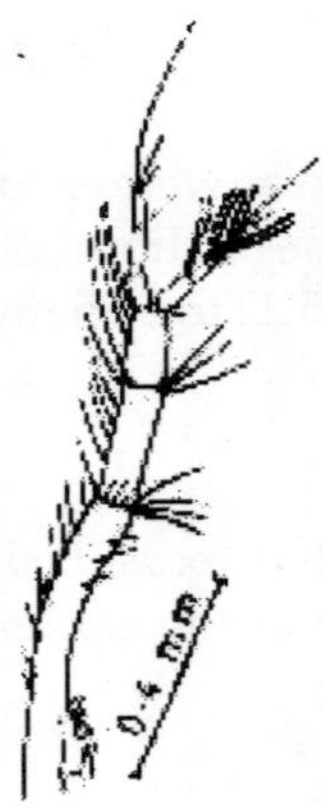	With statocyst at base of 1st segment, well developed ventro-median spine still present on basal segment; inner branch of distal segment 3 segmented, longer than outer and carries 4 setae apically, of which 1 is as long as the branch, outer branch 2 segmented carrying 8 aesthaetes and 3 setae.
Antennul–II		With endopod 6 segmented, distal segment apically bearing 3 long and 3 short setae; exopod with 27 setae and one anterolateral spine.

Characteristic	*Figure*	*Details*
Mandible	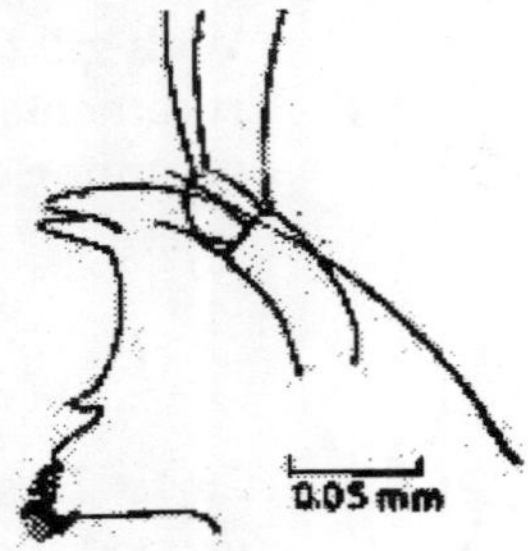	It has become almost symmetrical, free standing teeth lost, palp well developed and 2 segmented, carring setae.
Maxilla–I		With edopod much reduced, unsegmented and without setae; distal lobe of protopod larger than proximal, distal and proximal lobes with 13 to 18 and 7 to 8 setae respectively.
Maxillla–II		With much reduced protopod having 4 endites, proximal 2 endites with 2 setae, distal 2 endites vary 5 to 6 bristal like setae, endopod reduced unsegmented, without setae, scaphognathite vary conspicuous bearing 29 to 30 plumose setae.
Maxiliped–I		With endopod and exopod reduced in size without segments and setae, protopod has become wide with numerous setae, epipod well developed.
Maxiliped–II	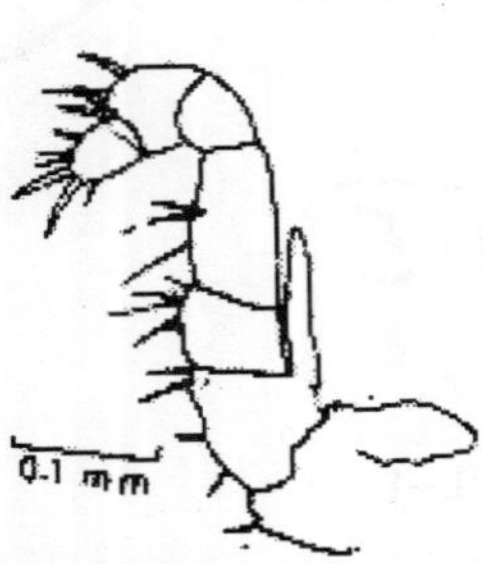	With exopod vestigial, endopod recurved, distal segment with 6 spine–like setae penultimate segment naked, protopod carries a gill.

Characteristic	*Figure*	*Details*
Maxiliped–III	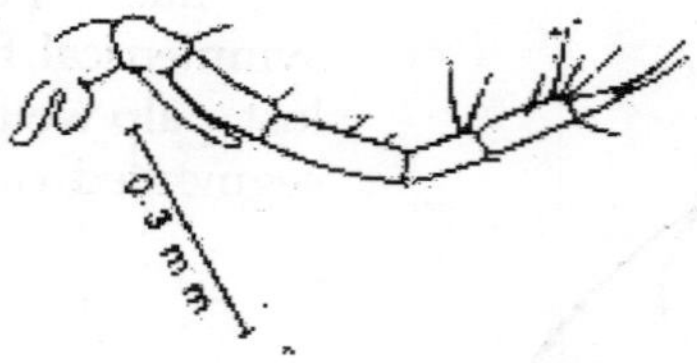	With gill on protopod, exopod rudimentary without setae, 4th segment of endopod with 5 setae.
Uropod	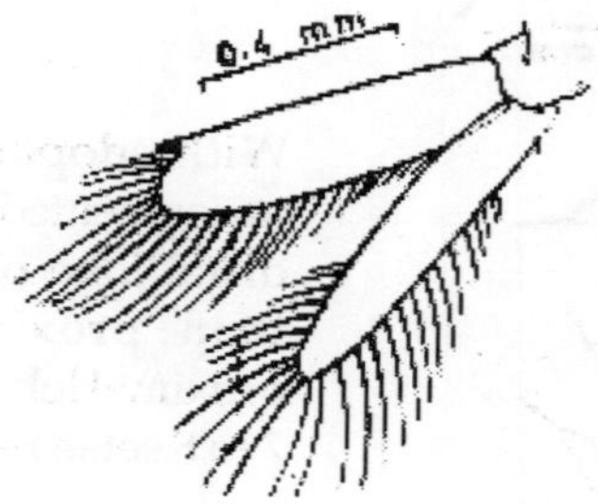	26 to 27 setae on margin of exopod and endopod of uropod.
Telson		Rectangular in shape varying 3 pairs of lateral and 5 pairs of terminal setae median notch practically absent.

Source: Muthu, M. S. *et al.*, 1978, Dholakia, A. D. 1994.

Post Larvae of *Penaeus merguiensis* de Man

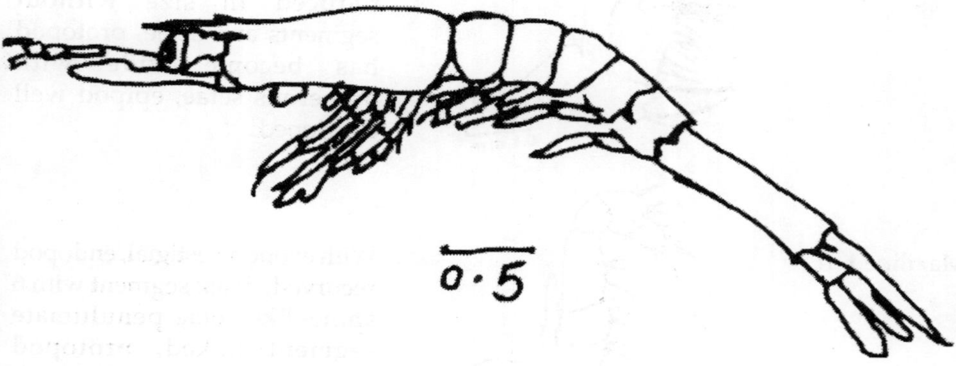

PL–1

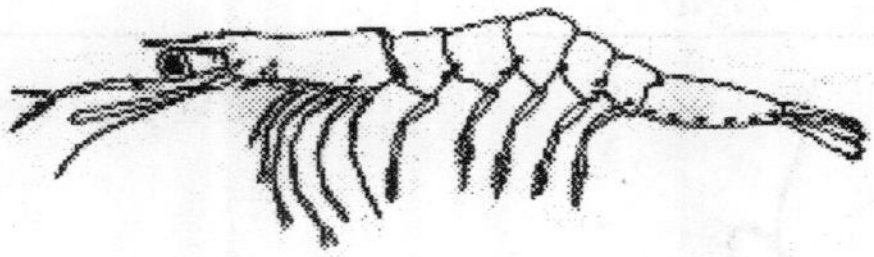

PL-20

Figure 3.4: ***Penaeus merguiensis***

Characteristic	*Figure*	*Details*
Carapace		First 3 legs became functional and also became pleopods. The carapace loses the pteygostomial spine but the suraraorbital and hepatic spines were persisting.
Antennual–I	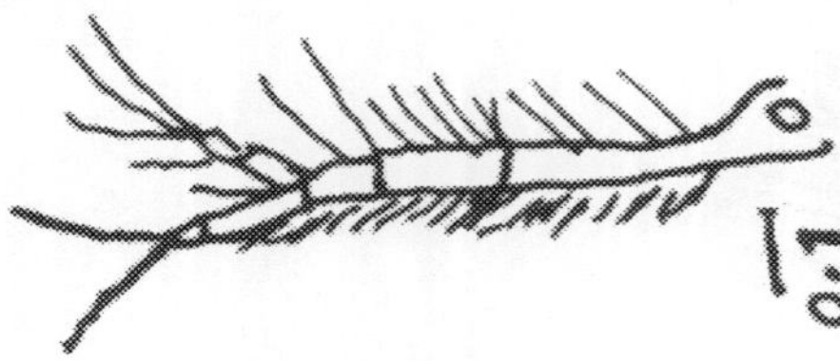	The 4th abdominal segment developed a small posterolateral spine.
Antennual–II	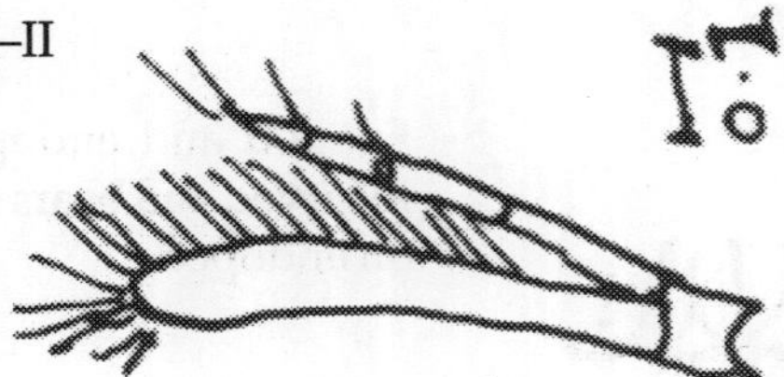	Exopod bears 24 setae.
Mandible	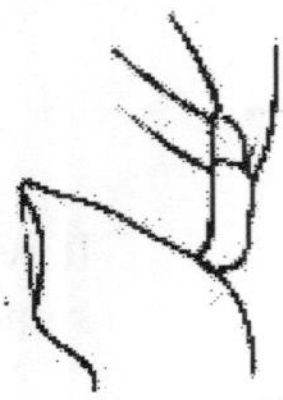	Palp of the mandible bears two setae on each segmentes.
Maxilla-I	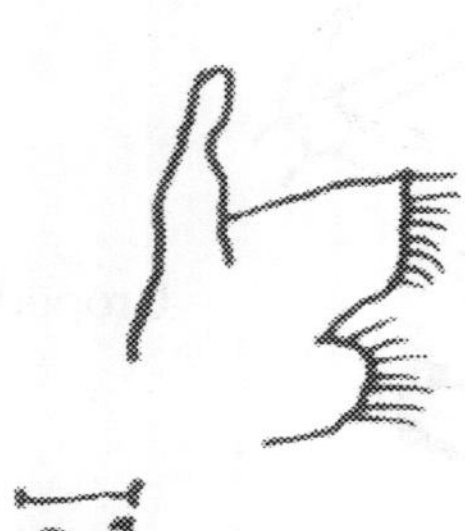	Endopod became vestigial and without setae.

Characteristic	*Figure*	*Details*
Maxillla–II	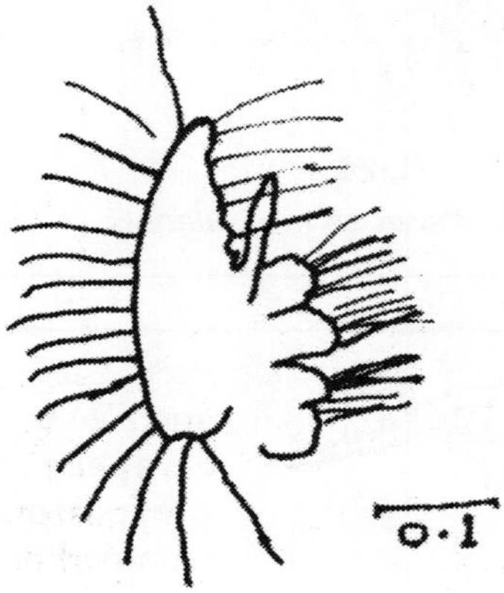	Exopod bears 22 setae.
Maxiliped–I	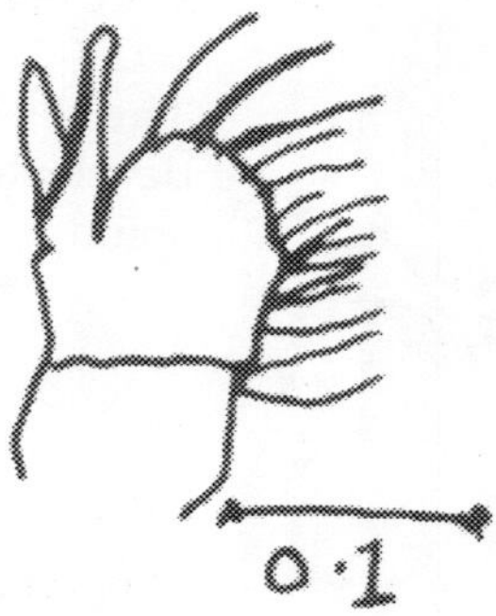	Exopod and endopod are unsegmented and without setae.
Maxiliped–II	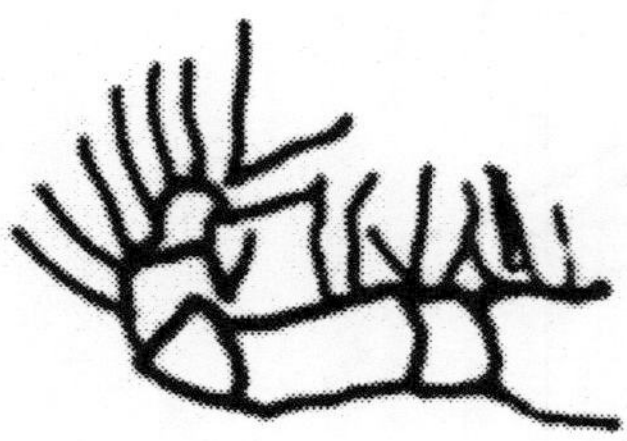	Exopod and endopod became recurved and bears several setae on endopod.
Maxiliped–III	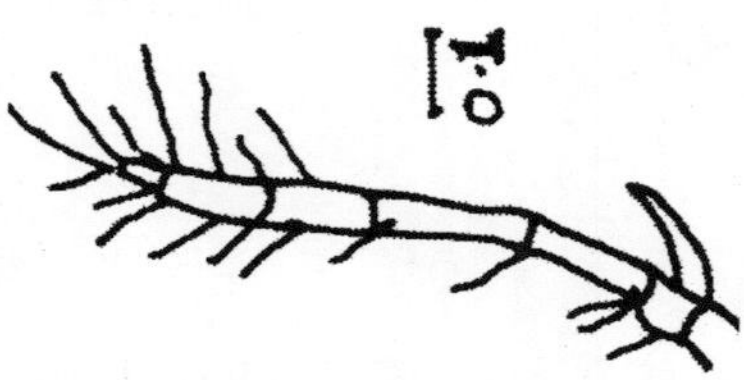	It was largest and bear several setae on exopod.
Uropod		Uropod bisegmented.

Characteristic	*Figure*	*Details*
Telson	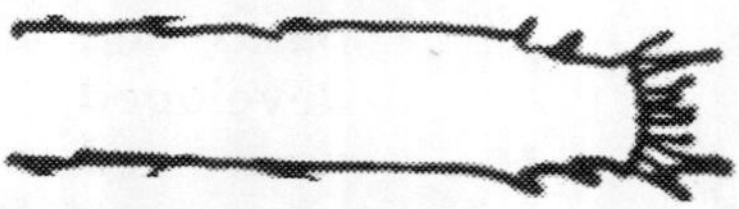	Telson with 8 + 8 spine formula.

Source: Dholakia, A. D. 1986.

Post Larvae of *Penaeus semisulcatus* DE HAAN

MCL: 1.50 mm (1.47–1.54 mm): MTL: 4.82 mm (4.76–4.87 mm).

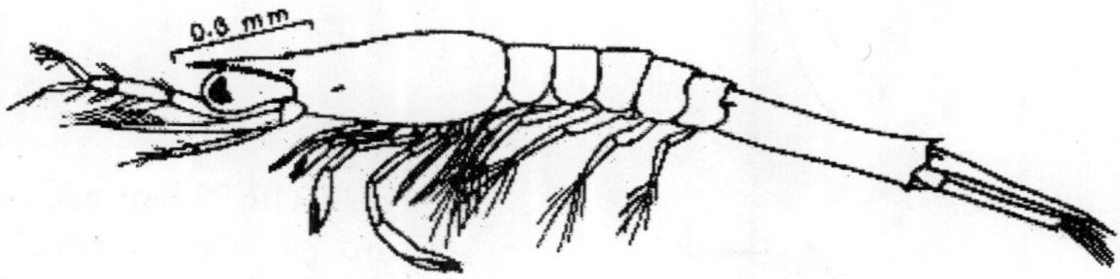

(PL-1)

(PL-20)

Characteristic	*Figure*	*Details*
Carapace	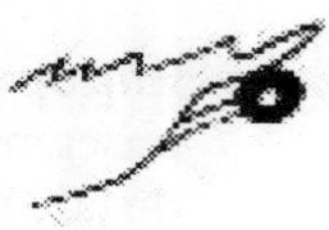	Rostrum longer than eye, bearing 1 dorsal spine, supraorbital, pterygostomial and hepatic spines present; 4th 5th and 6th abdominal segment with posteromedian dorsal spine, lateral spine present on 5th and 6th abdominal segments, anal spine persist; exopods of pareopod rudimentary pleopods, fully developed bearing 3 to 4 pairs of plumose setae.

Characteristic	*Figure*	*Details*
Antennual–I	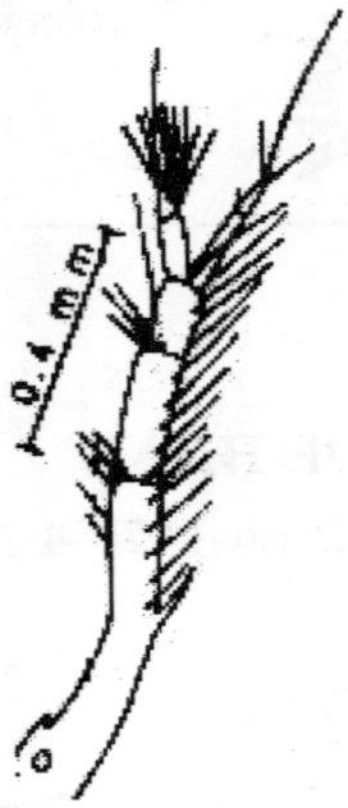	Basal segment with well developed statocyst, ventromedian spine persists, outer flagellum 2 segmented carrying 8 aesthaetes in 3 groups of 4 + 2 + 2, apically 2 slender setae present, inner flagellum longer than outer, 3 segmented with 3 apical setae.
Antennual–II		Scale with 22 long plumose setae and 1 distolateral spine, endopod 5 segmented almost same length as exopod.
Mandible	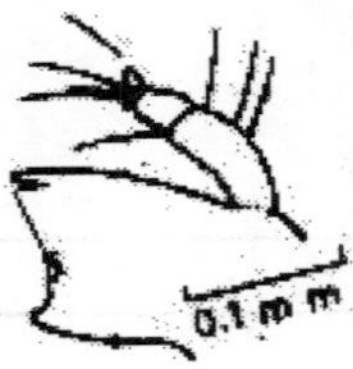	Symmetrical, standing teeth lacking palp 2 segmented, proximal segment with 3 setae on outer margin, 1 seta on distal inner border and distal segment with 3 apical setae.
Maxilla-I		Endopod rudimentary unsegmented and without setae, distal lobe longer than proximal and bears13 setae, proximal lobe with 7 to 8 setae.
Maxillla–II	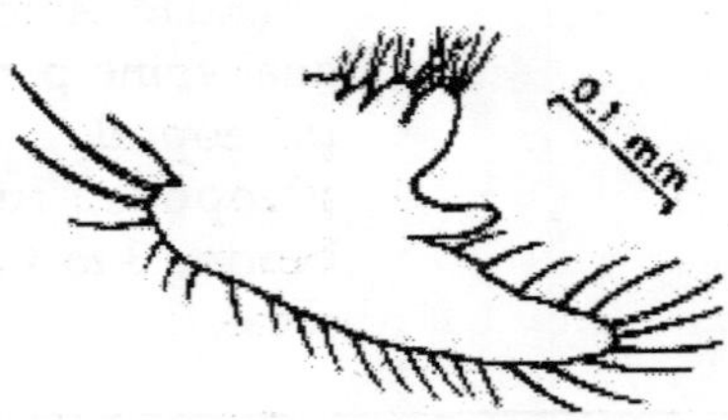	4 endites much reduced in size, proximal 2 endite each bearing 2 setae, distal 2 endites with 4 to 6 setae, endopod fairly large bears 28 plumose setae.

Characteristic	*Figure*	*Details*
Maxiliped–I		Endopod and exopod unsegmented, protopod broad carrying numerous setae.
Maxiliped–II		Acquires typical adult shape, exopod vestigial, endopod recurved, distal segment with 5 stout setae, 4th segment with 3 long setae, first 2 segments with 4 and 3 setae respectively.
Maxiliped–III	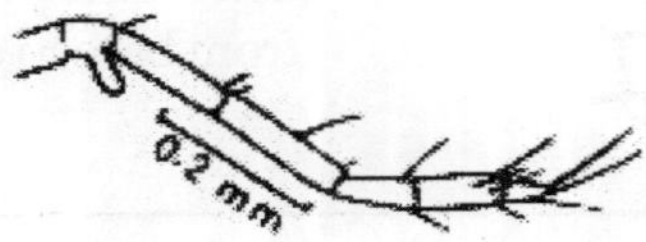	Protopod with gills, exopod vestigial and endopod 5 segmented, distal 2 segments with 3 to 4 long setae.
Uropod	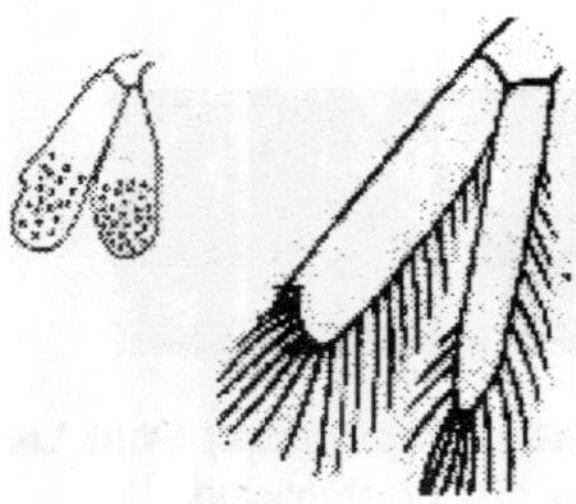	Exopod of uropod with 21 plumose setae and 2 short nonplumose setae, endopod with 20 plumose setae.
Telson	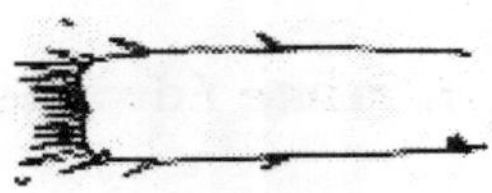	Telson with 5 pairs of distal and 3 pairs of lateral setae.

Source: Devrajan, K. *et al.*, 1978, Dholakia, A. D. 2004.

Juveniles of *Penaeus japonicus*

MTL 3.85 mm (3.68–4.09 mm); MCL 1.13 mm (1.12–1.16 mm).

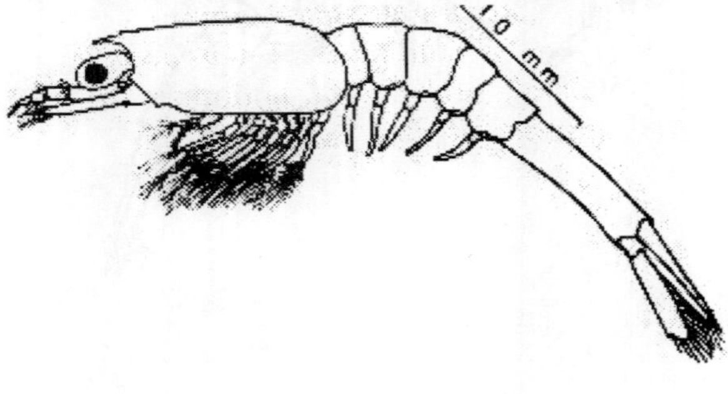

Characteristic	*Figure*	*Details*
Carapace		Short, reaching only half length of eye.
Sixth abdominal segment		10 to 11 ventral chromatophores are present on 6th abdominal segment. While no chromatophores is found on anterolateral segment.
Uropod		One Chromatophore is present on inner uropod ramus near base on median aspect.
Telson	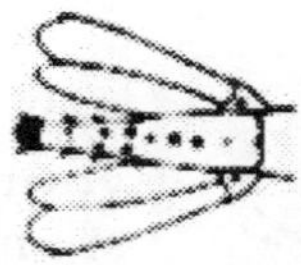	Chromatophores are present from base to distal end.

Source: Dholakia, A. D. 2004.

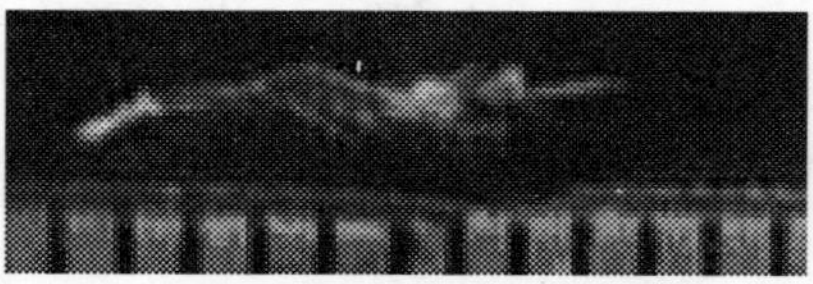

Figure 3.5: Kuruma Prawn (*Penaeus japonicus*) Post Larvae As "PL10" (10 Days old postlarvae)

Post Larvae of *Metapenaeus affinis* H. milne Edwards

TL 3.85–3.95 mm; CL 1.16–1.17 mm.

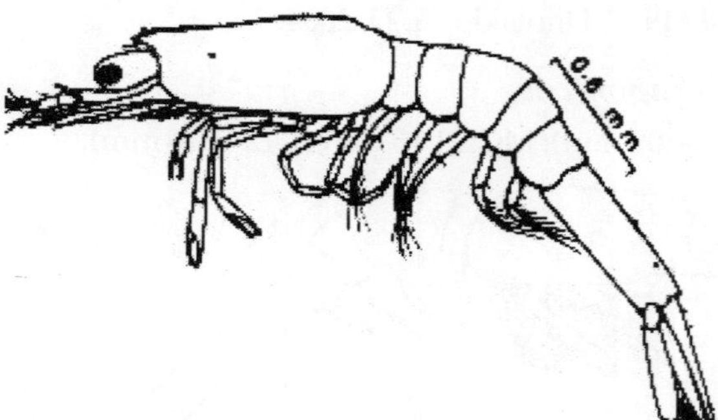

Characteristic	*Figure*	*Details*
Carapace		Carapace with 2 longer and 2 smaller spines and minute epigastric; Carapace with well developed antennal and hepatic spine; mouth parts do not show appreciable, change from those of mysis-III Rostrum short sharply pointed with 3 dorsal spines.
Antennual–I	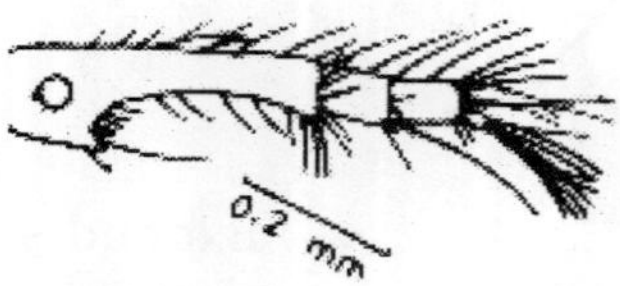	Inner flagellum longer than outer and faintly divided into 2, outer flagellum 2 segmented carrying 8 aesthaetes.
Antennual–II	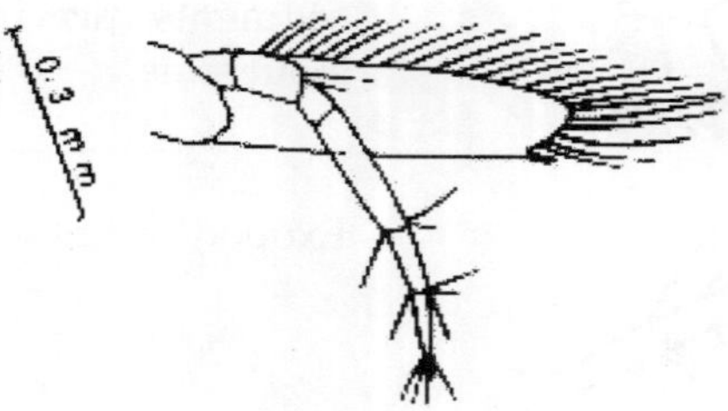	Endopod 6 segmented exopod with 23 to 24 plumose setae.
Mandible	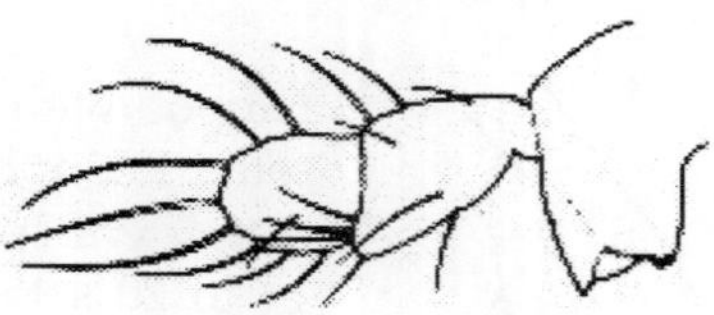	Palp bigger than mandible, number of setae increased.
Maxilla-I		Endopod small without segments, bearing terminally on inner side one seta, distal endite of protopod more flattened bearing 13 setae of which 1 is plumose.
Maxillla–II	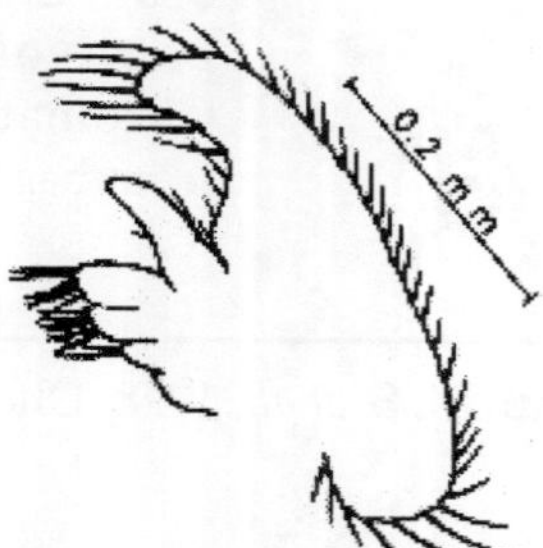	Exopod with more than 40 setae endopod reduced without segmentation, protopod with 4 endites, distal 2 endites with 6 setae.

Characteristic	*Figure*	*Details*
Maxiliped–I	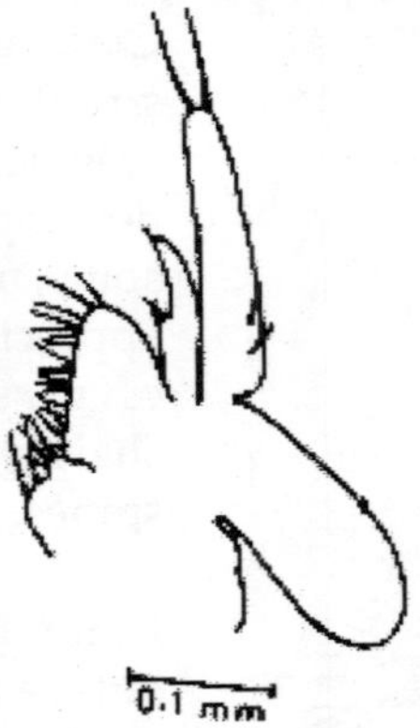	Exopod with 2 short plumose setae distally and proximally endopod reduced, without segmentation, with 2 short setae on inner side, gill large, protopod broader with a number of setae on inner margin.
Maxiliped–II	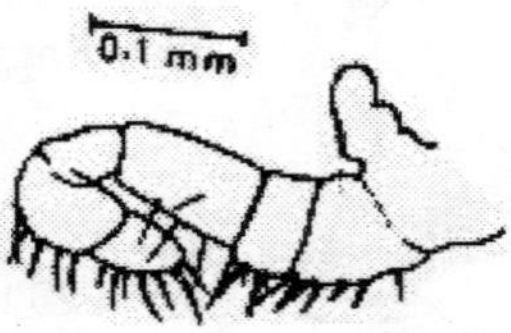	Exopod shrunken, endopod sharply recurved, distal 2 segments carrying number of stout setae.
Maxiliped–III	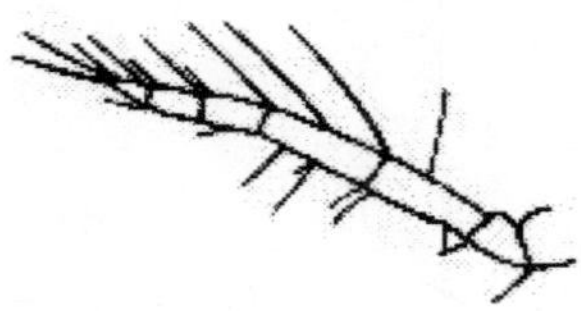	Exopod absent.
Uropod	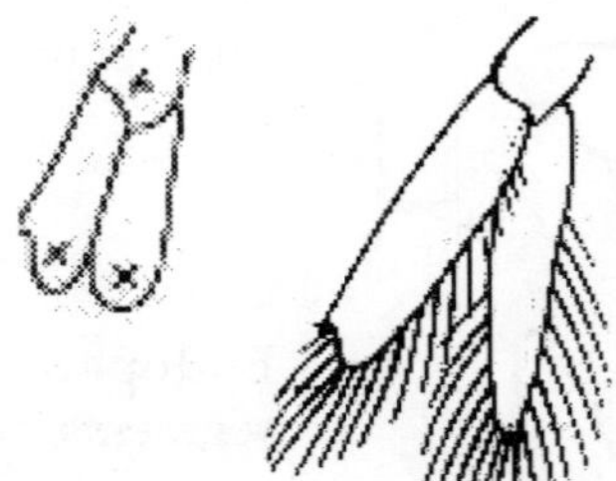	Exopod with 17 to 19 plumose setae, 1 nonplumose setae and 1 distolateral spine, endopod with 19 to 20 setae.One chromatophores is present on inner uropod ramus at distal end.
Telson	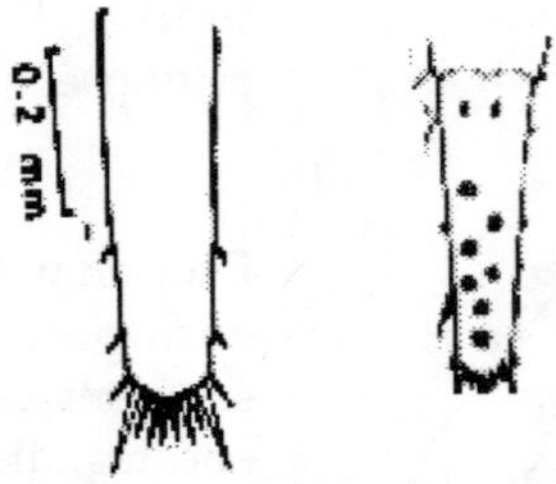	Telson convex posteriorly. It bears 2 pairs of lateral and 5 pairs of posterior spines. Chromatophores are present from base to distal end.

Source: P Vedvyas Rao 1973 and Muthu, M. S. *et al.*, 1978, Dholakia, A. D. 2004.

Metapenaeus dobsoni (Miers)

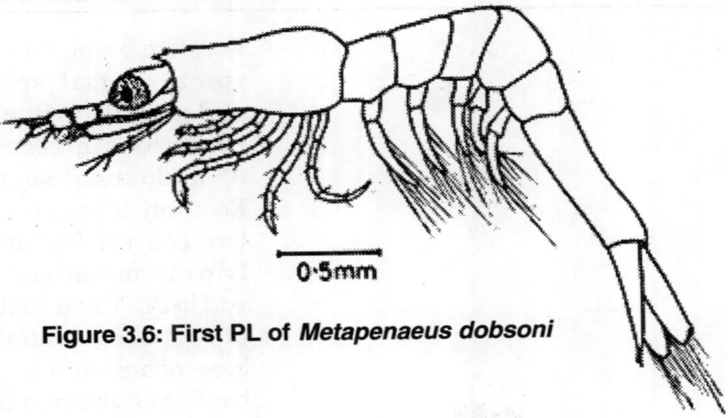

Figure 3.6: First PL of *Metapenaeus dobsoni*

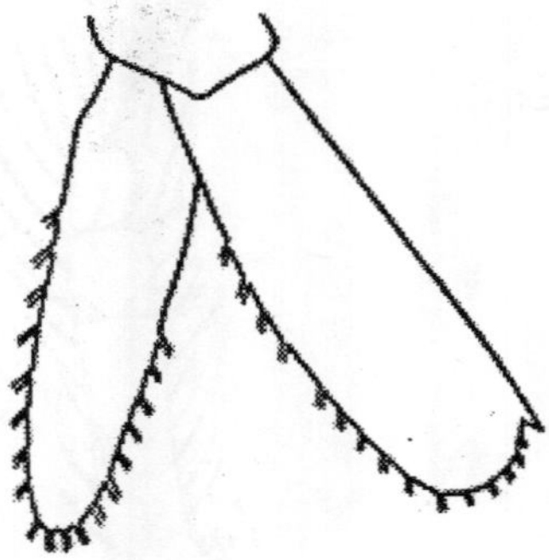

0.25 mm

Figure 3.7: Telson **Figure 3.8: Uropod**

Colouration

Tips of 2nd and 3rd segments of antennular peduncle slightly brownish. Eye stalks yellowish. The abdominal segments appear light yellowish and each segment is provided with a branched dark chromatophore. There is also a patch of this chromatophore in the middle region of the telson.

Post Larvae of *Metapenaeus dobsoni* (Miers)

TL- 3.05–3.15 mm; CL–0.75–0.80 mm.

Characteristic	*Figure*	*Details*
Carapace		Rostrum blunt with 2–3 dorsal spines dorsal spine on 5th abdominal segment and posteromedian ventral spine on 6th abdominal segment lost. Rostrum not extending beyond base of eye in PL-I and II. In PL-IV rostrum reaching one fourth and in PL- V and VI three fourth of eye; dorsoventral width of base of rostrum conspicuous, basal crest differentiating from PL- V onwards.
Antennual–I	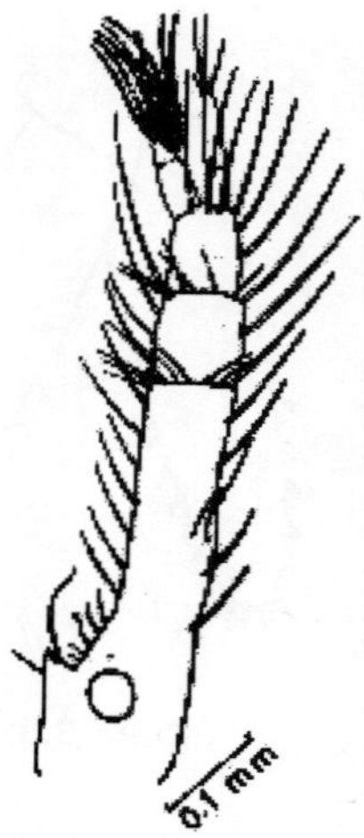	Both inner and outer flagella 2 segmented.
Antennul–II	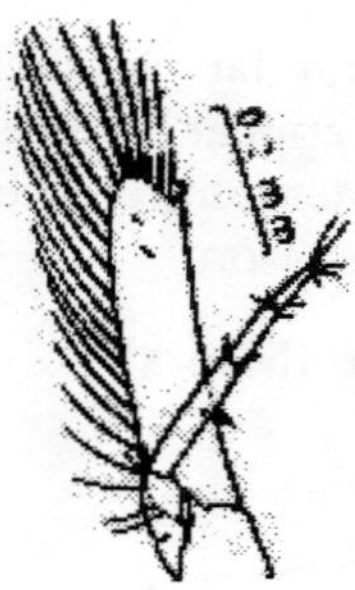	Is 5 segmented tipped with 7 short setae, bristal-like setae at junction of segments.
Mandible	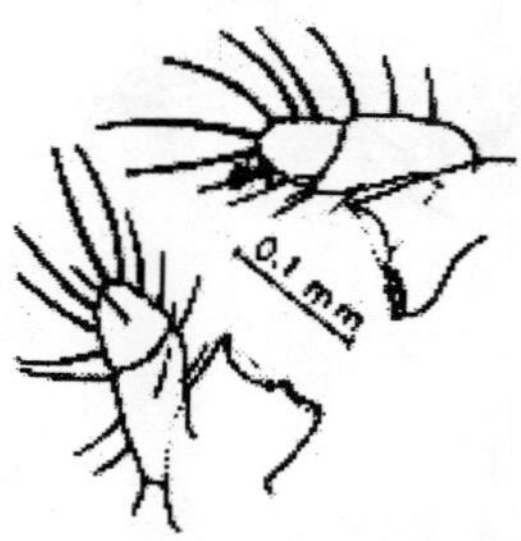	Palp clearly 2 segmented, standing teeth replaced by thin blade like cutting edge.

Characteristic	*Figure*	*Details*
Maxilla-I	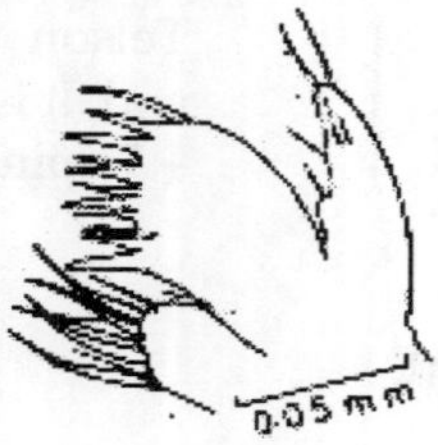	Endopod segmentation lost and setae hightly reduced in size, 12 setae on distal endite.
Maxillla–II		Endopod further reduced. Distal endite of protopod vestigial, exopod with 27 to 28 setae.
Maxiliped–I		Endopod segmentation lost, setae highly reduced. Exopod setae also reduced in size, protopod broader. Exopod still with long setae, gill rudiment long.
Maxiliped–II		Exopod without setae and reduced in size, endopod sharply recurved. Long plumose setae along inner margin replaced by stout setae, exopod still with long plumose setae.
Maxiliped–III		Setae on endopod reduced in length, exopod still with long plumose setae.
Uropod	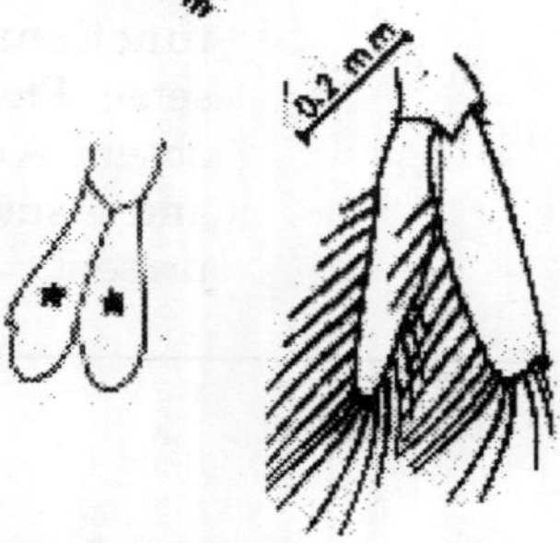	Endopod with 19 setae, no change in exopod. pleopods distal segment longer with 10 plumose setae basal segment broader with 2 distolateral short setae.

Characteristic	*Figure*	*Details*
Telson	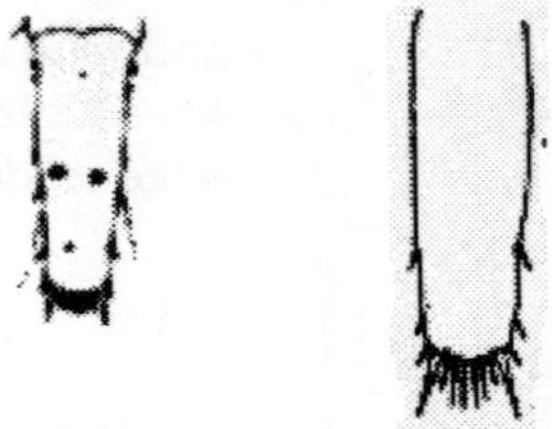	Telson with convex posterior end. It is almost truncate with 7 + 7 spines.

Source: Vedvyas Rao 1973 and Muthu, M. S. *et al.*, 1978, Dholakia, A.D. 2004.

Identification of First Post Larvae

Post Larvae of *Metapenaeus monoceros* (FABRICIOUS)

MTL: 3.02 MM (2.87–3.15 MM); MCL: 0.91 MM (0.91–0.92 MM).

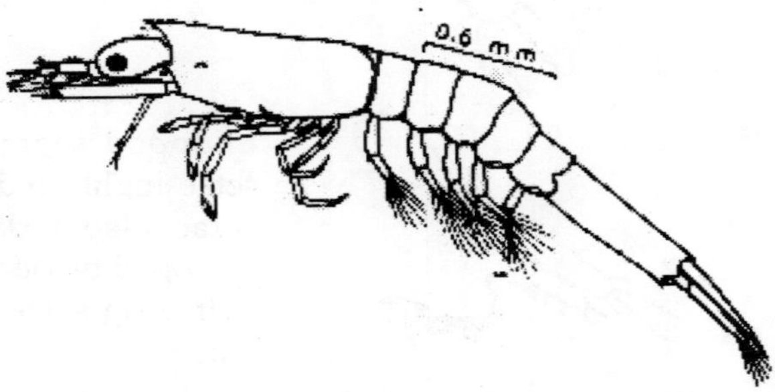

Characteristic	*Figure*	*Details*
Carapace		Rostrum less than half length of eye, with 3 dorsal teeth and with 2 principal and 2 smaller spines. The first has a smaller tooth in front and the second, larger tooth is followed by a much smaller spine, tip pointed. exopod of PL-1 to PL-5 reduced, without setae; pleopods functional bearing plumose setae. Pterygostomial spine is absent. A conspicuous antennal and a smaller hepatic spine is present.

Characteristic	*Figure*	*Details*
Antennual–I	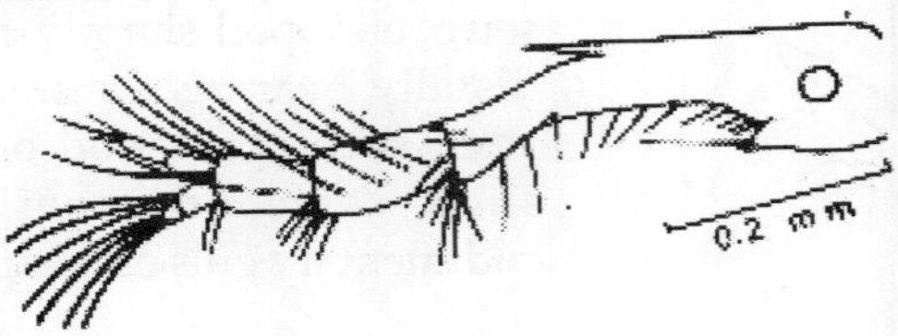	Inner flagellum longer than outer, 2 segmented bearing 3 setae apically, outer in 2 segmented bearing 7 to 8 aesthaetes, statocyst fully developed.
Antennual–II	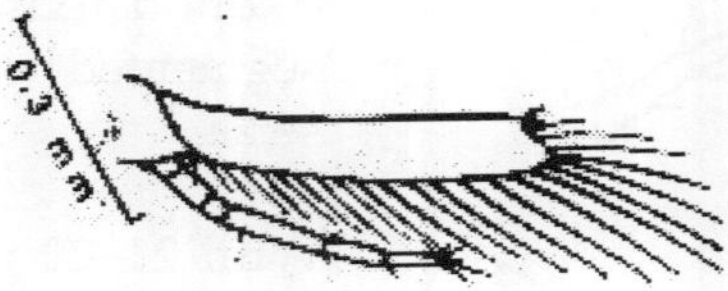	Exopod 5 to 6 segmented and distal segment with 6 short setae apically.
Mandible	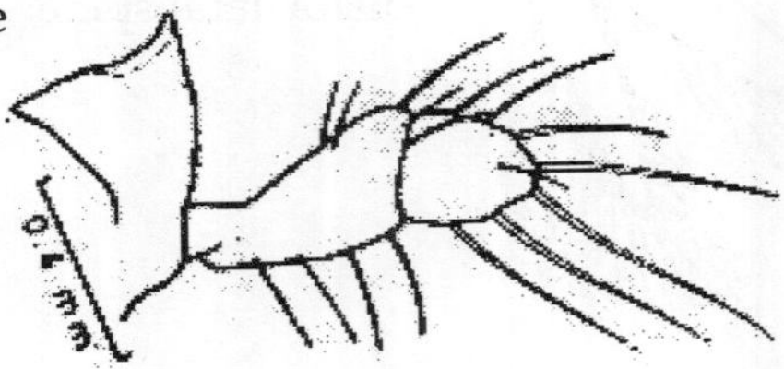	Palp big, flattened and 2 segmented carrying plumose setae, free standing teeth absent.
Maxilla-I	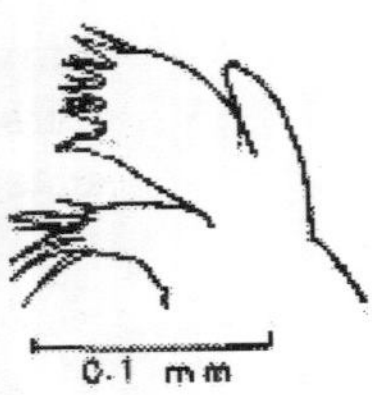	Endopod reduced, unsegmented and palp like, distal endites of protopod more flattened and bearing short stumpy setae proximal endite with long setae.
Maxillla–II	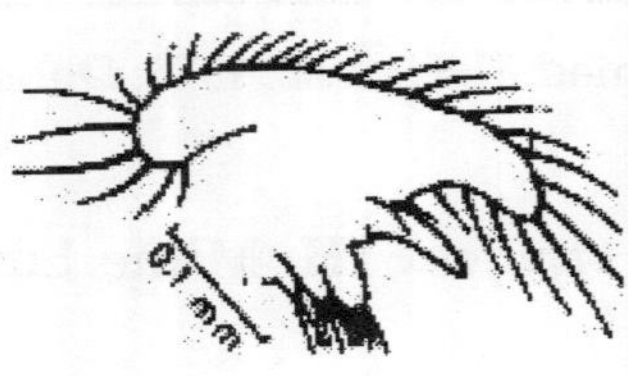	Exopod flattened leaf like bearing 33 to 40 plumose setae, endopod unsegmented, without setae, protopod with 3 endites bearing small bristal–like setae apically.
Maxiliped–I		Setae distally and 2 short setae on proximal outer margin, endopod unsegmented bearing 2 small setae, protopod flattened with 2 lobes carrying number of setae, gill well developed.

Characteristic	*Figure*	*Details*
Maxiliped–II	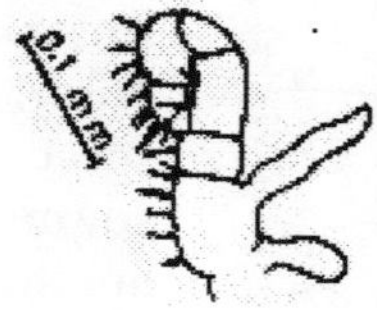	Exopod unsegmented without setae, endopod sharply curved distally bearing a number of stout setae, protopod bearing short setae on inner side, rudimentary gills developed.
Maxiliped–III		Exopod reduced, endopod 5 segmented bearing a number of setae.
Uropod	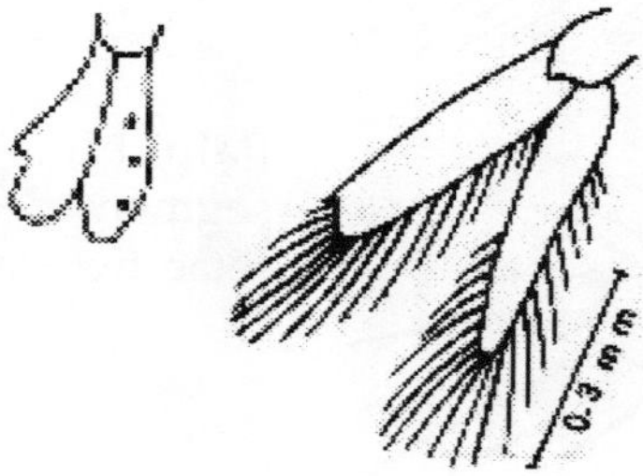	With 21–22 plumose setae, 2 short non-plumose setae and a distolateral spine.
Telson		Convex posteriorly bearing 3 pairs of lateral setae and 8 distal setae.

Source: Vedvyas Rao 1973 and Mohmad, K. H. *et al.*, 1978, Dholakia, A. D. 2004.

Post Larvae of *Parapenaeopsis stylifera* (H. Milne Edwards)

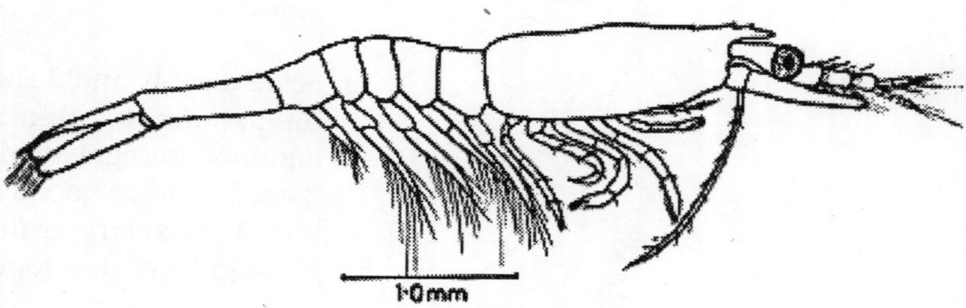

PL-10
Figure 3.9: *P. stylifera*

Figure 3.10: Telson

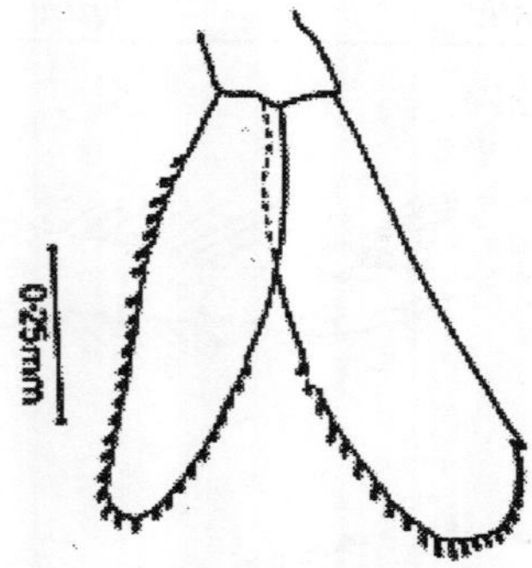

Figure 3.11: Uropod

Sizes ranged from 4.25 to 4.75 mm total length and 1.290 to 1.315 mm carapace length.

Parapenaeopsis stylifera (H. Milne Edwards)

MCL: 1.18 mm (1.13–1.20 mm); MTL: 3.80 mm (3.72–3.86 mm).

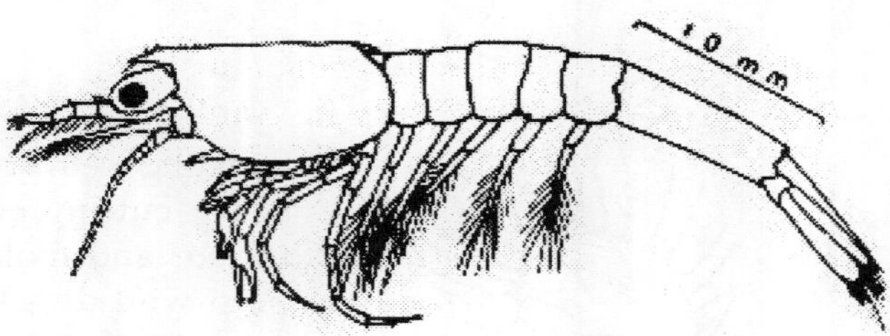

Characteristic	*Figure*	*Details*
Carapace		Rostrum reaching almost middle of eye stalk with 2 rostral teeth and an apigastric tooth, 2 long plumose setae originate subterminally on ventral side of rostrum, 2 distal rostral teeth, 3 spinules in between epigastric and penultimate rostral tooth and 1 minute spinule posterior to epigastric, hepatic spine present, dorsal spine on 5th abdominal segment has disappeared.

Characteristic	*Figure*	*Details*
Antennual-I	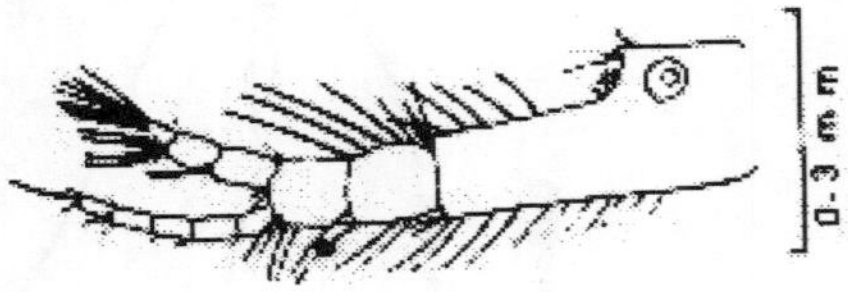	Outer flagellum shorter than inner and 3 segmented, inner flagellum 5 segmented.Basal segment remains always longer than other two segments; in PL-III, it develops a hollow to accommodate eye.
Antennual–II	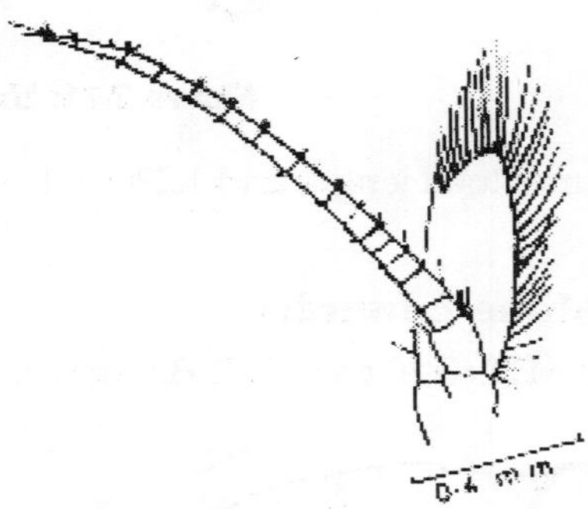	Scaphocerite broad with 33 plumose setae and a dostolateral spine, flagellum longer than scale with 15 segments, some segments faintly subdivided in 2.
Mandible		Standing teeth absent but with sharp cutting edge between incisor and molar processes, palp well developed with 2 segments, distal oval one longer than proximal segment, numerous plumose setae on both segments.
Maxilla–I	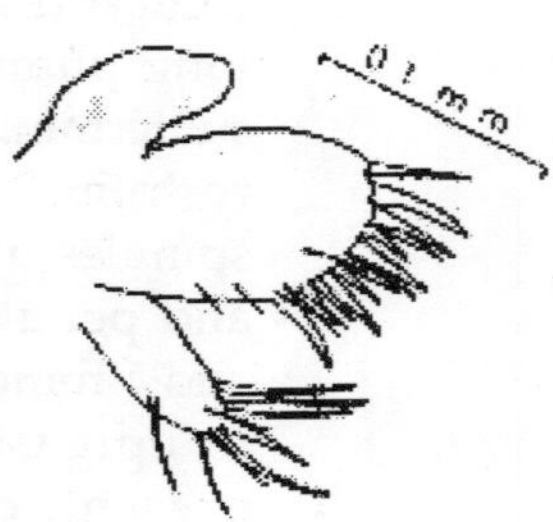	Endopod reduced to unsegmented naked palp, more setae to endites. Penuncular lobes unchanged with distal lobe being larger than proximal; endopod which is unjoined and without seta in PL-I, becomes broad at base and narrow at tip with a small seta and it remains in that condition until PL-X.

Characteristic	*Figure*	*Details*
Maxilla–II	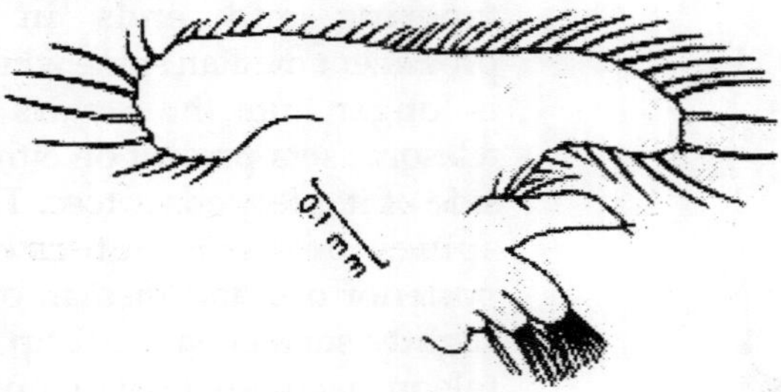	Exopod reduced with 48 to 50 plumose setae, endopod reduced to unsegmented palp bearing 3 outer lateral setae and 2 minute terminal setae, 4 endites seen, distal bearing 10 setae, 3rd 6 to 7, 2nd 2 and 1st none. It becomes more thinner and narrower; endopod appears slightly reduced especially with increase in size of PL.
Maxiliped-I	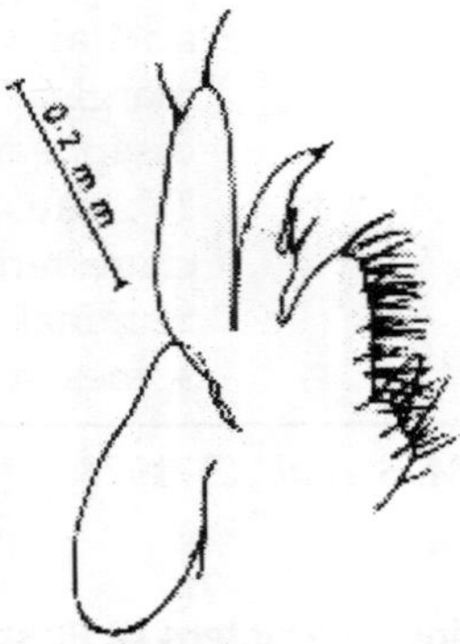	Protopod broad with more number of setae, endopod reduced to faintly 2 segmented palp with short terminal setae and 2 small lateral setae, exopod with only 2 setae, 1 terminal and 1 subterminal, gill rudiment large.
Maxiliped-II	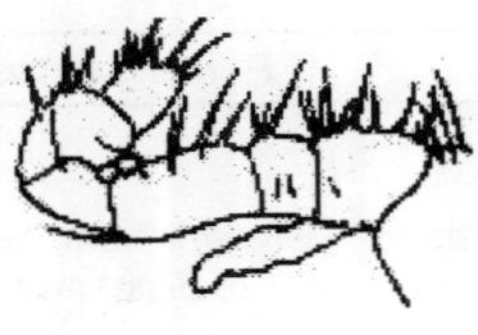	Endopod recurved, with stout bristles on distal and penultimate segments, exopod shrunken, without setae.
Maxiliped-III	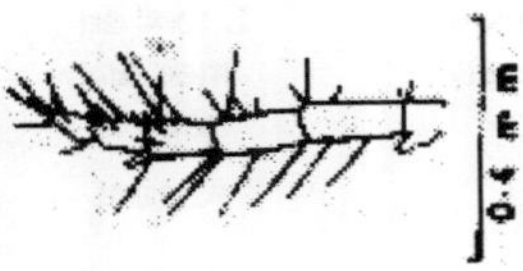	Exopod vestigial and endopod with more nonplumose setae.
Uropod	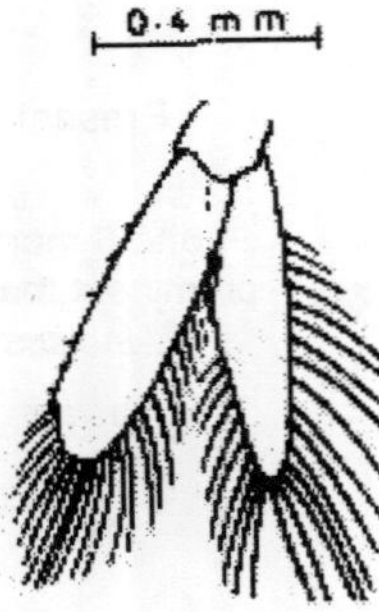	Exopod bud with 27 to 29 plumose setae and a distolateral tooth, endopod with 25 to 27 plumose setae. Uropod almost same in first two PL substage, with exopod being longer than endopod.

Characteristic	*Figure*	*Details*
Telson		Posterior margin of telson tapering and ends in a prominent median spine which is longer than the 4 pairs of telesonic seta present on either side of it; pleopods setose. The spines between outermost posterior one and median one slightly smaller in PL-II tip of telson becomes conspicuous and projects outward in PL-III and above mentioned small marginal spines become minute, disappearing completely in PL-IV, the later substage also characterized by arrangement of terminal and two inner lateral spines in "Tridentate" form.

Source: Vedvyas Rao 1973 and Muthu. M. S. *et al.*, (5) 1978.

Table 3.1: Comparison of Distinguishing Characters of Post Larvae of Penaeids

Sl.No.	*Characteristic*	*P. Indicus*	*P. merguiensis*	*P. monodon*
1	Chromatophores			
	i. Telson	Present in distal half only	Present from base to distal end	Present from base to distal end
	ii. Inner uropod ramus	May have a minute one, but usually absent	3 to 4 on median aspect	A raw of chromatophores along median aspect
	iii. Outer uropod ramus	Absent	Absent	3 to 5 in the middle on median aspect
	iv. 6th abdominal segment			
	a. Anterolateral chromatophores	Present	Present	Absent
	b. Ventral chromatophores	6	5 to 6 more prominent than in *P. indicus.*	13 to 18
	c. Dorsal chromatophores	Absent	Absent	Absent
2.	Colour of chromatophores in live specimens	Red	Red	Maroon

Contd....

Table 3.1–Contd...

Sl.No.	Characteristic	P. Indicus	P. merguiensis	P. monodon
3.	Posteromedian dorsal spines in 5th abdominal segment	Absent after two rostral spine stage	Present up to 5 rostral spine stage	Absent after two rostral spine stage
4.	Rostrum	Reaches end of eye	Reaches end of eye	Reaches end of eye
5.	Spinules on dorsal carina of 6th abdominal segment	Absent	Absent	Absent
6.	Anternnal spine on carapace	Absent	Absent	Absent

Sl.No.	Characteristic	P. semisulcatus	P, japonicus
1.	Chromatophores		
	i. Telson	Present from base to distal end	Present from base to distal end
	ii. Inner uropod ramus	Numerous on distal half only	One near base on median aspect
	iii. Outer uropod ramus	Numerous on distal half only	Absent
	iv. 6th abdominal segment		
	a. Anterolateral chromatophores	Present	Absent
	b. Ventral chromatophores	9 to 10	10 to 11
	c. Dorsal chromatophores	Absent	Absent
2.	Colour of chromatophores in live specimens	Brown	Crimson
3.	Posteromedian dorsal spines in 5th abdominal segment	Present up to 4 rostral spine stage	Present up to 4 rostral spine stage
4.	Rostrum	Long, surpasses end of eye	Short, reaching only half length of eye
5.	Spinules on dorsal carina of 6th abdominal segment	Absent	Present all along length of carina
6.	Anternnal spine on carapace	Absent	Prominent

Sl.No.	Characteristic	M. dobsoni	M. monoceros	M.affinis
1.	Chromatophores			
	i. Telson	Two prominent ones in the middle	Present from base to distal end	Present from base to distal end
	ii. Inner uropod ramus	One prominent in the middle	2–3 present	One present at distal end

Contd...

Table 3.1–Contd...

Sl.No.	*Characteristic*	*M. dobsoni*	*M. monoceros*	*M.affinis*
	iii. Outer uropod ramus	One prominent in the middle	Absent	One present at distal end
	iv. 6th abdominal segment			
	a. Anterolateral chromatophores	Absent	Present	Present
	b. Ventral chromatophores	Single	4 to 5	2
	c. Dorsal chromatophores	Usually absent	Single	Single
2.	Colour of chromatophores in live specimens	Brown	Reddish brown	Brown
3.	Posteromedian dorsal spines in 5th abdominal segment	Absent	Absent	Absent
4.	Rostrum	Very short blunt	Very short pointed	Very short pointed
5.	Spinules on dorsal carina of 6th abdominal segment	2 to 3 may be present near posterior end	1 to 2 may be present near posterior end	1 to 2 may be present near posterior end
6.	Anternnal spine on carapace	Present	Present	Present

Source: Muthu, M. S. 1978.

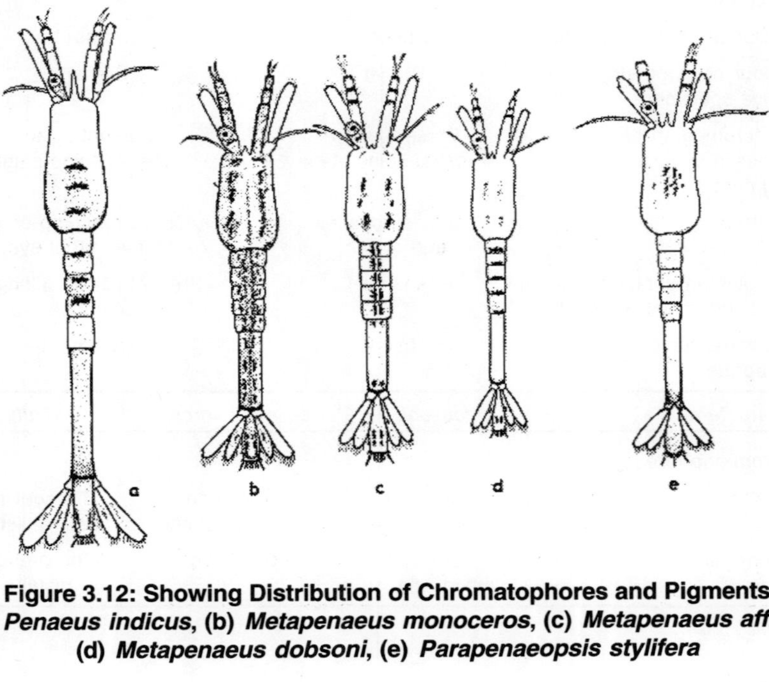

Figure 3.12: Showing Distribution of Chromatophores and Pigments (a) *Penaeus indicus*, (b) *Metapenaeus monoceros*, (c) *Metapenaeus affinis* (d) *Metapenaeus dobsoni*, (e) *Parapenaeopsis stylifera*

Table 3.2: Comparison of First Post Larval Stages of Five Species

Structure	*Penaeus indicus*	*Metapenaeus monoceros*	*Metapenaeus affinis*	*Metapenaeus dobsoni*	*Parapenaeopsis stylifera*
Total length (mm)	5.22–5.95	3.75–3.95	3.85–3.95	3.00–3.50	4.25–4.75
Carapace length (mm)	1.28–1.35	1.104–1.112	1.168–1.170	0.928–0.975	1.290–1.315
Rostrum	Exceeding eye; 1 dorsal tooth	Small, not quite reaching middle of eye; 2 pairs of dorsal spines	Small; just projecting beyond frontal margin of carapace; 2 pairs of dorsal spines and 1 epigastric	Small; 2 dorsal spines	Reaching middle of eye; 2 dorsal and 1 epigastric spines
Carapace	With supraorbital and hepatic spines	With antennal and hepatic spines	With antennal and hepatic spines	With antennal and hepatic spines	With antennal, hepatic and branchiostegal spines
Antenna I	Inner branch of distal peduncular segment 3-jointed with 2 setae at tip; outer branch 2-jointed	Inner branch of distal peduncular segment 2-jointed with 4 setae; outer branch faintly segmented	Inner and outer branches of distal segment faintly jointed; inner branch with 3 setae	Inner branch of distal peduncular segment 2-jointed with 4 setae at tip; outer branch faintly segmented	Inner branch of distal peduncular segment 5-jointed with 6 setae at tip; outer branch 3-jointed
Antenna II	Flagellum 4-jointed; scale with 28 setae	Flagellum 4–5 jointed; scale with 26 setae	Flagellum 4-jointed; scale with 26 setae	Flagellum 4-jointed; scale with 24–25 setae	Flagellum 10-jointed; scale with 39 setae
	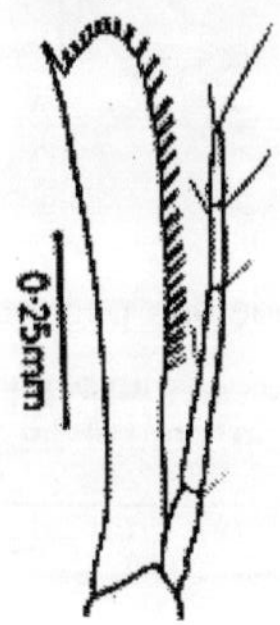	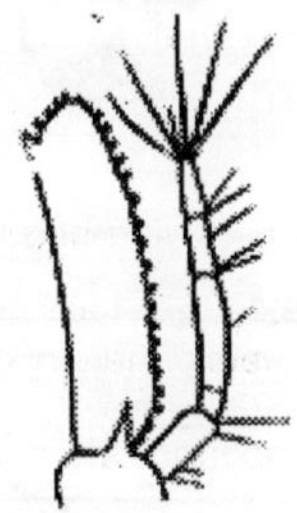	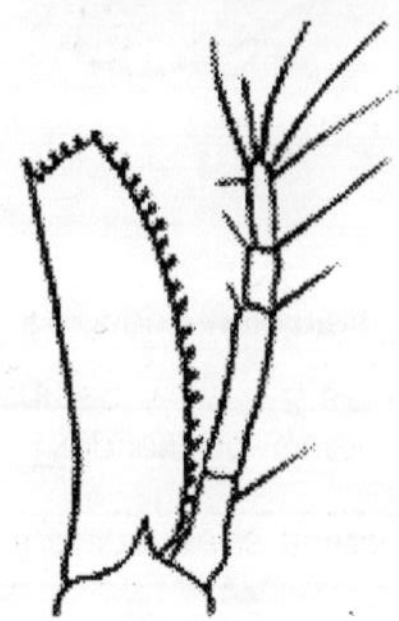	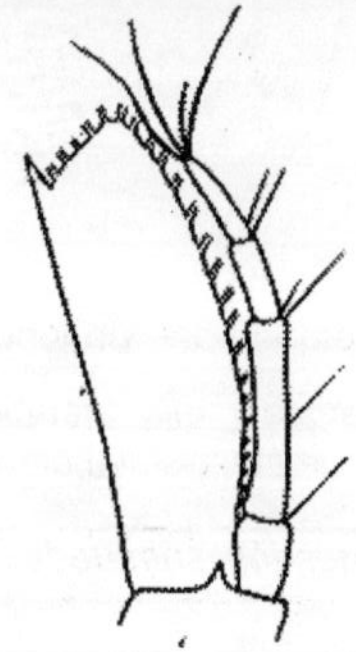	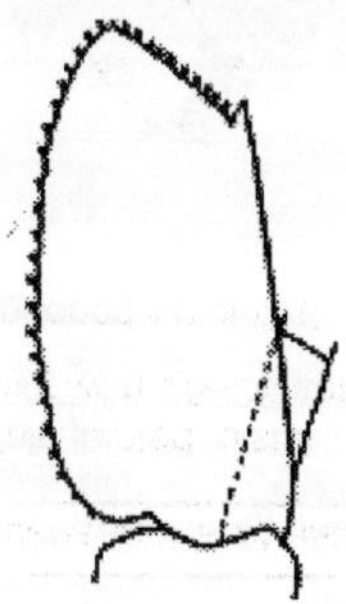

Contd...

Table 3.2–Contd...

Structure	*Penaeus indicus*	*Metapenaeus monoceros*	*Metapenaeus affinis*	*Metapenaeus dobsoni*	*Parapenaeopsis stylifera*
Mandible	Palp jointed, distal joint with 3–4 setae	Palp jointed, distal segment with 5–6 setae	Palp jointed, distal segment with 7–8 setae	Palp jointed, distal segment with 7 setae	Palp jointed, distal segment with 25–26 setae
Maxilla I	Endopod segmented 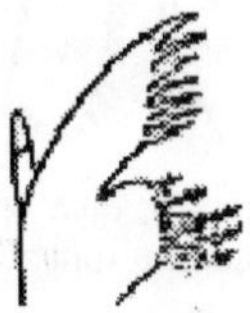	Endopod unjointed 	Endopod unjointed 	Endopod segmented 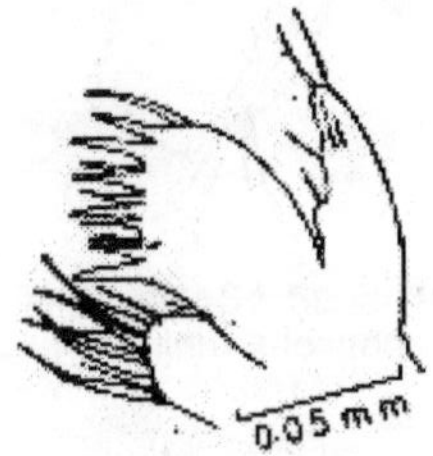	Endopod unjointed 
Maxilla II	4 endites; palp faintly segmented, scaphognathite with 27 setae 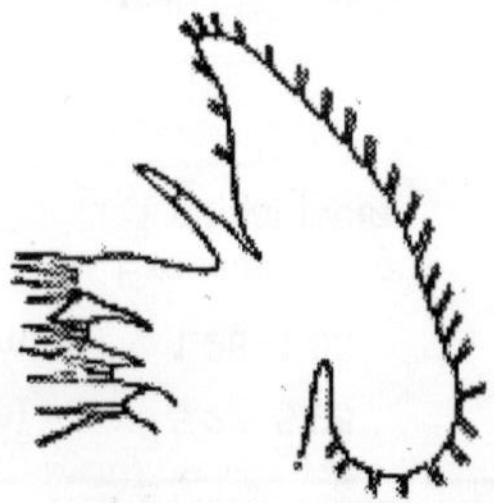	3 endites; palp unjointed scaphognathite with 32 setae 	3 endites; palp unjointed, scaphognathite with 39 setae 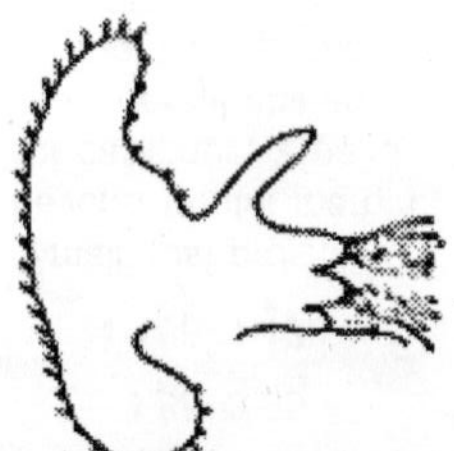	3 endites; palp unjointed, scaphognathite with 26 setae 	4 endites; palp unjointed, scaphognathite with 54–55 setae 

Contd...

Table 3.2–Contd...

Structure	*Penaeus indicus*	*Metapenaeus monoceros*	*Metapenaeus affinis*	*Metapenaeus dobsoni*	*Parapenaeopsis stylifera*
Pereiopods	1st 3 pairs without spines	1st pair with basial and ischial spines; 2nd and 3rd with basial spine only	No spines	1st 3 pairs with basial spines	No spines
Abdomen	Median dorsal spine on 5th and 6th segments	Median dorsal spine on 6th segment	Median dorsal spine on 6th segment	Median dorsal spine on 6th segment	None of the segments with median dorsal spine
Pleopod	Uniramous, 3-jointed; distal segment with 10 setae	Same as in *P.indicus*	Same as in *P.indicus*	Same as in *P.indicus*	Uniramous, 4-segmented; distal segment with 18–21 setae
Telson	Spine formula 8 + 8; posterior margin straight	Spine formula 7 + 7; posterior margin straight	Spine formula 7 + 7; posterior margin straight	Spine formula 7 + 7; posterior margin straight	Spine formula 8 + 1 + 8; margin of telson tapering between postero-lateral spines
	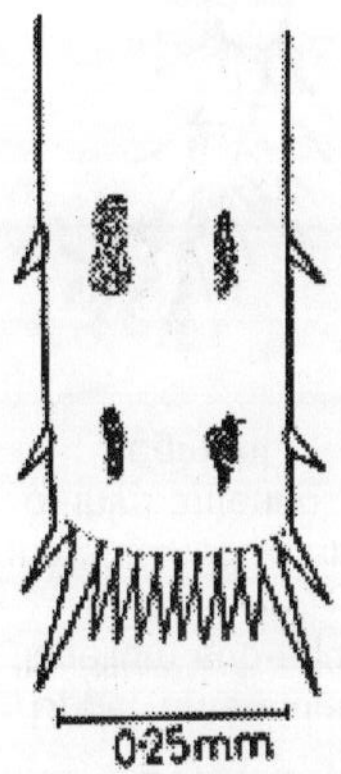				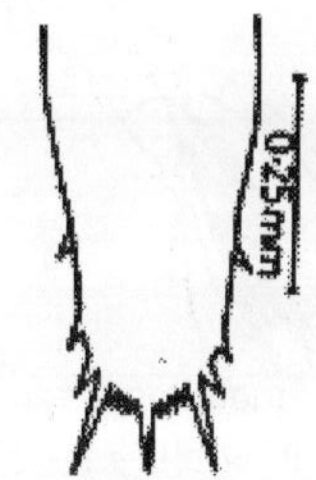

Contd...

Table 3.2–Contd...

Structure	*Penaeus indicus*	*Metapenaeus monoceros*	*Metapenaeus affinis*	*Metapenaeus dobsoni*	*Parapenaeopsis stylifera*
Antenule	Upper antennular flagellum sub-equal to the lower one attached to apex of third antennular segment. 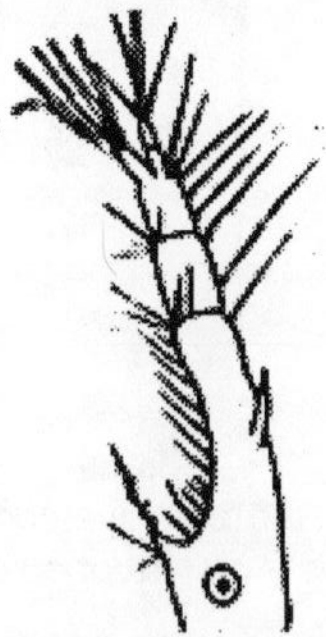	Peduncle 3-segmented 	Peduncle 3-segmented with a conspicuous statocyst at the base 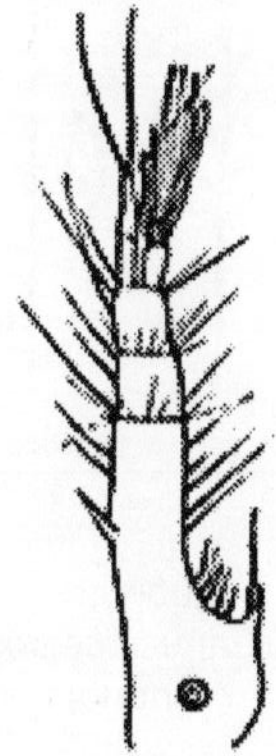	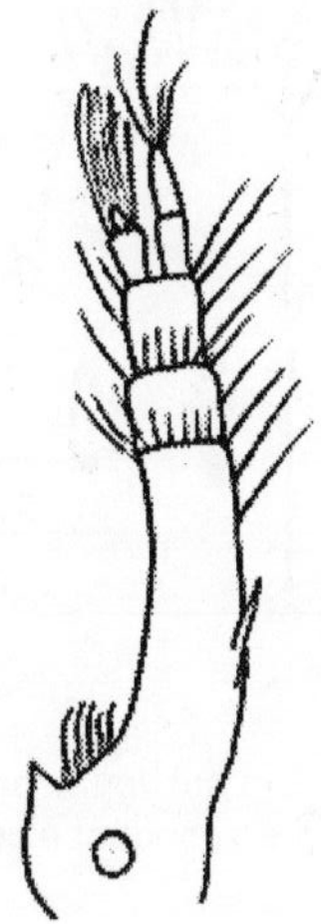	Peduncle 3-segmented as in other penaeids and with a median ventral spine on the basal segment. 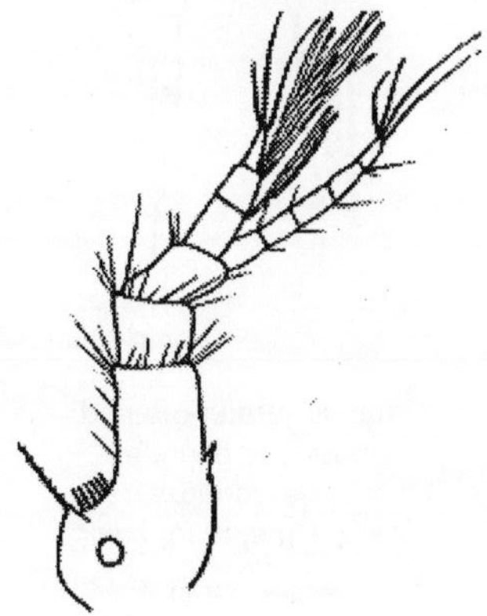

Source: Mohmad K.H, P. Vednyas Rao, and M. J. George, 2006.

Chapter 4

Field Identification

When I was giving lectures to students of Fisheries Faculty, I observed that the students were facing difficulty in identifying prawns, even separating families. During field visit by the following feature one can separate marine and freshwater prawns/ shrimps. Similarly they can separate families at one glance. However, for exact identification you have to follow the key provided.

Distinguish Penaeus, Parapenaeopsis and *Metapenaeus* Species

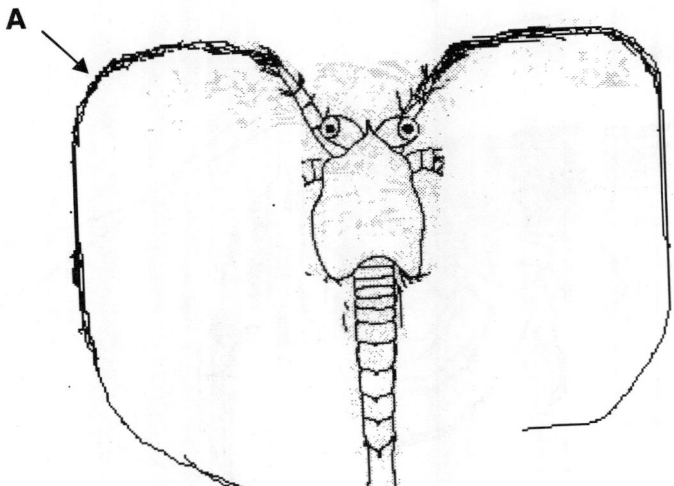

Figure 4.1: Antinnuls of Penaeus and Parapenaeopsis (A = Round shape when it is swimming)

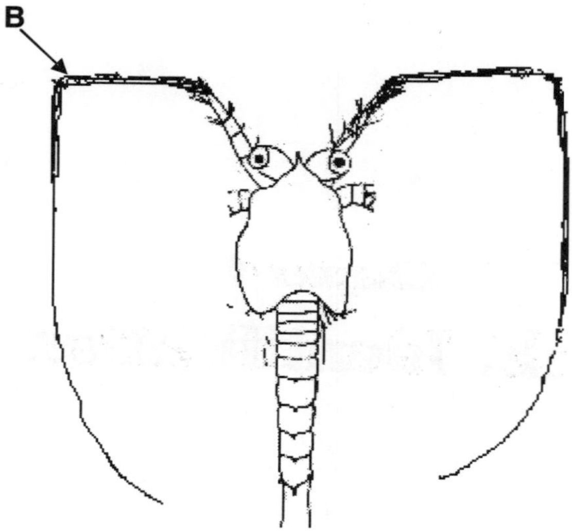

Figure 4.2: Antinnules of Metapenaeus
(B = Forms angle when swimming)

Distinguish Characters of Marine and Freshwater Prawn

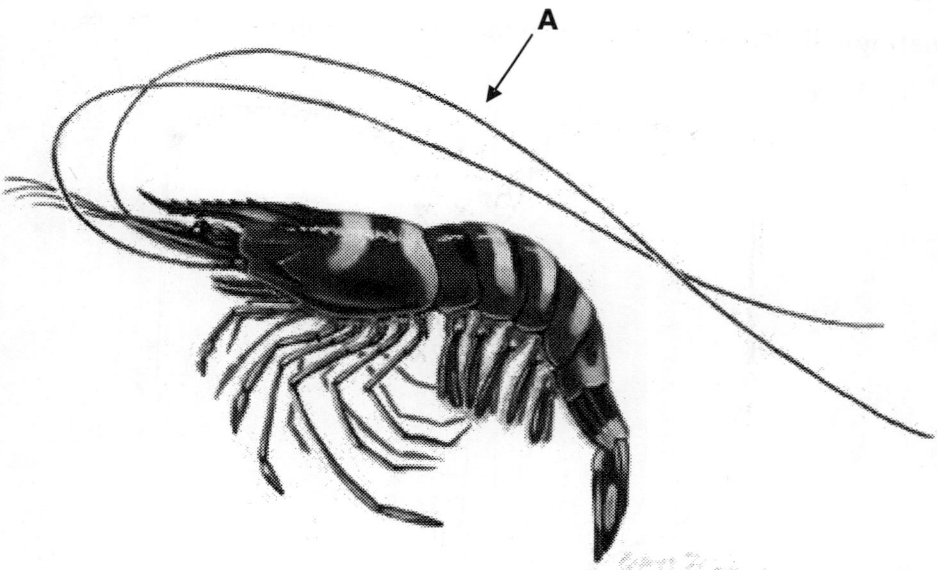

Figure 4.3: Marine Water Prawn
(A = Two Antinnuals)

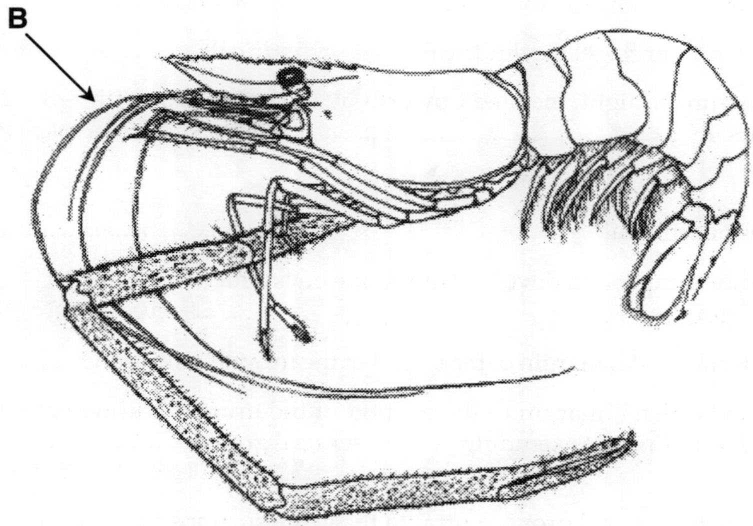

Figure 4.4: Freshwater Shrimp
(B = Four Antinnuals)

The Central Marine Fisheries Research Institute (CMFRI) Cochin had prepared Identification key for prawn. On the basis of it, Kurian and Sebastian (1976) had published the same in 'Prawn and Prawn Fisheries of India'. It is taken as main source of classification given below. However, classifications given by some other scientists are also included in this chapter.

KEY TO THE COMMERCIALLY IMPORTANT PRAWNS OF INDIA

FAMILY PENAEIDAE

Post orbital spine present Sub family *Solenocerinae* .. 4

Post orbital spine absent 2

2(1) Carapace with a median dentate crest extending nearly or quite to the posterior margin Sub family *Sicyoninae* 8

— Carapace without a median dentate crest except occasionally over the eyes 3

3(2) Distinct median tubercle on ocular peduncle; upper antennular flagellum inserted near posterior border of third anetennular segment, strikingly shorter than lower one Sub family *Aristeinae* 9

— No distinct median tubercle on ocular peduncle; upper antennular flagellum sub-equal to the lower one attached to apex of third antennular segment Sub family *Penaeinae* 12

4(1) Antennular flagella folisceous *Solenocera* 5

— Antennular flagella cylindrical or sub-cylindrical; .... *Hymenopenaeus*

— Rostrum straight, inclined upwards at an angle of 20° with 7–8 + 2 teeth dorsally .... *Hymenopenaeus acqualis* (Bate)

Solenocera

5(4) Telson trifurcate 6

— Telson simple and devoid of any spine on lateral margin ... *Solenocera indica* (Natraj)

6(5) Externo–distal margin of the expod of the uropod with spine 7

— Externo-distal margin of the exopod of the uropod without spine; post rostral carina not extending beyond cervical groove .... *Solenocera pectinata* (Bate)

7 6) Spine on cervical grove ventral to the posterior-most spine of the rostral series present; 'L' shaped grove on either branchiostegal region present .... *Solenocera hextii* Wood Masson

— Spine on the cervical groove ventral to the posterior-most spine of the rostral spine absent; 'L' shaped groove either branchiostegal region absent .... *Solenocera choprai* Natraj

8(2) Post rostral carina armed with 5 teeth; abdominal pleura of 1st and 2nd segment unispinose and 3rd, 4th and 5th with 3 spines .... *Sicyonia lancifer* (Oliver)

9(3) Rostrum three-toothed dorsally; hepatic spine absent *Aristeus* 10

— Rostrum with many teeth on upper border; hepatic spine present; teeth of ptery-gostomian region more than 2.5 times its greater breadth .... *Aristeomorpha woodmasoni* Calman

10(9) Integument glabrous 11

— Integument pubescent .... *Aristus virilis* (Bate)

11(10) Pleurobranchiae on segments X–XIII on distinct filaments provided with pinnules .... *Aristeus alcocki* Ramdan

12(3) Rostrum without ventral teeth 13

— Rostrum with ventral teeth Penaeus 19

13(12) A distal fixed pair of spines on the telson and 1–3 pairs of mobile spines 14

No distal fixed pair of spines on the telson; lateral mobile spines may be present 16

14(13) Petasma symmetrical; 3rd maxilliped without basial spine 15

— Petasma symmetrical; 3rd maxilliped with basial spine. *Metapeneaopsis* 26

15(14) Carapace with longitudinal sutures extending from post orbital margin to almost posterior margin *Parapenaeus* 30

— Carapace without longitudinal sutures; branchiostergal spine present .... *Peneopsis*

— Telson with 3 pairs of movable marginal spines in addition to the fixed pair .... *Penaeopsis rectacuta* (Bate)

16(13) No exopod on 5th pereopod; pleurobranch on 7th thoracic somite present .... *Metapenaeus*

— Exopod on 5th pereopod; pleurobranch on 7th thoracic somite absent 17

— Hepatic spine present; petasma not constricted distally; anterior plate of thelycum rounded posteriorly .... *Atypopsenaeus strenodactylus* (Stimpson)

18(17) 3rd pereopod with epipodite *Tracihypenaeus* 41

— 3rd pereopod without epipodite Parapenaeopsis 43

Penaeus

19(12) Adrostral carina reaching almost to posterior border of carapace; gastro-frontal carina present 20

— Adristral carina not reaching behind middle of carapace; gastro–frontal carina absent 22

20(19) Telson armed usually with 3 pairs of spinules 21

— Telson unarmed; rostrum with ventral tooth

21(20) Adrostral sulcus narrower than post rostral carina; anterior plate of thelycum rounded at the apex .... *Penaeus japonicus* Bate

— Adrostral sulcus as wide as post rostral carina; anterior plate of thelycum bifid at the apex .... *Penaeus latisulcatus* Kishionuye

22(20) Hepatic carina present 23

— Hepatic carina absent 24

23(22) Hepatic carina horizontally straight; 5th pereopod without exopodite .... *Penaeus monodon* Fabricius

— Hepatic carina inclined at an angle of 20° anteroventrally; 5th pereopod with small exopodite .... *Penaeus semisulcatus* de Hann

24(22) Gastro-orbital carina occupying the posterior 2/3 distiance between hepatic spine and orbital angle; rostral crest may be elevated but nit triangular in profile .... *Penaeus indicus* Milne Edward.

— Gastro-orbital carina absent or not reaching hepatic splne and occupying the middle 1/3 distance between hepatic spine and orbital angle 25

25(24) Dectyle of 3rd maxilliped of adult male ½ size of propods;adrostral carina not reaching as far as epigastric tooth; rostral crest triangular in profile .... *Penaeus merguiensis* de Man

— Dectyle of 3rd maxilliped of adult male much longer than propodus; adrostral carina reaching just beyond epigastric tooth; rostral crest markedly elevated .... *Penaeus penicillatus* Alcock

Parapenaeus

30(15) Branchiostegal spine present; 5th pereopods not reaching tip of antennal scale 31

— Branchiostegal spine absent; 5th pereopods extending antennal scale by dactyl .... *Parapenaeus longipes* Alcock

31(30) Branchiostegal spine on anterior margin of carapace; 6th abdominal somite less than twice length of 5th, process 'a' of petasmal bifurcate, directed laterally; thelycum with anterior, intermediate and posterior plates .... *Parapenaeus fissures* (Bate)

— Branchiostegal spine a little behind anterior margin of carapace; 6th abdominal somite more than twice length of 5th; rostrum reaching distal end of 1st segment of antennular peduncle ....
.... *Parapenaeus investigatoris* Alcock

Metapenaeus

32(16) Distomedian petasmal projection with fully developed or vestigial apical filament; thelycum of impregnated females usually with white conjoined pad 33

— Distomedian petasmal projection without apical filament; thelycum of impregnated females without white conjoined pads.

33(22) Rostrum wide and short, not reaching to distal end of basal antennular segment; thelycum with ovoid anterior and lateral plates of sub-equal size; conjoined pads usually set askew; apical filaments of petasmal vestgeal represented by a pair of rounded bosses ....
.... *Metapenaeus lysianassa* (de Man)

34(33) Posterior part of rostrum is with distinctly elevated crest; basial spine on male 3rd pereopod simple; apical petasmal filaments slender slightly covering; thelycum with a large anterior and small lateral plates. ....
Metapenaeus brevicornis (Milne Edward)

— Posterior part of rostrum is without distinctly elevated crest; basial spine on male 3rd pereopod long and barbed; apical petasmal filaments not really visible; anterior thelycal plate tongue-like.

35(32) Branchiocardiac sulcus are distinct in at least posterior 1/3 carapace; distomedian petasmal projection flap-like 36

— Branchiocardiac sulcus almost completely absent; distomedian projections anteriorly filiform each with a serrate venteral margin ....
....*Metapenaeus stebbingii* (Nobili)

36(35) Ischial spine on 1st pereopod distinct 37

— Ischial spine on 1st pereop small or absent 40

37(36) Distomedian petasmal projections directed anterioral; lateral thelycum plates with raised lateral ridges each with posterior inwardly curved triangular plate ....*Metapenaeus ensis* (de Haan)

— Distomedian petasmal projections directed anterolaterally; anterior thelycum plate tongue-like 38

38(37) Lateral thelycal plates with salient and parallel ear shaped lateral ridges; distomedian petasmal projection hood-like ....
.... *Metapenaeus monoceros* (Fabricius)

— Lateral thelycal plates without lateral ridges; distomedian petasmal projection not hood-like 39

39(38) Posterior extension of the anterior median thelycal plates bound laterally by an oval flat plate on each side; distomedian petasmal projections overlaying lateral projections and distally tri-lobed ....
.... *Metapenaeus alcoki* George and Rao

— Posterior extension of the anterior median thelycal plates not bound laterally by an oval flat plate on either side; distomedian petasmal projections not overlaying lateral projections ....
.... *Metapenaeus kutchensis* George and Rao

40(36) Branchiocardiac distinctly, extending from posterior margin of carapace almost to hepatic spine; anterior thelycal plate longitudinally grooved, wider posteriorly than anteriorly; distomedian petasmal projections crescent–shaped .... *Metapenaeus affinis* (Miln.Edward)

— Branchiocardiac carina feeble or ill-defined, anterior end not extending posterior 1/3 of carapace; distal margin of anterior thelycal plate convex to indistinctly triangular; petasmal with laminose and strongly diverging distomedian projections .... *Metapenaeus brevicornis* Kubo

Trachypenaeus

41(18) Epipodites present on 1st and 2nd pereopods 42

— Epipodite absent on 1st and 2nd pereopods; distolateral projections of petasmal with sharp tip reaching coxae of 4th pereopods; anterolaterally with large wing-like flaps on outer curvature .... *Trachypenaeus pescadorensis* Schmidt

42(41) The plates of thelycum with raised anterior and lateral margins .... *Trachypenaeus sedili* Hall

— The anterior plates of thelycum may have a raised enterior margin but laterally the margins are not raised, an excavation present between the anterior plate and the transverse sternal ridge .... *Trachypenaeus curvirostris* (Stimpson).

Parapenaeopsis

Family Penaeopsis

Carapace is with longitudional and transverse sutures. Male are with symmetrical petasmal, a single arthrobranch on last throracic segent. No trace of a second arthrobranch.

43(18) Epipodite present on 1st and 2nd peropods 44

— Epipodite absent on 1st and 2nd peropods 49

44(43) 2nd pereopods with basial spines 45

— 2nd pereopods without basial spines ...... *Parapenaeopsis uncta* (Alcock)

45(44) Telson with pair of fixed subapical spines; at least distal ½ free portion of rostrum unarmed ... *Parapenaeopsis stylefera* (Milne Edward)

— Telson without pair of fixed subapical spines; with or without lateral movable spines;1/3 or less free portion of rostrum unarmed 46

46(45) Petasma with a pair of disto-lateral projections directly laterally or distolaterally usually short and spout–like 47

— Petasma with a pair of long slender caliper like disto-lateral projection directed forwards; thelycum with median ruft of long setae behind posterior edge of last thoracic sternite; third pereopod of female with basial spine .. ... *Parapenaeopsis maxillipedo* (Alcock)

47(46) Post-rostral carina reaching almost t posterior border of carapace; petasmal with pair of short spout like disto lateral projections and pair of cap like distal projections 48

— Post-rostral carina reaching ¾ carapace; petasmal with pair of disto lateral projections directed laterally, cap like distal projections absent .... *Parapenaeopsis nama* (Alcock)

48(47) Antennular flagella 0.5–0.6 length of carapace; thelycum with median tuft of setae on posterior plate .... *Parapenaeopsis sculptilis* (Heller)

— Antennular flagella 0.7 length of carapace or longer; thelycum without a median tuft of setae on posterior plate ..... *Parapenaeopsis hardwikii* (Miers)

49(43) Anterior plate of thelycum with V-shaped posterior edge; and 2 accessory ridge on anterior edge of posterior plate; rostrum with proximal 1.3 rising from carapace; remainder more or less horizontal

— Anterior plate of thelycum with more or less straight transverse posterior edge; No accessory ridge on anterior edge of posterior plate; rostrum inclined up wards at an angle to carapace for whole of its length. .... *Parapenaeopsis acclivirostris* (Alcock)

Source: Kurian, C. V. and V. O. Sebastian 1976.

Some deep sea prawns are found in India and they are considered as potentially important commercial species. Identification of such species, coming under family PANDALIDAE are given below.

Key for Identification of Prawns of Family PANDALIDAE

1. Carapace with longitudinal carina on the lateral surface, integument very firm; pereopods of 2nd pair very unequal *Heterocarpus* 4

— No longitudional carina on the carapace except for the post rostral crest; 3rd maxilliped with exopod 2

2(1) At least the first 2 pereopods with epipods; posterior lobe of scaphognathite truncate; stylocerita pointed anteriorly *Plesionika* 3

— Pereopod without epipods; upper margin of rostrum finely and evenly serrate along its whole length; carpus of 5th leg shorter than propodus; minimum thickness of 6th abdominal somite, when looked at dorsally, 2/5 length of this somite; telson almost 1 ½ as long as 5th somite .... *Parapandalus spiripes* (Bate)

3(2) Posterior border of 3rd abdominal tergum acutely produced into a sharp tooth that overlaps the next tergum .... *Plesionika ensis* (A. Milne Edw.)

— Posterior border of 3rd abdominal tergum coovex is not acutely produced; rostrum 45 to 67 per cent of the length of the body from orbit to tip of the telson. .... *Plesionika martia* (A. Milne Edw.)

4(1) Abdominal terga, through carinated, never produced posteriorly into overhanging spines; post ocular carinae present; supper margin of rostrum proper, armed with 2 or 3 teeth. .... *Heterocarpus gibbosus* Bate.

— Third abdominal tergum armed with an acute spine arising from the anterior half; post ocular carina completely wanting .... *Heterocarpus wood masoni* Alcock.

Source: Kurian, C. V. and V. O. Sebastian 1976.

Key to the Identification of Prawn of the Family SERGESTIDAE

In India only Genus *Acetes* is found in commercially important prawns in the family *Sergestidae.* The identification of the same species is given as under.

Genus Acetes

1. Procured spine is present 1st pair of pleopods 2

— No procured spine is present 1st pair of pleopods 3

2(1) Trochanter (basis) of the 3rd pereopod with tooth on inner free margin; petasmal without membraneous coupling folds .... *Acetes indicus* H. Milne Edw.

— Trochanter (basis) of the 3rd pereopod without tooth on inner free margin; petasmal with a pair of folded coupling membranes armed with hooks .... *Acetes erythraeus* Nobili.

3(1) External antennular flagellum in male with two clasping spines; apex of telson rounded or truncated. 4

— External antennular flagellum in male with single clasping spines; apex of telson triangular .... *Acetes sibogae* Hansen

4(3) Segment preceding the one bearing the clasping spine with angular process pointing back wards; apex of telson truncated and with a tooth at each corner .... *Acetes serrulatus* Kroyer

— Segment preceding the one bearing the clasping spine without any process; apex of telson round and third thoracic sterite produced posteriorly as large plate in female 5

5(4) Ciliated and non-ciliated portion of external border of exopod of uropod not separated by a tooth; distal portions of pars externa with tubercles. .... *Acetes japonicus* Krishinouye

— Ciliated and non-ciliated portion of external border of exopod of uropod separated by a tooth; distal portions of pars externa with tubercles .... *Acetes cochinensis* Rao

Source: Kurian, C. V. and V. O. Sebastian 1976.

Identification of other genes of *Sergesters* are given under:

1. Body and appendages (when fresh) with scattered red chromatophores, not uniform red, transparent so that organs of Pesta (pigmented and luminous modifications of the gastrohepatic gland) visible; hepatic spine and usually supraorbital spine present. Subgenus *Sergestes* s.s. Relatively small species. 7

2. The two distal segments of the fifth legs setose on only one margin; supraorbital spine present; third maxillipeds subequal with third legs; third

segment of antennular peduncle shorter than first. *S. arcticus* Kroyer, 1855. Relatively common between 50 and 600 fms. in our waters. Length up to 2.5 in 3

3(2) The two distal segments of the fifth legs setose on both margins 4

4 . Supraorbital spine absent. *S. seminudus* Hansen, 1919 5

5(4) Supraorbital spine present. Third maxillipeds subequal with third legs *S disjunctus* Burkenroad, 1940 6

6. Supraorbital spine present. Third maxillipeds greatly enlarged, considerably longer than third legs .... S. index Burkenroad, 1940

7(1) Body and appendages (when fresh) uniform opaque red in colour; organs of Pesta absent; hepatic and supraorbital spines absent. Subgenus *Sergia*. Relatively large species 8

8. Exoskeleton rigid; rostrum acute anteriorly and with acute apical tooth; cornea considerably wider than peduncle; numerous purple lens-less photophores on body and appendages. * *S. potens* Burkenroad, 1940 9

9(8) Exoskeleton membranous; rostrum obtuse; cornea barely wider than peduncle; no dermal photophores. *S. japonicus* Bate, 1881. (In literature synonyms as *S. mollis*) 10

10. Exoskeleton membranous; rostrum obtuse; cornea considerably wider than peduncle and later bearing a median tubercle; no dermal photophores .... *S. kroyeri* Bate, 1881

It is difficult to differentiate prawn species in early stages especially when species are very nearer to each other. It is, therefore, some keys are given based on field collection and field identification.

Table 4.1: Field Key for the Identification of Commercially Important Adult Penaeid Prawns of India

	Observation	*Conclusion/Next State*
I	Antennule with 2 flagella; peura of 2nd abdominal segment overlaps only 1st abdominal segment.	PENAEIDEA (1)
II	Antennule with 3 flagella; pleura of 2nd abdominal segment overlaps 1st and 3rd abdominal segments	CARIDEA
1	Antennular flagellum very long and foliaceaous	(2)
2	Telson simple, without any lateral spinules	*S. CRASSICORNIS*
	PENAEUS (Ventral teeth present)	
3	Rostrum with ventral teeth	(PENAEIDAE)..5
4	Rostrum without ventral teeth	17
5	Rostrum with 1–3 ventral teeth	7

Contd...

Table 4.1–Contd...

	Observation	Conclusion/Next State
6	Rostrum with 1 ventral teeth body with yellow and brown cross bands	8
7	Rostrum with4–6 ventral teeth	13
8	Telson without lateral spinules; one teeth on ventral side of rostrum	*P. CANALICULATUS*
9	Telson with 3 pairs of lateral spinules.	*P.JAPONICUS*
10	Rostrum with 3 ventral teeth	11
11	Hepatic carina straight, horizontal; white and brown cross bands on the body; 6th abdominal segment is having parallel colour band.	*P. MONODON*
12	Hepatic carina inclined; greenish brown and white bands on body; three teeth on ventral side; 6th abdominal segment is having slanting colour band	*P. SEMISULCATUS*
13	Rostrum long, without prominent crest; gastro-orbital carina occupy posterior 2/3 distance from hepatic spine and orbital angle 4–6 teeth present on ventral side rostrum crest feeble.	*P. INDICUS*
14	Gastro-orbital carina occupy posterior 1/3 distance from hepatic spine and orbital angle	15
15	Rostrum shirt, with much elevated triangular, basal crest; dactylus of 3rd maxillipeds of adult only ½ the length of propodus; 4–6 teeth present on ventral side rostrum crest prominent.	*P. MERGUIENSIS*
16	Rostrum long, without elevated basal crest; dactylus of 3rd maxillipeds of adult much longer than propodus	*P. PENICILLATUS*
17	Expodite present on 5th pereopod	26
18	Exopodite absent on 5th pereopod	19
19	Rostrum straight	21
20	Rostrum curved	23
	METAPENAEUS (Ventral teeth absent on rostrum)	
21	Rostral crest absent; 1st abdominal segment with dorsal carina larger in size; rostrum extends beyond 2nd segment of antennular peduncle	*M. MONOCEROS*
22	Rostral crest present; no carina of 1st abdominal segment; rostrum does not extend beyond 2nd segment of antennular pedncle	*M. BREVICORNIS*
23	Slightly curved rostrum, without crest; last pair of perepod surpass the antennal scale, 5th leg of thorax surpassing the antennual scale.	*M. AFFINIS*
24	Rostrum more curved; with basal crest; last pair of pereopods fall short of middle of antennal scale, 5th leg of thorax do not surpassing the antennual scale.	*M. DOBSONI*
25	Rostral tip more straight than in *M. affinis*; posterior extension of anteromedian thelycal plate not bound by oval plate on either side; dorsomedian abdominal carina from 4th segment, size small.	*M. KUTCHENSIS*

Contd...

Table 4.1–Contd...

	Observation	*Conclusion/Next State*
	PERAPENAOPSIS (Ventral teeth absent on rostrum and exopodite present on 5th leg of thorax)	
26	Telson with fixed subapical spines; at least distal ½ free portion of rostrum without teeth and styliforms, Hepatic carina do not run on to brancheostegal teeth.	*P. STYLIFERA*
27	Telson without fixed subapical spines 1/3 or less free portion of rostrum without teeth	28
28	Patasma with long, slender forwardly directed distolateral projections; thelycum with median tuft of long setae behind posterior edge of last thoracic plate	30
29	Petasma with short, spout-like distolateral projections directed towards the sides	33
30	3rd pereopods of female with basal spoine	*P. MAXILLIPEDO*
31	3RD pereopod of female without basal spine	*P. CORNUTA*
32	Antennular flagellum longer; thelycum without tuft of setae on posterior	*Parapenaeopsis hardwikii* (Miers)
33	Antennular flagellum ½ or little more length of carapace; thelycum with median tuft of setae on posterior plate, Hepatic carina running over brancheostegal tooth.	*Parapenaeopsis sculptilis* (Heller)
34	Rostrum inclined upwards at an angle to carapace for whole of its length	*Parapenaeopsis acclivirostris (Alcock)*
35	Small prawns with very short rostrum having 2–3 teeth on the dorsal side of carapace; eyes elongated, about 1/3 the carapace length; 5th pereopos absent; red spot on the base of uropod, representing the "tan organ"	*ACETES INDICUS*

Note: Identification characters of prawns are based on field observations.

Key to the Species of *Metapenaeus* (Modified from Racek and Dall (1965)

1 Telson armed with 3 or 4 pairs of conspicuous spines 2

— Telson armed with a single row of very minute mobile spinules, with or without 1–2 pairs of somewhat larger distal spines 4

3(2) Branchial region with small pubescent areas; coxal projection of 4th pereiopod long and curved, dagger-like; thelycum with rounded median boss posterior to lateral plates; distomedian petasmal projections without an anterolateral spinous process .... *M. intermedius* (Kishinouye)

Branchial region with 2 large pubescent areas; coxal projection of 4th pereiopod a straight conical spine; thelycum without a rounded boss posterior to lateral plates; distomedian petasmal projections with a distinct anterolateral spinous process .... *M. endeavouri* (Schmitt)

4(1) Distomedian petasmal projection with fully developed or vestigial apical filament; thelycum of impregnated females usually with white conjoined pads 5

Distomedian petasmal projection without apical filament; thelycum of impregnated females without white conjoined pads 9

5(4) Rostrum wide and short, not reaching to distal end of basal antennular segment; thelycum with ovoid anterior and lateral plates of subequal size; conjoined pads usually set askew; apical filaments of petasma vestigial, represented by a pair of rounded bosses .... *M. lysianassa* (de Man)

Rostrum projecting beyond basal antennular segment, with a marked edentate distal portion 6

6(5) Posterior part of rostrum with distinctly elevated crest; basial spine on 3rd pereiopod simple 7

Posterior part of rostrum without distinctly elevated crest; basial spine on 3rd pereiopod long and barbed 8

7(6) Ischial spine on 1st peroiopod subequal to basial spine; telson usually with 1 distal pair of slightly larger spinules; distolateral petasmal projections directed outwards; apical filaments of distomedian projections slender, slightly converging; thelycum with a large anterior and small lateral plates .... *M. brevicornis* (H. Milne Edwards)

Ischial spine on 1st pereiopod much smaller than basial spine; telson usually with 2 distal pairs of slightly larger spinules; distolateral petasmal projections pointing anteriorly; apical filaments of distomedian projections lobelike; thelycum with a small anterior and very large lateral plates ...... *M. tenuipes* Kubo (*M. spinulatus* Kubo)

8(6) Apical petasmal filaments not readily visible; anterior thelycal plate tongue-like .... *M. dobsoni* (Miers)

Apical petasmal filaments large and lobe-like, curved dorsally; anterior thelycal plate styliform .... *M. joyneri* (Miers)

9(4) Branchiocardiac sulcus distinct in at least posterior 1/3 carapace; distomedian petasmal projections flap-like 10

Branchiocardiac sulcus almost completely absent; distomedian petasmal projections anteriorly filiform, each with a serrate ventral margin .... *M. stebbingi* (Nobili)

10(9) Ischial spine on 1st pereiopod distinct 11

Ischial spine on 1st pereiopod small or absent 15

11(10) Ischial spine subequal to basial spine; petasmal apices turned at 30° towards midline, semicircular; anterior thelycal plate spoon-like; lateral plates with raised ventral ridges, each with anterolateral and posteromedian spinous process .... *M. suluensis* Racek

and Dall Ischial spine much smaller than basial spine; lateral thelycal plates without spinous processes 12

12(11) Distomedian petasmal projections directed anteriorly; lateral thelycal plates with raised lateral ridges, each with a posterior inwardly-curved triangular plate .... *M. ensis* (de Haan)
(= *M. mastersii* (Haswell) = *M. incisipes* (Bate))

Distomedian petasmal projections directed anterolaterally; anterior thelycal plate tongue-like 13

13(12) Lateral thelycal plates with salient and parallel earshaped lateral ridges; distomedian petasmal projections hood-like .... *M. monoceros* (Fabricius)

Lateral thelycal plates without lateral raised ridges; distomedian petasmal projections not hood-like 14

14(13) Posterior extension of anterior median thelycal plate bound laterally by an oval flat plate on each side; distomedian petasmal projections overlying lateral projections and distally trilobed .... *M. alcocki* George and Rao

Posterior extension of anterior median thelycal plate not bound laterally by oval plate on either side; distomedian petasmal projections not overlying lateral projections ...... *M. kutchensis* George, George and Rao

15(10) Ischial spine minute and blunt 16

Ischial spine absent 19

16(15) Rostral teeth more or less evenly spaced; thelycal structure posteriorly open 17

Rostral teeth unevenly spaced, anterior 2 teeth separated from each other and from the rostral apex by a much wider space; thelycal structure posteriorly closed .... *M. demani* (Roux)

17(16) Distomedian petasmal projections not superficially separated into 2 lobes, almost completely overlying distolateral projections; lateral thelycal plates kidney-shaped, with strongly raised ventrolateral ridges ....
.... *M. conjunctus* Racek and Dall

18(17) Distomedian petasmal projections more or less superficially separated into 2 lobes, not overlying distolateral projections, lateral thelycal plates ear-shaped, with salient lateral ridges 18

— Distomedian petasmal projections parallel and directed anteriorly, longitudinal sulcus ill-defined; posterior end of salient ridges on lateral thelycal plates curved outwards; spine on merus of 5th pereiopod slightly bent inwards .... *M. papuensis* Racek and Dall

Distomedian petasmal projections diverging and directed anterolaterally, longitudinal sulcus distinct; posterior end of salient ridges on lateral thelycal plates curved inwards; spine on merus of 5th pereiopod slightly bent outwards .... *M. elegans* (de Man)
.... (= M. *singaporensis* Hall)

19(15) Rostrum with a marked edentate distal portion; anterior thelycal plate bluntly pointed, lateral plates large, separated by a narrow fissure ....
.... *M. eboracensis* Dall

Rostrum without edentate distal portion 20

20(19) Branchiocardiac carina distinct, extending from posterior margin of carapace almost to hepatic spine; anterior thelycal plate longitudinally grooved, wider posteriorly than anteriorly; distomedian petasmal projections crescent-shaped ....
.... *M. affinis* (H. Milne Edwards)
(= *M. mutatus* (Lanchester) =
M. necopinans Hall)

Branchiocardiao carina feeble or ill-defined, anterior end not exceeding posterior 1/3 of carapace 21

21(20) Anterior thelycal plate tongue-like, with a pair of anterolateral rounded tubercles; lateral plates with characteristic patch of dense setae; distomedian petasmal projections strongly diverging, each forming a broad out-wardly-curved tooth .... *M. insolitus* Racek and Dall

22(21) Anterior thelycal plate flask-shaped, with a longitudinal median ridge; distomedian petasmal projections finger-shaped 22

— Anterior margin of anterior thelycal plate with 3 tubercles 23

Anterior margin of anterior thelycal plate with 2 fang-like teeth and a median indistinct tubercle; petasma with slightly diverging tubular distomedian projections .... *M. dalli* Racek

23 (22) Median tubercle more prominent than lateral ones; distal margin of anterior thelycal plate distinctly triangular; petasma with almost parallel tubular distomedian projections, their distal half twisted dorsoventrally ....
.... *M. bennettae* Racek and Dall

All tubercles of equal size; distal margin of anterior thelycal plate convex to indistinctly triangular; petasma with laminose and strongly dirverging distomedian projections .... *M. burkenroadi* Kubo

Main Characters of Family *Metapenaeopsis*

Carapace without lateral keels outing portion of mandible short and massive. Rostrum toothed on a dorsal margin only. Telson, usually without fixed spinners. No spine on inner border of 1st article of antennular pedencle. Rostrum longer than the eye inner border of 1st of antennular pedencle bearing a spine. Carapace without longitudional and transverse sutures Males with asymmetrical petasmal. 2 arthrobranchs present on last therastic segment one of them well developed the other vestigial ....(Satyal Nadlal and Timothy Pickering)

Further Classification

26(14) Stridulating organ present on posterior branchiostegite 27

— Stridulating organ absent from posterior branchiostegite 28

27(26) Dorsal carina of 3rd pleonic somite sulcate; stridulating organ almost straight; anterior edge of thelycum plate entire; left petasmal lobe sharply pointed and triangular .... *Metapenaeopsis stridulans* Alcock

28(26) A pair of tooth- like platelets behind thelycal plate, posterior tubercles lacking .... *Metapenaeopsis mogiensis* (Rathbum)

— No teeth like platelets immediately posterior to thelycal plate 29

29(28) Posterior extension of thelycal plate with indistinct median sulcus and angular postero-lateral corners .... *Metapenaeopsis andamanensis* (Wood-Mason and Alcock)

— Posterior extension of thelycal plate with distinct median sulcus and evenly rounded postero-lateral corners .... *Metapenaeopsis philippii* (Bate)

Key to the Identification of Different Genera of Family Penaeudae

Rostrum tooth are well developed.carapace without postorbital spine and with a short carvical groove ending well below dorsal midline. Last two pairs of pereopods well developed. Third two pairs of pereopods well developed.

Endopods of second pair of pleopods in males bearing appendix masaulina only (lacking of appendix interna and muscular projection) Last three pleurae are keeled dorsally. Telson sharply pointed.

Rostral teeth 7-10/3-7; chelae of second legs without teeth; the 2 pairs of dorsal telsonic spines clearly not placed at lateral margin. *P. (H.) batei Holthuis*, 1950. They are endemic, abundant throughout New Zealand in sheltered shallow harbours and on shelf from low tide to about 30 fathoms. Transparent, lightly scattered with green and red pigment. Length is up to 1.5 inch. possibly oviparous throughout year. (In our literature identified as *Palaemon* or *Brachycarpus audouini*).

Table 4.2: A Comparison of the Diagnostic Characters of Three Species of *Metapenaeus*

Sl.No.	*Item*	*M. kutchensis*	*M. monoceros*	*M. affinis*
1.	Pabescence	Body only partly covered with harsh and very short to minimum	Body fully covered with harsh and very short tomentum	Carpace finely setose; abdomen may have some giabrous areas.
2.	Rostrum	Straight and with a small crest	Nearly straight and uplifted	More curved and less uplifted.
3.	Mid-dorsal carination of the abdominal segments	Carination commences from the fourth segment	Carination commences from the second segment	Carination commences from the second segment
4.	Ischial spine on 1st pereiopod	Present	Present	Generally absent; if present small denticle only
5.	Length of 5th pereiopod	Reaches a little beyond the middle of antennal scale	Reaches a little beyond the middle of antennal scale	Surpasses the tip of the antennal scale by dactylus
6.	5th pereiopod of adult male	With a shallow notch and feeble tooth at the base of merus	With a notch and hook-like spine at the base of merus	With a notch and tooth at the base of the merus
7.	Petasma	Distomedian lobs more transversely placed with proximal and narrow and distal end broad	Distomedium lobes hook-like	Distomedium lobes ending in a pair of two-lipped spouts resembling a pair of short horns
8.	Thelycum	Posterior plate concave, without ear like lobes, cut transversely into two unequal segments, with no apparent clusters of setae between them.	Posterior plate concave, bounded laterally by elevated ear-like lobes	Posterior plate laterally flat, cut transversely into two unequal segments, with conspicuous clusters of setae between them.

Source: George P. C. *et., al.* 1963.

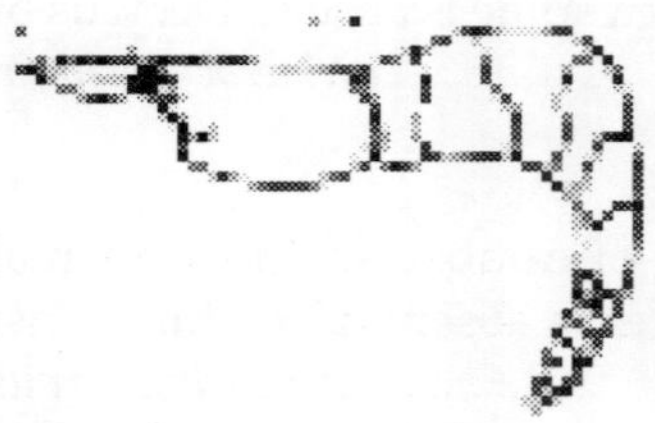

Figure 4.5: *Palaemon affinis*

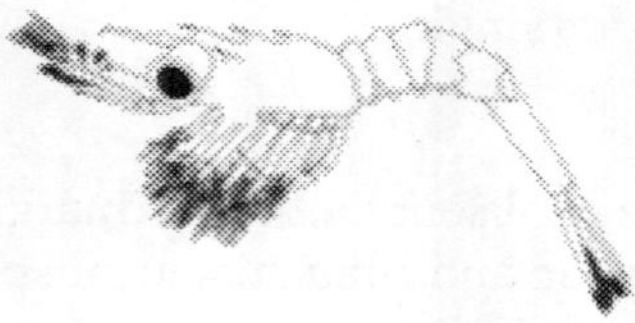

Figure 4.6: *Palaemon serratus*

Figure 4.7: *Croagon croagon* (Variety of *Palaemon serratus*)

Key to the Families of Penaeidea

1(2) First three pairs of legs chelate; fourth and fifth pairs well developed; rostrum usually extends beyond eye; gills numerous (more than 8 on each side). Family. .... PENAEIDAE.

2(1) First legs characteristically non-chelate; fourth and fifth pairs obviously shorter than anterior legs, or even absent; rostrum short or absent; gills few (not more than 8 on each side) or absent.

3(4) Head not elongated; second and third legs with minute chelae. Family. .... SERGESTIDAE

4(3) Head greatly elongated; first two pairs of legs non-chelate; third minutely chelate; fourth and fifth legs absent; gills absent. Family .. LUCIFERIDAE

(Contains the single planktonic genus *Lucifer*, one species known here. *L. typus* M.-Edwards, 1837, with relatively long eyestalks together with eyes subequal to elongated 'head', taken off North Auckland Peninsula. Length up to 0.5 in. In our literature as *L. batei*).

Family *Penaeopsis*

Carapace is with longitudional and transverse sutures. Male are with symmetrical petasmal, a single arthrobranch on last throracic segent. No trace of a second arthrobranch.

Family *Trochpeneopsis*

Carapace without lateral keels outing portion of mandible short and massive.

Rostrum toothed on a dorsal margin only. Rostrum toothed on dorsal as well as on ventral margin. Telson tridentate, with a fixed spine on each side of tip. Rostrum shorter than the eyes no spines on inner border of first article of antennular pendunce.

Family *Plaemonidae*

Branchiostegal spine is absent, anoterior margin of carapace below antennal spine unarmed. Both hepatic and branchiostegal spines absent .... *Leptocarus*

Branchiostegal spine is absent, anoterior margin of carapace below antennal spine unarmed. Hepastic spine is present. Branchiostega spine is absent. Decylus of last three pairs of pereopods single .... *Macrobranchium*

Genus *Brachyocarpus*

Branchiostegal spine is absent, anoterior margin of carapace below antennal spine unarmed. Both hepatic and branchiostegal spines is absent. Dectylus of last three pairs of pereopods are bifid. .... *Brachyocarpus*

Genus *Palaemon*

Rostral teeth 7-10/3-7; chelae of second legs without teeth; the 2 pairs of dorsal telsonic spines clearly not placed at lateral margin. *P. (H.) batei Holthuis*, 1950. Transparent, lightly scattered with green and red pigment. Length is up to 1.5 in. Possibly ovigerous found throughout year. (In our literature as *Palaemon* or *Brachycarpus audouini*.)

Branchiostergal spine present; mandible with a palp; eyes distinctly pigmented; first pleopod of male with or without a rudimental appendix interna on the endopod; branchiostergal spines present as a sharp line; propodus of 5th pereopod with transverse rows of setae on the distal part of the posterior margin; the two median hairs of the posterior margin of telson are slender Palaemon.

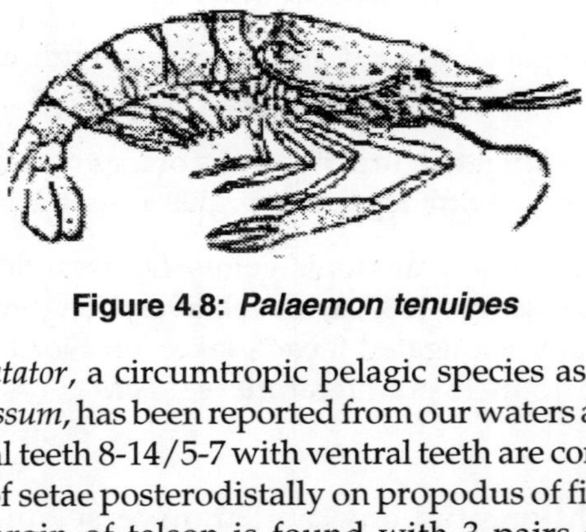

Figure 4.8: ***Palaemon tenuipes***

Palaemon or *L. natator*, a circumtropic pelagic species associated with floating algae, especially *Sargassum*, has been reported from our waters and should be watched for in the north. Rostral teeth 8-14/5-7 with ventral teeth are concealed by double row of setae and no rows of setae posterodistally on propodus of fifth leg. Length is up to 2 inch.) Posterior margin of telson is found with 3 pairs of spines. (Subfamily PONTONIINAE.) l They are rage group of small, secretive, often commercial forms, mainly Indopacific in distribution. Many genera, those in New Zealand with rostrum compressed and toothed; all maxillipeds with exopods; pleura of first to fifth abdominal segments rounded or bluntly pointed; dactyli of fifth legs without basal protuberance (mandibular palp absent).

Figure 4.9: Rostrum of ***Palaemon affinis***

Palaemon affinis M.-Edwards, 1837 with rostral teeth 5-10/2-5 and propodus of fifth leg with transverse rows of setae posterodistally, is our endemic, common, intertidal shrimp. Found throughout New Zealand, being extremely abundant in the brackish waters and mangrove swamps of the north. It is transparent with longtudinal wavy red and green bands and a prominent diagnostic orange and black spot laterally on sixth abdominal segment. Length is up to 3 inch.

Genus *Palaemonetes*

Branchiostegal spine present, situated on or slightly behind anterior margin of carapace. Endopods of first pleopods of male without well developed appenmdix interna, fifth pair of pereopod with transverse row of hairs on distal part of posterior margin. Rostrum is without an elevated basal dorsal crest of teeth, pleura of fifth abdominal segment ending in a tooth. Mandibular palp absent .... *Palaemonetes*

Genus *Nemaptopalaemon*

Branchiostegal spine present, situated on or slightly behind anterior margin of carapace. Endopods of first pleopods of male without well developed appendix interna, fifth pair of pereopod with transverse row of hairs on distal part of posterior margin. Rostrum with an elevated basal dorsal crest of teeth, pleura of fifth abdominalo rounded. Dectyls of last 3 pairs of pereopods very strongly lengthen, longer than carpus and propodus together, branchiostegal groove absent from carapace. .... *Nemaptopalaemon*

Genus *Exopalaemon*

Branchiostegal spine present, situated on or slightly behind anterior margin of carapace. Endopods of first pleopods of male without well developed appendix internal, fifth pair of pereopod with transverse row of hairs on distal part of posterior margin. Rostrum with an elevated basal dorsal crest of teeth, pleura of fifth abdominal rounded. Dectyls of last 3 pairs of pereopods always shorter than produs branchiostegal groove present on carapace .... *Exopalaemon*

Chapter 5

Identification of Adult Prawns/Shrimp

A thorough knowledge of morphological character is highly essential for the identification of prawns/shrimp. These characters are illustrated clearly in the schematic drawing of penaeid prawn. The Taxonomic names of each part are mentioned here, which will be used in identification in this book.

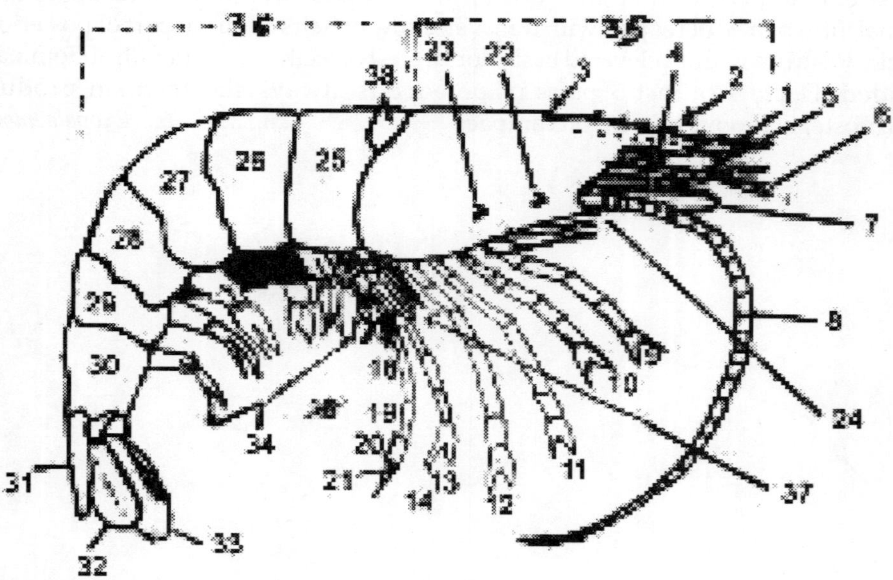

Figure 5.1: Drawing and Taxonomic Names of Each Part of Penaeid Prawns

Diagramatic View of Penaeid Prawn Showing Taxonomic Characters

1. Rostrum	2. Rostral spine
3. Epigastric spine	4. Eye
5. Antennule	6. Antennular flagellum
7. Antennal scale	8. Antennal flagellum
9. Third maxilliped	10. First pereopods
11. Second pereopods	12. Third pereopods
13. Fourth pereopods	14. Fifth pereopods
15. Coxa	16. Basis
17. Ischium	18. Merus
19. Carpus	20. Propodus
21. Dectylus	22. Post orbital spine
23. Hepatic spine	24, Pterygostomian spine
25. First Abdominal segment	26. Second Abdominal segment
27. Third Abdominal segment	28. Fourth Abdominal segment
29. Fifth Abdominal segment	30. Sixth Abdominal segment
31. Telson	32. Endopodite
33. Exopodite	34. Pleopods
35. Cephalothorax	36. Abdomen
37. Petasma	38. Arthroidal membrane

Source: Dholakia, A. D. 2004

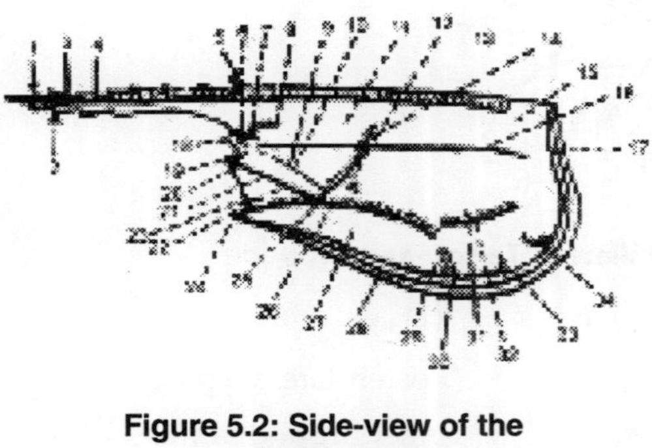

Figure 5.2: Side-view of the Carapace

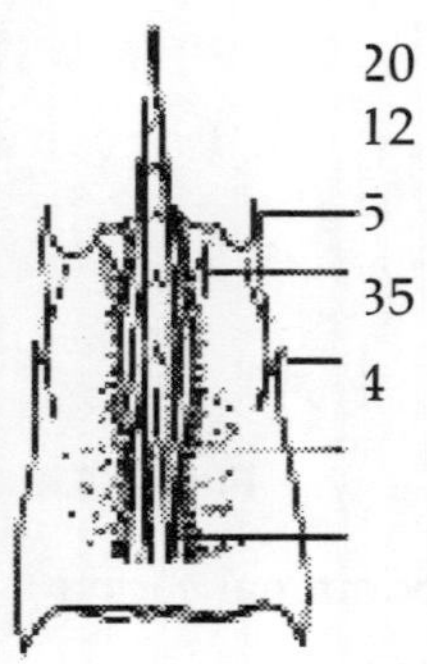

Figure 5.3: A View from the Above of the Carapace

Identification Characters Using Carapace of Penaeid Prawn

1. Rostral tooth
2. Ventral rostral tooth
3. Adrostral carina
4. Adrostal sulcus
5. Epigastric spine
6. Gastrofrontal carina
7. Gastrofrontal sulcus
8. Postocular sulcus
9. Orbito-antennal sulcus
10. Gastroorbital carina
11. Gastric region
12. Hepatic spine
13. Cervical sulcus
14. Cervical carina
15. Cardiac region
16. Longitudinal suture
17. Ptery-gastomian sulcus
18. Orbital or supra orbital spine
19. Antennal
20. Postorbital spine
21. Antennal carina
22. Postantannal spine
23. Branchiostegal spine
24. Ptery-gastamian spine
25. Hepatic carina
26. Hepatic sulcus
27. Ptery-gastpmain region
28. Marginar region
29. Inferior carina
30. Inferior sulcus
31. Branchiocardinal carina
32. Branchiocardinal sulcus
33. Transverse suture
34. Strridulatin organ
35. Postrostral carina
36. Postrostral or median sulcus

Source: Dholakia, A. D. 2004

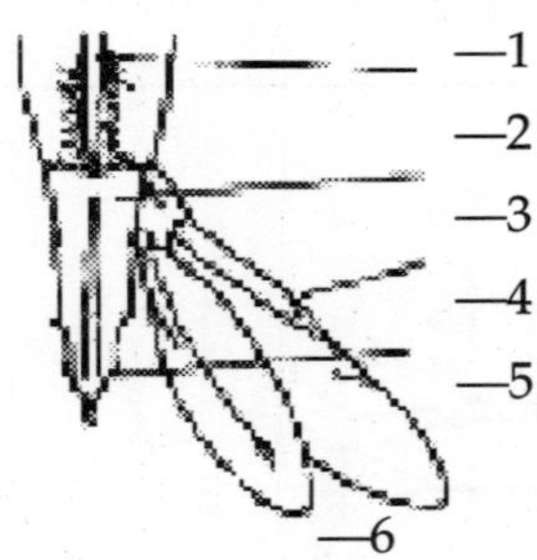

Figure 5.4: Dorsal View of Telson and Uropod

1. Abdominal carina
2. Telson
3. Uropods
4. Posterolateral spine
5. Outer uropod ramus
6. Inner uropod ramus

IDENTIFICATION OF ADULTS

Classification

Phylum	Arthropoda
Class	Crustacea
Sub class	Malacostraca
Order	Decapoda
Sub order	Natantia
Family	Penaeidae
Genus	Penaeus
Species	Monodon

Penaeus monodon (Fabricius)

New name: Penaeus (Penaeus) monodon Fobricius, 1798

Common name: Giant tiger prawn.

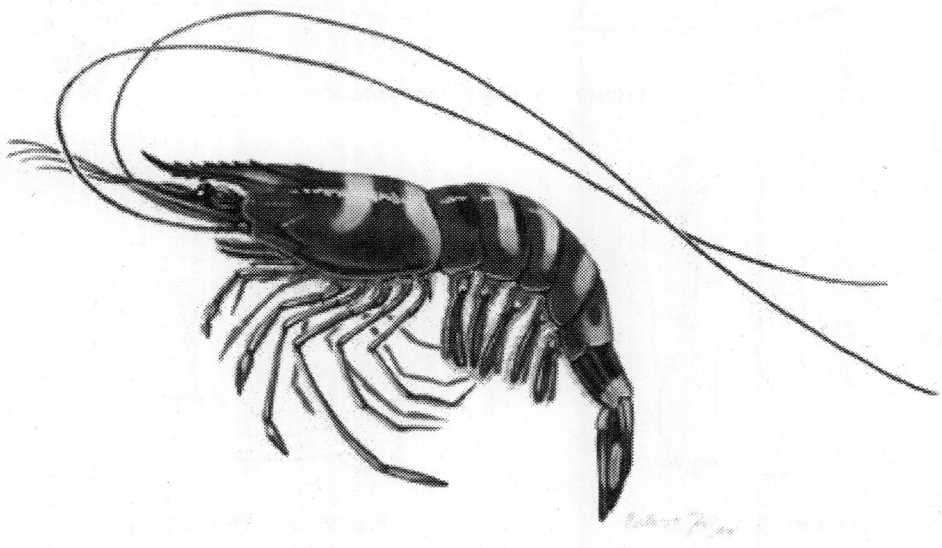

Figure 5.5

Distinctive Characters

Carapace is smooth. Rostrum armed with 7 or 8 teeth on dorsal, and 3 or 4 (rarely 2) teeth on ventral margin; adrostral crest and groove extending as far as, or slightly ahead, of epigastric tooth; postrostral crest well developed, almost reaching posterior margin of carapace, with or without a feeble median groove; gastrofrontal crest absent; antennal crest very prominent, ending above middle of hepatic crest; gastro-orbital crest extending over posterior half, or less, of distance between hepatic spine and orbital margin; hepatic crest straight, almost horizontal, distinctly separated from

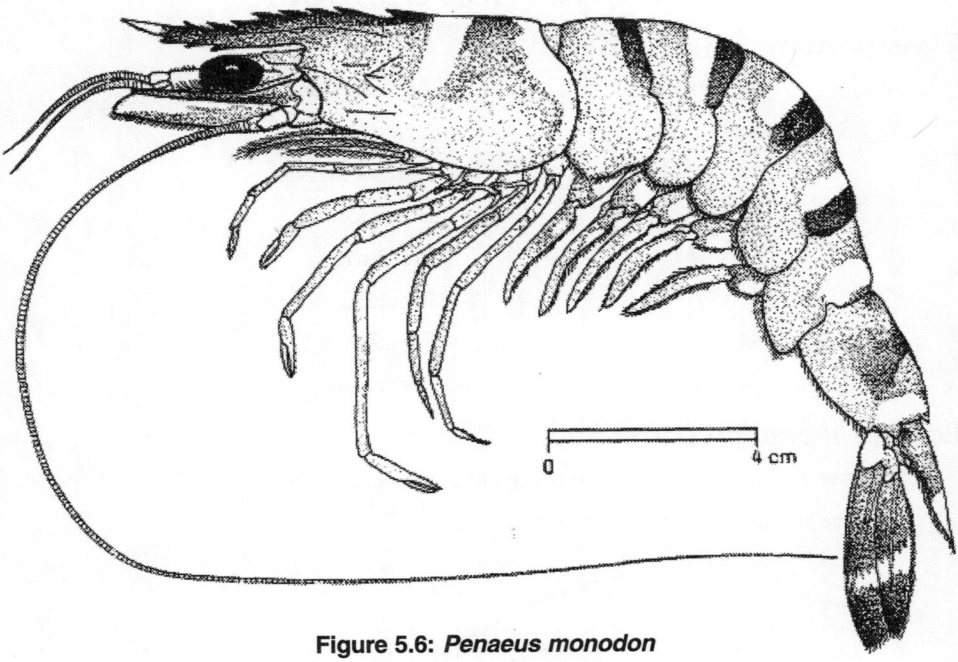

Figure 5.6: ***Penaeus monodon***

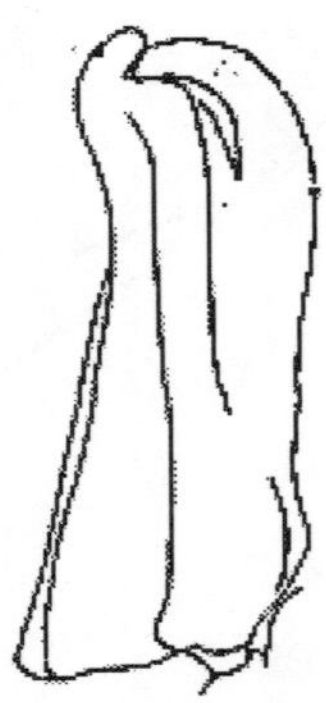

Figure 5.7: Petasma

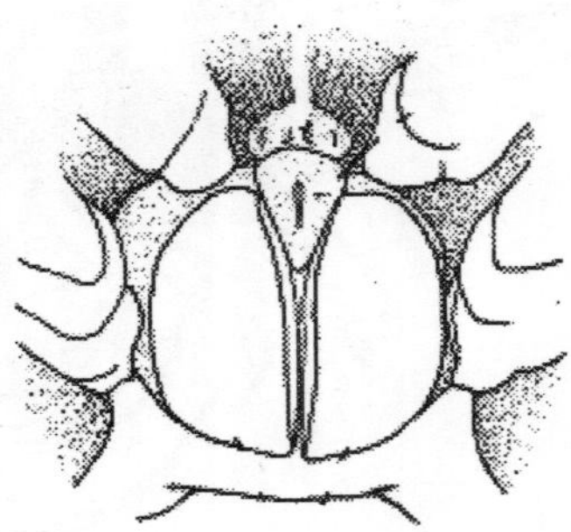

Figure 5.8: Thelycum

base of antennal crest; fifth pereopod without exopod. Petasma (in males) with distomedian projections slightly overhanging distal margin of costae; ventral costae generally unarmed, sometimes minutely serrate at tip; outer surface of lateral lobes generally unarmed; inner surface of lateral lobes armed with spinules. Thelycum (in females) with lateral plates, their median margin sometimes forming tumid lips; anterior process concave, rounded distally; posterior process subtriangular, partly inserted between thelycal plates.

Post orbital spine absent; carapace without a median denate crest occasionally over the eyes. No distinct median tubercle on ocular peduncle; upper antennular flagellum sub-equal to the lower one attached to apex of third antennular segment. Rostrum is with ventral teeth. Adrostral carina is not reaching behind middle of carapace; gastro frontal carina absent. Hepatic carina is horizontally straight; 5th pereopod without expodite. Maximum size of this species is recorded as 320 mm.

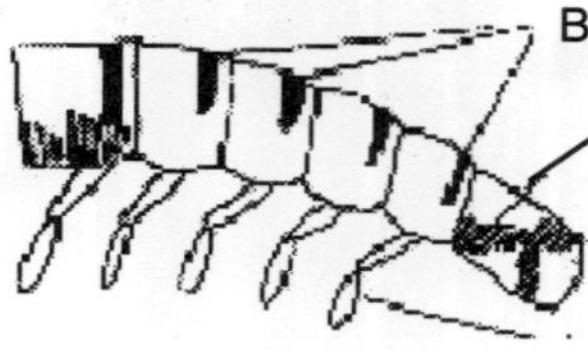

Figure 5.9: Colouration of Tail of *P. monodon*

Colour

Body green-grey to brown, sometimes reddish or bluish; dorsoposterior margin of carapace generally cream-yellow, often a transverse band of the same colour near middle of carapace; abdomen with dark brown to dark grey and pale yellow dorsal transverse bands (exceptionally absent); antennae uniform pink-brown; pereopods and pleopods of same colour as body or darker, with cream-coloured spots; uropods brown, green-grey or bluish, with a pale yellow to pink median transverse band. Juveniles are pale green, with dark transverse bands on first, third and last abdominal segments.

Size

Maximum total length: males, 26.8 cm; females, 33.7 cm; this is the largest penaeid species known.

Distribution

Within the area, from south and east Africa to the Gulf of Aden and the Red Sea, the west coast of Madagascar, Mauritius and La Réunion; also present from Pakistan to south India and Sri Lanka.

Carapace of similar looking species.

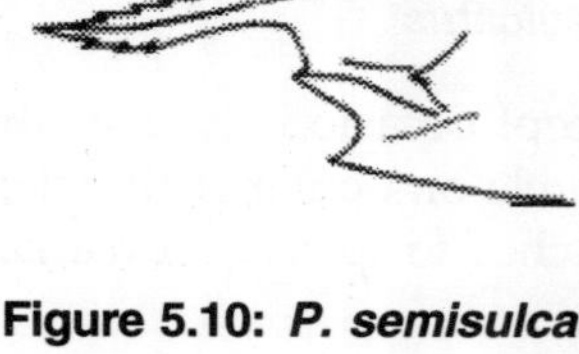

Figure 5.10: *P. semisulcatus*

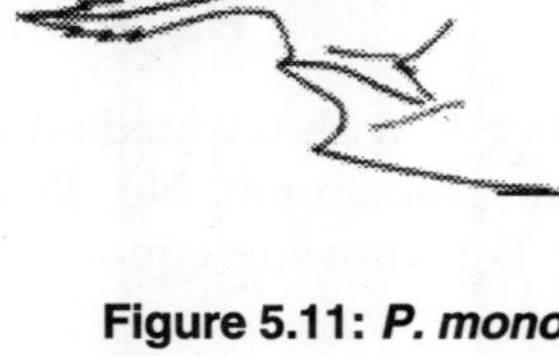

Figure 5.11: *P. monodon*

Penaeus semisulcatus (de-Haan)

Penaeus (Penaeus) semisulcatus De Haan, 1844

Common name: Green tiger prawn

Figure 5.12

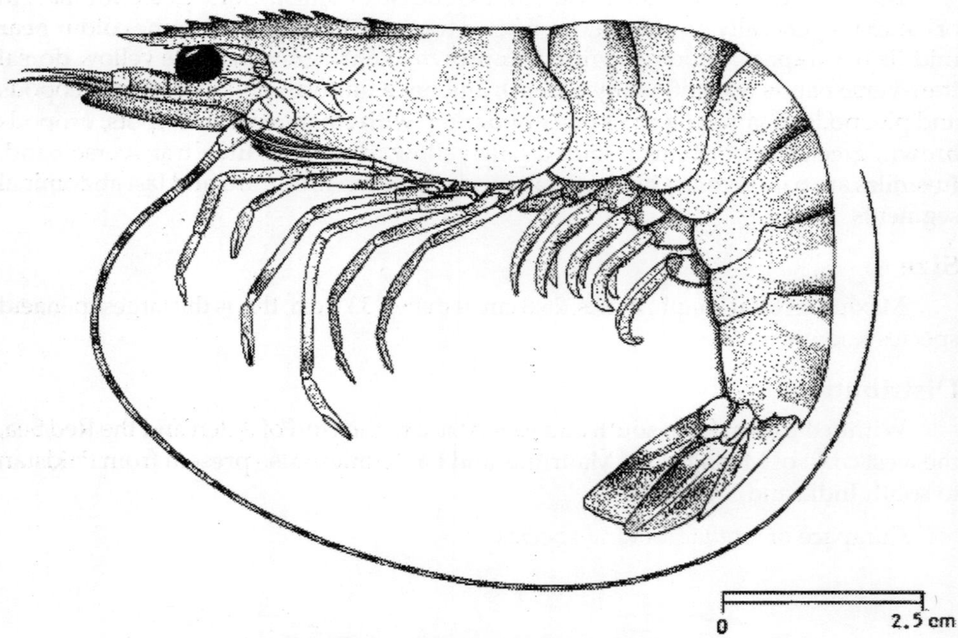

Figure 5.13: *Penaeus semisulcatus*

Carapace is without median dentate crest except occasionally over the eyes. Post–orbital spine absent. No distinct median tubercle on ocular peduncle; upper antennular flagellum sub-equal to the lower one attached to apex of third antennular segment.

Figure 5.14: Petasma

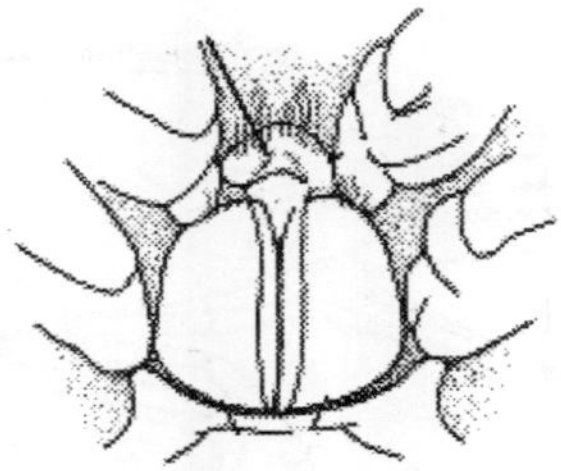

Figure 5.15: Thelycum

Adrostral cerina not reaching behind middle of carapace; gastro-frontal carina absent; Hepatic carina inclined at an angle of 20° anteroventrally; 5th pereopod with small expodite Maximum size of this species is recorded as 250 mm.

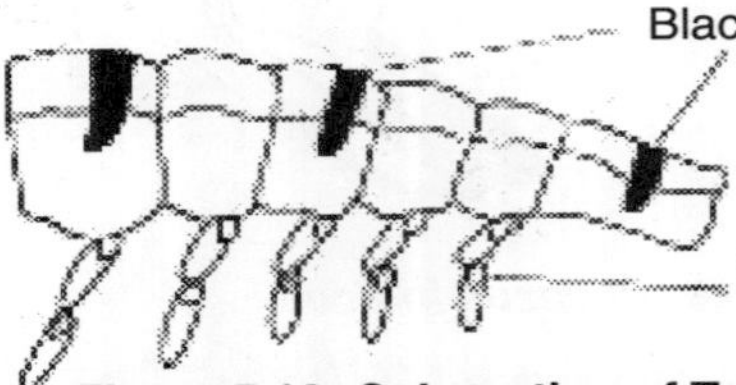

Figure 5.16: Colouration of Tail of *P. semisulcatus*

Penaeus indicus Milne Edw.

New name: Penaeus (Fenneropenaeus) indicus H. Milne Edwards, 1837

Common name: Indian white shrimp

Figure 5.17

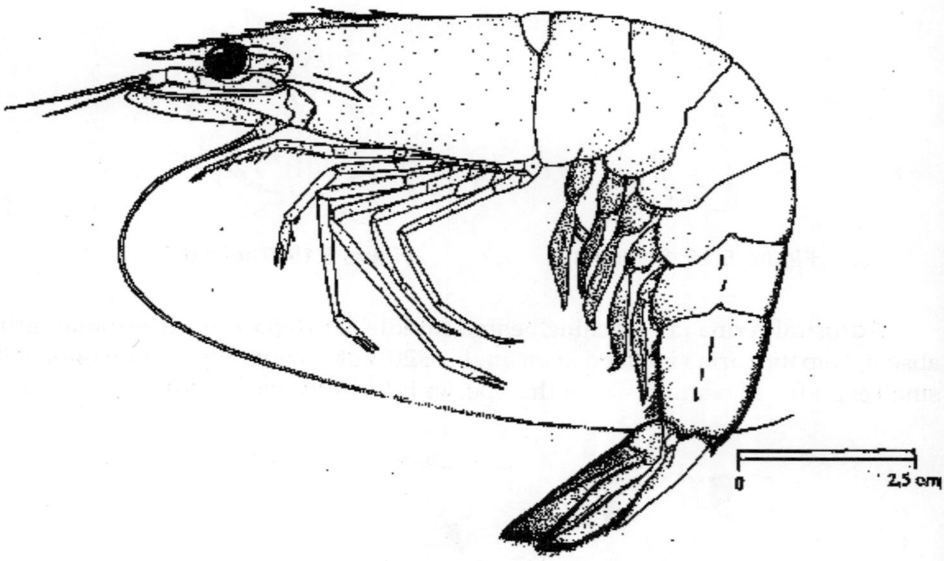

Figure 5.18: *Penaeus indicus*

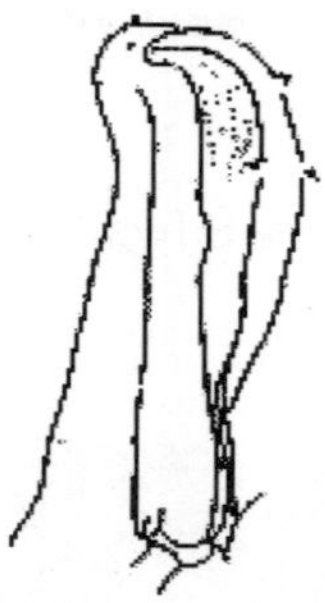

Figure 5.19: Petasma

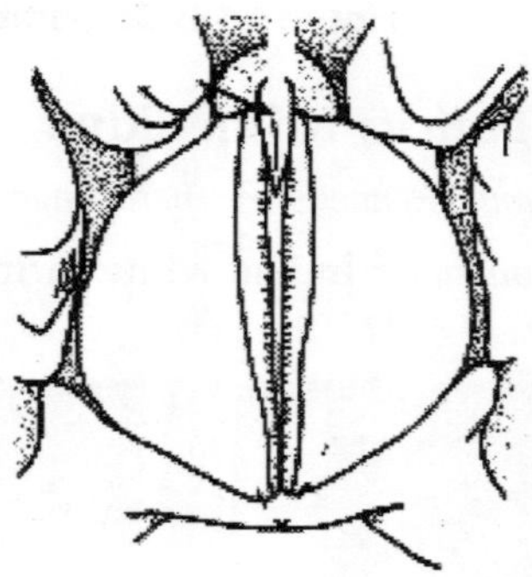

Figure 5.20: Thelycum

Distinctive Characters

Carapace is hairless. Rostrum slender and long, with 7 to 9 teeth on dorsal, and 4 to 6 teeth on ventral margin; blade of rostrum becoming moderately high in large specimens; adrostral crest and groove extending as far as, or just beyond epigastric tooth, the groove shallow; postrostral crest ending distinctly before posterior margin of carapace; gastrofrontal and hepatic crests absent; gastroorbital crest extending over posterior 213 of distance between hepatic spine and orbital margin; in adult males dactyl of third maxilliped about as long as propodus (0.85 to 1.0 times). Petasma in males with distomedian projections overhanging distal margin of costae; ventral costae unarmed; outer surface of lateral lobes with a few rows of minute tubercles. Thelycum (in females) with lateral plates, their median margins forming tumid lips beset with papillae on their inner surface; anterior process rounded distally; posterior process ill-calcified, almost completely inserted between lateral plates.

Adrostral carina is not reaching behind middle of carapace; gastro–frontal carina absent. Hepatic carina is also absent. Gastro–orbital carina occupying the posterior 2/3 distance between hepatic spine and orbital angle; rosteral crest must be elevated but not triangular in profile.

Rostrum is with ventral teeth. On dorsal side of rostrum 7 to 9 spine are present while on ventral side they are 4 to 6; rostrum crest feeble. Post orbital spine absent; carapace without a median dentate crest except occasionally over the eyes; No distinct median tubercle on ocular peduncle; upper antennular flagellum sub–equal to the lower one attached to apex of third antennular segment.

Size

Maximum size of this species found in India is 230 mm.

Colour

Body is pale pink to yellowish, semi-translucent, with olive-green to grey-blue speckles; rostral and middorsal abdominal crests mostly brown, but reddish at base; pereopods generally of same colour as body; pleopods pink or red; distal part of uropods green or red, fringe of setae usually red. Juveniles are whitish, with specks of same colour as adults; rostral crest semi-translucent; pleopods whitish.

Distribution

Within the area, from south and east African coast, to India and Sri Lanka including Madagascar and the Red Sea, but possibly absent or very rare in the Gulf of Oman and the "Gulf". Further east it extends to south China, the Philippines and northern Australia. Inhabits shelf areas from the coastline to depths of about 90 m; most abundant in shallow waters in less than 30 m on sand or mud (slight preference for sandy bottoms); a very euryhaline species since in the Red Sea it tolerates salinities of more than 45°Jm; caught by day as well as at night.

Present Fishing Grounds

In India it is one of the most abundant, and at the same time, is among the most highly priced shrimp; also abundant in Sri Lanka but not so from the Gulf of Aden to Pakistan; juveniles contribute to a good fishery in estuarine regions. It is cultivated in India.

Penaeus merguiensis (de–Man)

New name: *Penaeus (fenneropenaeus) merquiensis* De Man, 1888

Common name: Banana shrimp

Distinctive Characters

Carapace is smooth. Rostrum armed with 6 to 9 teeth on dorsal, and 3 to 6 on ventral margin; blade of rostrum high, broadly triangular in shape; adrostral crest and groove not reaching as far as epigastric tooth, the groove shallow; postrostral

Figure 5.21

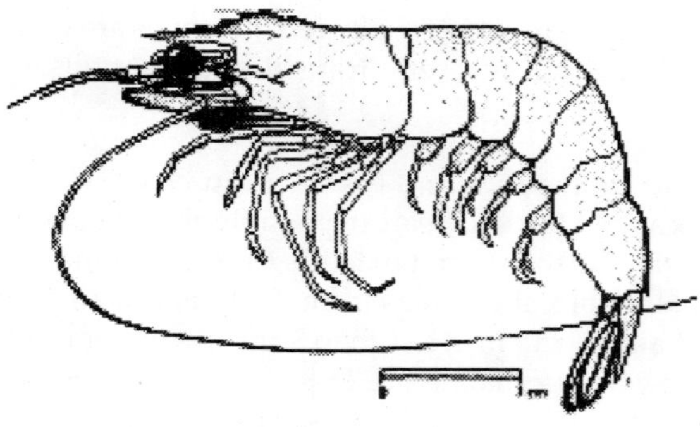

Figure 5.22: *Penaeus merguiensis*

Figure 5.23: Petasma

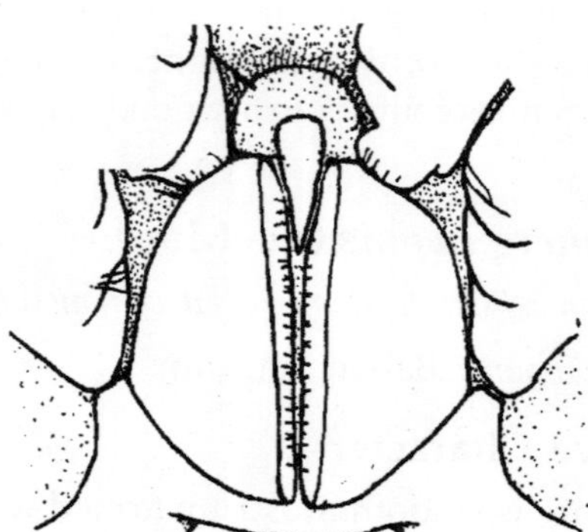

Figure 5.24: Thelycum

crest ending distinctly before posterior margin of-carapace; gastrofrontal and hepatic crests absent; gastro-orbital crest extending over middle third to half of distance between hepatic spine and orbital margin; in adult males, dactyl of third maxilliped half (0.5 to 0.6 times as long as propodus). Petasma in males is with distomedian projections overhanging distal margin of costae; free border of ventral costae serrate near apex; outer surface of lateral lobes with several rows of minute tubercles. Thelycum (in females) with lateral plates, their median margins forming tumid lips beset with papillae on their inner surface; anterior process slightly concave, rounded distally; posterior process ill calcified, almost completely inserted between lateral plates. No distinct median tubercle on ocular pedicles upper antennular flagellum sub-equal to the lower one attached to apex of third antennular segment. Maximum size of this species is recorded as 240 mm.

Colour

Body pinkish to pale yellow, sometimes green-greyish, with green-grey to grey-blue speckles; rostral and middorsal abdominal crests brown-red, sometimes grey; pereopods of same colour as body; pleopods yellowpink to red; distal part of uropods green or red, fringe of setae red. Juveniles are whitish, with specks of same colour as adults; rostral crest translucent; pleopods whitish; distal half of uropods translucent, their tips red.

Size

Maximum total length: males, 19.5 cm; females, 24 cm.

Distribution

Inhabits shelf areas from the coastline to depths of about 55 m, but is most abundant in shallow waters of less than 20 m depth, on mud or sandy-mud; prefers turbid waters (this is not a burrowing species) and forms large shoals when the density of the population is high; caught by day as well as at night.

Present Fishing Grounds

In Pakistan it is commercially very important and, together with *P. penicillatus*, outnumbers all the other species in catches from coastal waters; it also contributes to

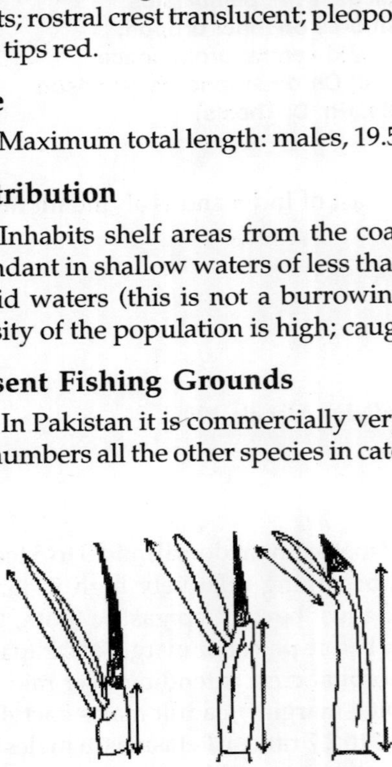

Figure 5.25: Third Maxilliped

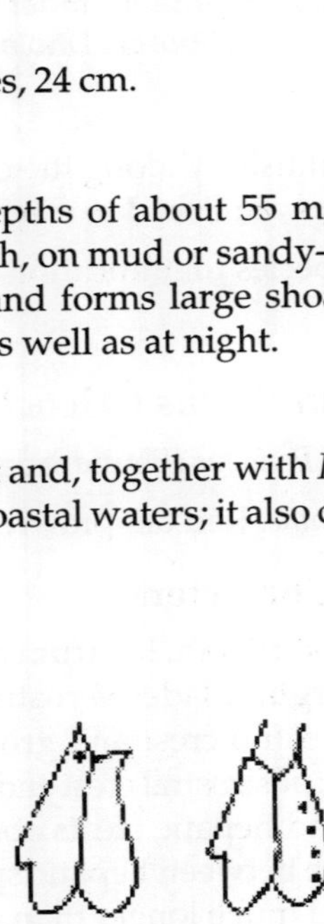

Figure 5.26: Chromatophores

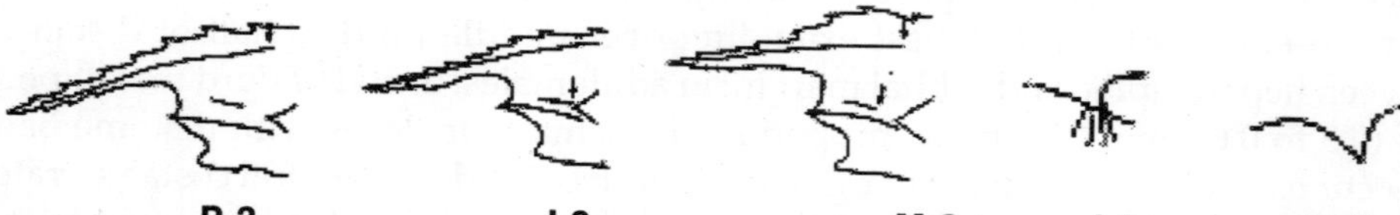

Figure 5.27: Position of Gastroorbital Crest

Figure 5.28: Identification of Difference Between *Penaeus merguiensis, P. indicus* and *P. penicillatus*

M: *Penaeus merguiensis;* I: *P. indicus;* P: *P. penicillatus*
(1) Third maxilliped; (2) Chromatophores on inner uropod; (3) Position of Gastroorbital crest; (4) Mid-ventral prominence; (5) Position of anterior most dorsal tooth; (6) Chromatophores on telson (*Source*: Dholakia, A. D. 1994, Ph. D. Thesis)

the commercial fishery along the central west coast of India and is abundant in Sri Lanka; juveniles are fished in estuarine regions.

Source: FAO Species Identification Sheets Pen 14 1983.

Penaeus penicillatus (Alcock)

New name: *Penaeus* (Fenneropenaeus) *penicillatus* Alcock, 1905

Common name: Red tail prawn

Distinctive Characters

Carapace is smooth. Rostrum armed with 7 to 9 teeth on dorsal, and 3 to 5 teeth on ventral margin; blade of rostrum convex, becoming relatively high in large specimens; adrostral crest and groove extending just beyond epigastric tooth, the groove shallow; postrostral crest ending distinctly before posterior margin of carapace; gastrofrontal and hepatic crests absent; gastro-orbital crest extending over middle third of distance between hepatic spine and orbital margin; in adult males dactyl of third maxilliped much longer than propodus (1.5 to 2.7 times). Petasma (in males) is with very short distomedian projections, generally not reaching distal margin of costae. Thelycum in females is with lateral plates, their median margin forming tumid lips; anterior process slightly concave, rounded distally; posterior process ill-calcified, almost completely inserted between lateral plates.

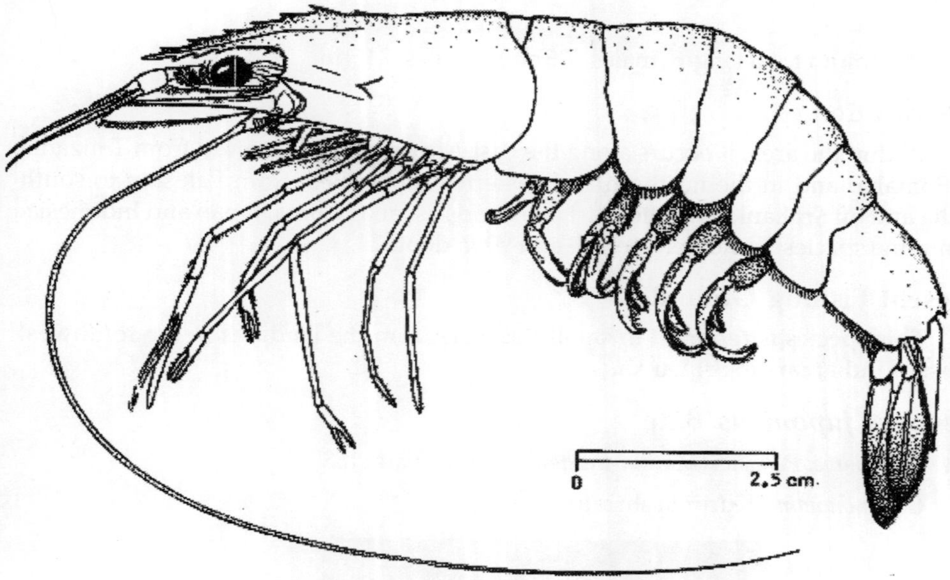

Figure 5.29: ***Penaeus penicillatus***

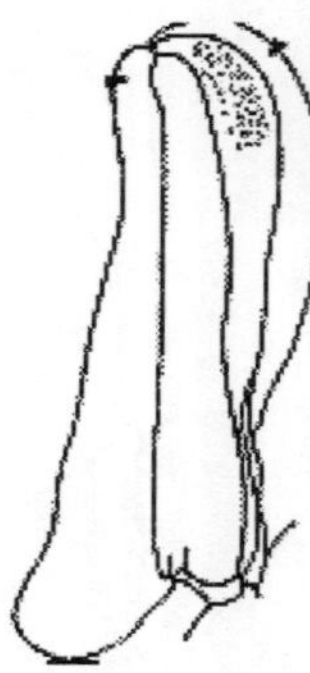

Figure 5.30: Petasma

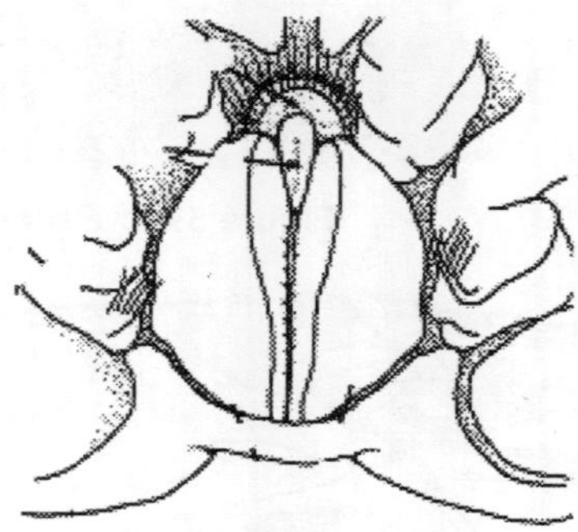

Figure 5.31: Thelycum

Carapace of similar looking species

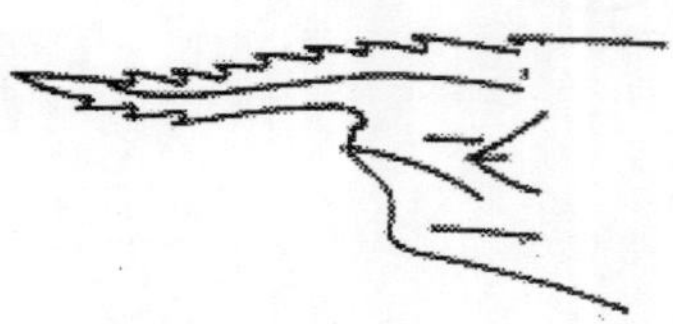

Figure 5.32: Carapace of *P. monodon*

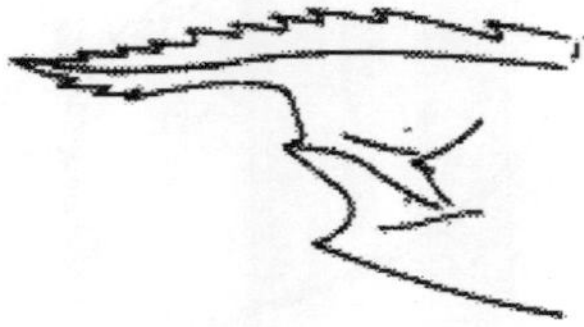

Figure 5.33: Carapace of *P. semisulcatus*

Size

Maximum total length: males, 7.6 cm; females, 7.9 cm.

Distribution

Within the area, it occurs along the east African coast (reported from Tanzania to Somalia) and in the northern and eastern Arabian Sea, from Pakistan to south India and off Sri Lanka. Further east, it extends as far as China, Japan and Indonesia. A marine species found between 10 and 90 m depth.

Present Fishing Grounds

This species is reported in small numbers from the landings in the southwest coast of India; rare in Sri Lanka.

Penaeus japonicus Bate

New name: Penaeus (Marsupenaeus) japonicus Bate, 1888

Common name: Kuruma shrimp

Figure 5.34: ***Penaeus japonicus***

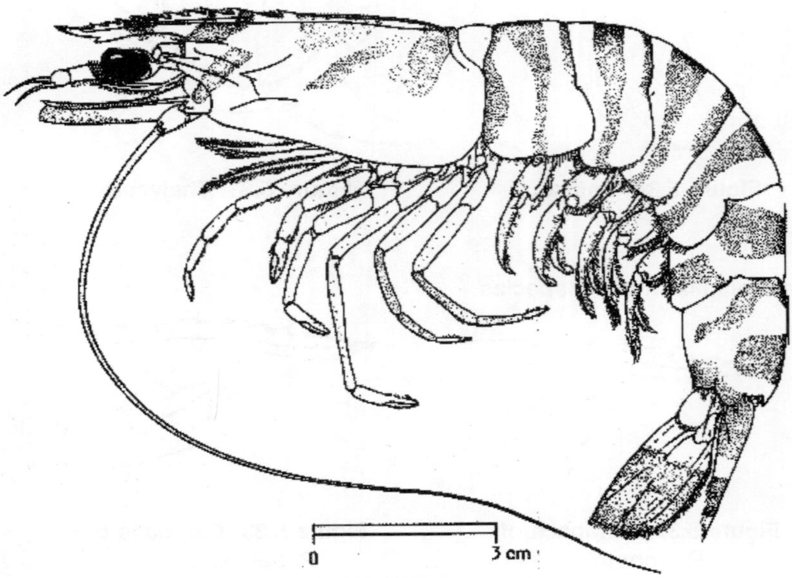

Figure 5.35: ***Penaeus japonicus***

Figure 5.36: Petasma

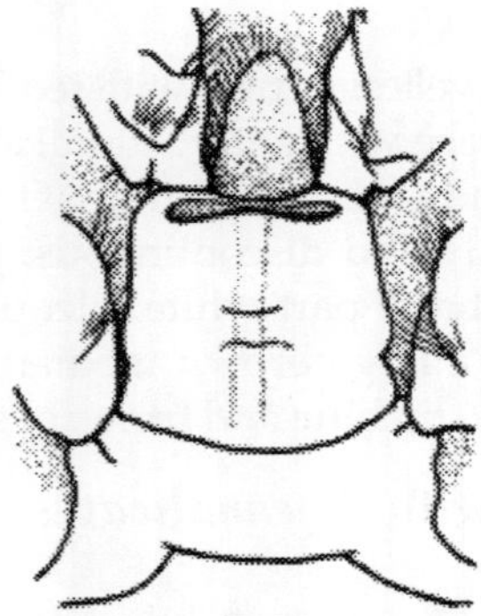

Figure 5.37: Thelycum

Marsupenaeus japonicus

Kuruma Shrimp (Penaeus japonicus): Native to the Indian Ocean and the Southwestern Pacific Ocean from Japan to Australia, kuruma shrimp are farmed in Japan and Australia. Live kuruma shrimp bring outrageously high prices in Japan, as high as $100 a pound! It's relatively easy to ship live animals without water, they mature and spawn in ponds, and they tolerate low water temperatures better than any other farmed species, down to 10 degrees Celsius. They require clean, sandy bottoms and high protein diets (55 per cent). Markets are limited to Japan.

Distinctive Characters

Carapace is smooth. Rostrum is short, reaching only half length of eye. Rostrum armed with 9 to 11 teeth on dorsal, and a single tooth on ventral margin, with an accessory crest on the blade; adrostral crest and groove long, extending almost to posterior margin of carapace, the groove wide; postrostral crest well developed as far back as adrostral groove, with a deep median groove throughout its length; gastrofrontal crest present; gastrofrontal groove bifurcate posteriorly; hepatic crest almost horizontal to base of antennal crest and from there sloping anteroventrally; telson armed with 3 pairs of movable spines; no ischial spine on first pereopod. Chromatophores on telson are present from base to distal end. Inner rami of uropods are with a single chromatophore near base on medial aspect (usually hidden by telson). 10 to 11 Ventral chromatophores are present. Colour of chromatophores in live specimens at PL size is Crimson. Petasma (in males) with long distomedian projections distinctly unhanding distal margins of costae, tips enlarged; ventral costae slightly broadened epically and unarmed; outer surface of lateral lobes not tuberculate. Posteromedian dorsal spines in 5th abdominal segment present up to 4 rostral spine stage. Spinules on dorsal carina of 6th abdominal are present all along length of carina. Anternnal spine on carapace is prominent. Thelycum (in females) without lateral plates but with a pouch widely open anteriorly; anterior and posterior processes forming a triangular, concave plate.

Colour

Body pale yellow to pink with red-brown to dark brown transverse bands; rostrum banded; carapace with anterolateral and dorsal patches (the latter circular in dorsal view and 2 bands, the anterior one at middle of carapace and leaning anteroventrally; last abdominal band discontinuous; pereopods yellow proximally, blue or bluish distally, their basal part white; pleopods yellow, tips bluish, white spots at bases; uropods with a large brown median transverse band, poximally white-creamish, distally yellow, tip blue and fringe of setae red.

Coparision with *P. canalicatus*

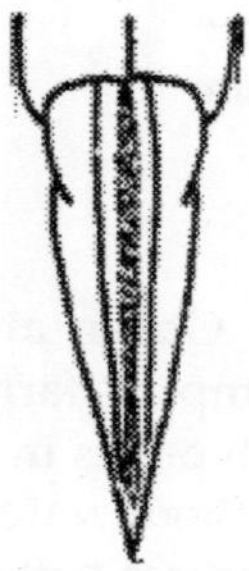
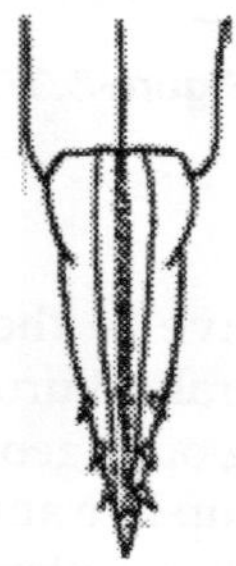

End of Telson

Last Abdominal Segment

Figure 5.38: ***P. canilicatus***

Figure 5.39: ***P. japonicus***

Figure 5.40: ***P. canilicatus***

Figure 5.41: ***P. japonicus***

Penaeus vannamei

New Scientific Name: *Litopenaeus vannamei*

Common name: West Coast white shrimp, Camaron blanco, Langostino.

F.A.O. names: White leg shrimp, Crevette pattes blanches, Camaron patiblanco

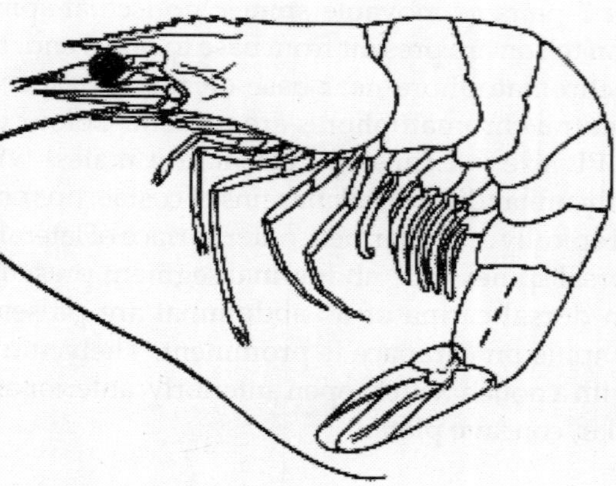

Figure 5.42: ***Litopenaeus vannamei***

Figure 5.43: ***Litopenaeus vannamei***

Penaeus vannamei Boone, 1931

They are with maximum total length 230 mm; maximum carapace length 90 mm. They are found at depth 0 to 72 m. bottom mud in Marine (adults) and estuarine (juveniles).

White shrimp can be stocked at small sizes, have a uniform growth rate and reach a maximum length of 230 millimeters. They breed in captivity better than *monodon,* but not as readily as many of the other penaeids Hatchery survivals are high, from 50 to 60 per cent.

Distinguishing Features

The rostrum is armed with dorsal and usually, 2-4 (occasionally, 5-8) ventral teeth, which are moderately long, and in young distinctly surpassing antennular peduncle. They are shorter in adults, sometimes reaching only to the middle-length of second antennular segment. Carapace has pronounced antennal and hepatic spines, and lacks orbital and pterygostomian spines. The postocular sulcus is absent. The postrostral carina is of variable length, sometimes almost reaching posterior margin of carapace. The adrostral carina and sulcus are short, extending to or only slightly beyond epigastric tooth. Gastrofrontal carina are absent, whereas the gastro-orbital carina is relatively short, usually extending (at most) anteriorly about two-thirds of distance between hepatic spine and orbital margin. The orbito-antennal sulcus is well marked, with sharp cervical and hepatic carinae, and deep accompanying sulci. Branchiocardiac carina is lacking and longitudinal and transverse sutures absent. The sixth abdominal somite bears three cicatrices, dorsolateral sulcus extremely narrow or absent. The telson is unarmed. Antennules lack a parapenaeid spine and antennular flagella are much shorter than the carapace. The palp of first maxilla is elongate, consisting of 3 or 4 articles, with distal ones together flagelliform. The basal article is produced into setose proximal lobes on the lateral and mesial margins, which bear 1 or 2 long distomesial spines, and distolateral row of spinules. Basial and ischial spines are present on first pereopod, and a basial on second.

In mature males the petasma is symmetrical, semi open, not hooded, lacking distomedian projections, and has short ventral costae, not nearly reaching distal margin and distinctly gaping. The spermatophores are extremely complex, consisting of a sperm mass encapsulated by a sheath and bearing various attachment structures (anterior wing, lateral flap, caudal flange, dorsal plate), as well as adhesive and glutinous materials.

The mature female has an open thelycum and sternite XIV bearing ridges, prominences, depressions, or grooves.

Larval Stages

This species has six nauplii stages, three protozeal stages, and three mysis stages in its life history. The CL of *L. vannamei* postlarvae are in range from 0.88 to 3.00 mm. The larval stages (1.95–2.73 mm CL) can be recognized by the lack of a thoracic spine on the 7th sternite, and relative rostral length against the length of eye plus eye stalk ranges from 2/5–3/5, rarely 4/5. The most distinguishable morphological character is the development of supraorbital spines in the second and third protozoea.

Coloration

Translucent white, thus it is most commonly known as the "white shrimp". The body of the species often has a bluish hue that is due to a predominance of blue chromatophores which are concentrated near the margins of the telson and uropods.

Size

It grows to about 230 mm.

Biology

Habitat: This marine shrimp likes muddy bottoms at depths from the shoreline down to about 72 meters [235 feet]. Physio-Ecology: Food and Feeding Habits: Reproduction: In *L. vannamei* the carapace is translucent, permitting the color of the ovaries to been seen. In females, the gonad which is first whitish turns golden brown or greenish brown on the day of spawning. The males deposit the spermatophores only on hard-shell females which will spawn a few hours later. The courtship and mating behavior begins in the afternoon in relation to light intensity. Regression of developing ovaries is very rare and development of the ovaries leads almost every time to spawning. The spawning process begins by sudden jumps and active swimming of the female and the whole process lasts about one minute. The cortical reaction is very rapid and first segmentation occurs in a few minutes (Ogle, 1992). The numbers of eggs varies according to individual size. Female of *L. vannamei* of 30 to 45 g size gives 100.000 to 250.000 eggs. Eggs are approximately 0.22 mm in diameter. Cleavage to the first nauplius stage occurs approximately 14 hours after spawning (Aquacop, 1979). Toxicity: None. Timing and Method of Introduction: The pacific white shrimp was imported in 1985, as postlarvae, by many shrimp farms in South Carolina (Sandifer *et al.*, 1988).

Penaeus esculentus

Figure 5.44: *Penaeus esculentus*

Penaeus esculentus are generally brown with dark banding. Their rostrum and antennae are also banded. *Penaeus esculentus* Prawns are endemic to Australian waters. Adult Tiger Prawns are found to depths of 200 metres, but are mostly trawled in 10-20 metres of water over coarse sediments. A female tiger prawn with a 39mm carapace length produces about 364,000 eggs. *Penaeus esculentus* the brown tiger shrimp looks a lot like the giant tiger shrimp *(monodon)*, only smaller and browner. It is fished year-round and is often caught along with the green tiger shrimp *(semisulcatus)*.An aggressive detrital feeder; *esculentus* has potential in bacterial based systems.

Juvenile Tiger Prawns are found in shallow waters associated with sea grass beds, sometimes on top of coral reef platforms.

Tiger prawns feed primarily at night. Their diet consists of molluscs, crustaceans and polychaete worms.

Eastern King Prawn (*Penaeus plebejus*)

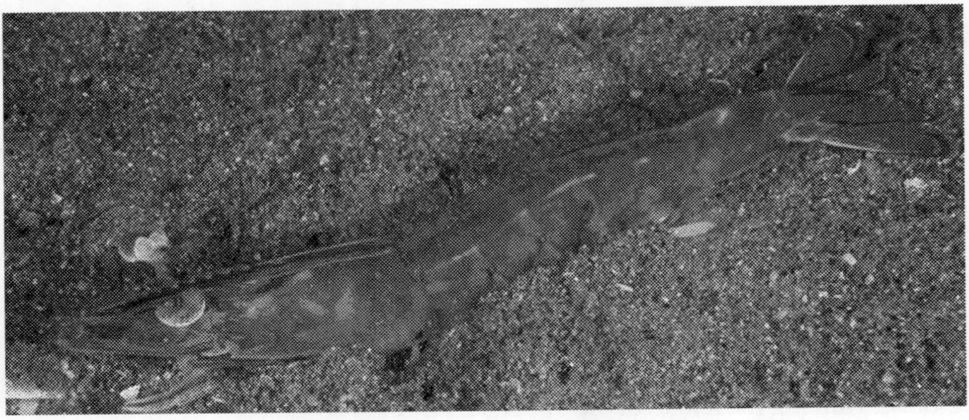

Figure 5.45: *Penaeus plebejus*

This species prefers to live in estuarine water that has sadly bottoms and they are also found in oceanic waters to depths of 220 meters. They are brightly coloured tail with blue and red lines. The body is cream to yellow in colour. Penaeus plebejus grows to approximately 30 cm total length for female and 19 cm for male.

Red Spot King Prawn (*Penaeus longistylus*)

Figure 5.46

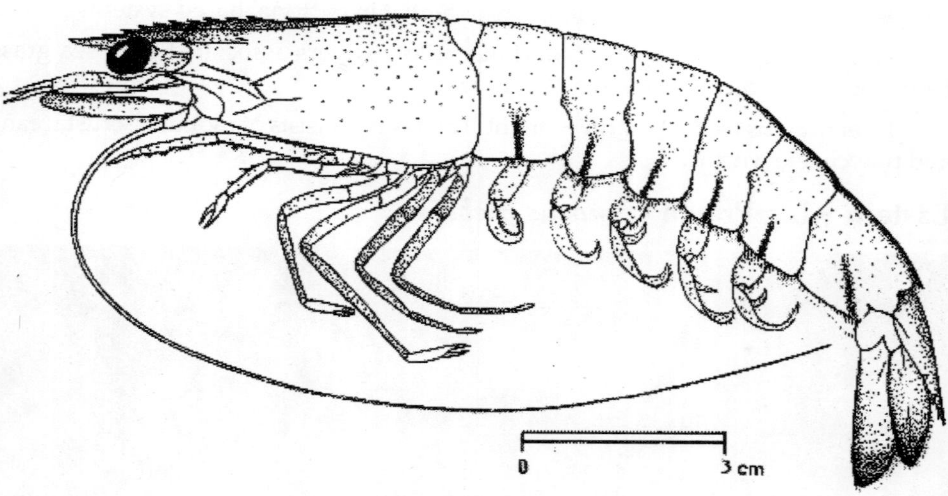

Figure 5.47: ***Penaeus longistylus***

Red Spot King Prawns are grooved prawns that are they have a pair of parallel grooves running the length of the upper surface of the carapace.

Red Spot King Prawns can be distinguished from Western King Prawns by having a red abdominal spot and blue-red over cream body markings (blue over cream in western king prawns).

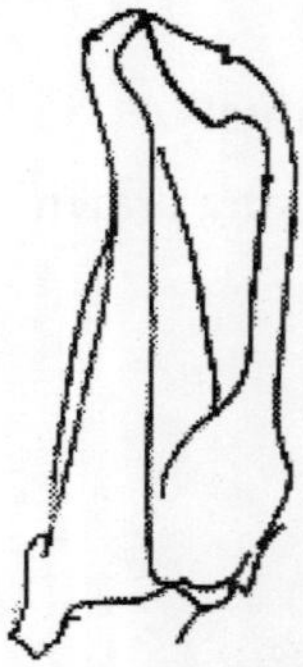

Figure 5.48: Petasma

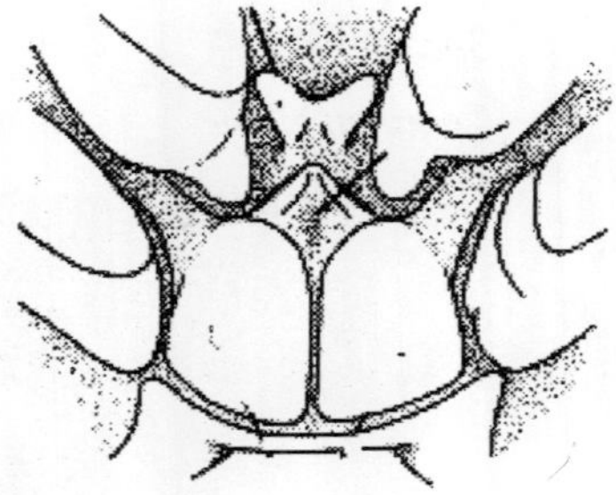

Figure 5.49: Thelycum

In Western Australia, Red Spot King Prawns are found with sponges, soft corals, coarse sand or shell grit substrates, coral reefs and coral rubble. On the Queensland east coast they are rarely found more than 30km from coral reef systems. Through much of Queensland, Red Spot King Prawn juveniles inhabit coral reef lagoons in depths of 1-3 metres, yet in north-eastern Queensland the juveniles inhabit estuaries and reef tops. Adults inhabit inter-reef channels and adjacent waters in depths from about 18m to 60m, often on sand-mud substrates.

Red Spot king prawns have an extended spawning period between May and October. Not much information is available on the prawn's larval form and maximum age. The Red Spot King Prawn appears to be unusual in that, unlike some other prawns, they do not require estuarine or coastal environments for the successful completion of their life cycle.

Commercial fishing for Red Spot King Prawns is primarily in the eastern Queensland part of its range. The main fishing area is between about Lucinda and Bowen on the central Queensland coast, over a 20-30 wide strip to the west of the Great Barrier Reef.

Red Spot King Prawns are caught by demersal otter trawlers operating at night. The main fishing area is open water in depths of 40-60 metres in the vicinity of coral reefs.

Penaeus latisulcatus Kishinouye

New name: *Penaeus* (Melicertus) *latisulcatus* Kishinouye, 1896

Classification:

Kingdom	Animalia
Phylum	Arthropoda
Class	Malacosta
Order	Decapoda
Sub order	Dendrobranchiata
Family	Penaedae

Genus	Penaeus
Species	Latisulcatus
Common Name	Western king prawn

Figure 5.50

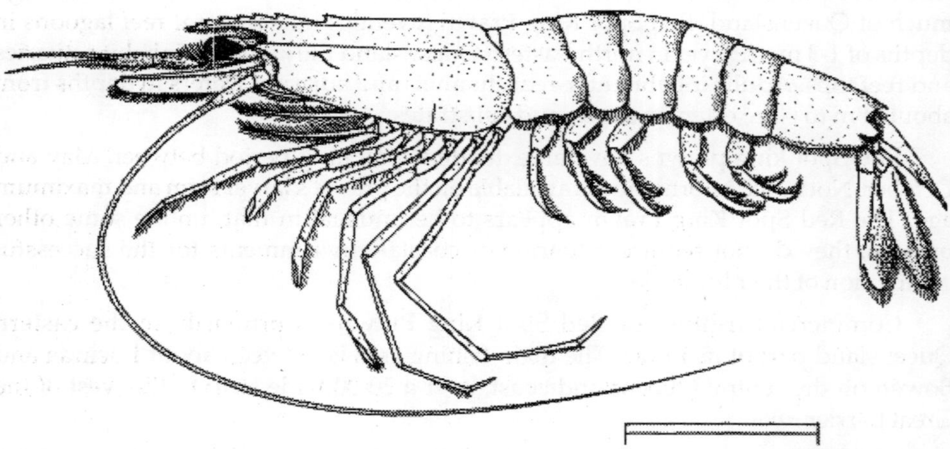

Figure 5.51: *Penaeus latisulcatus*

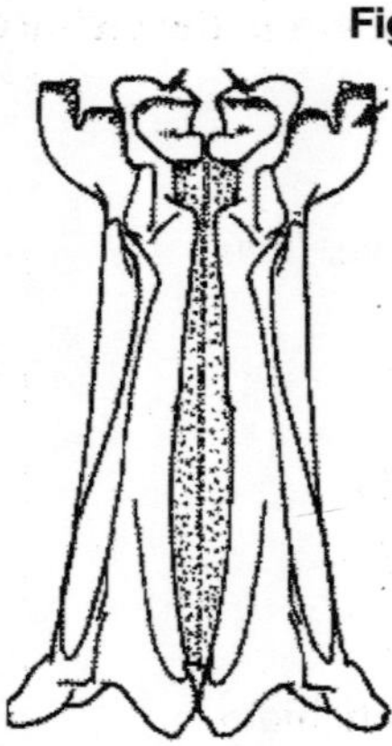

Figure 5.52: Petasma

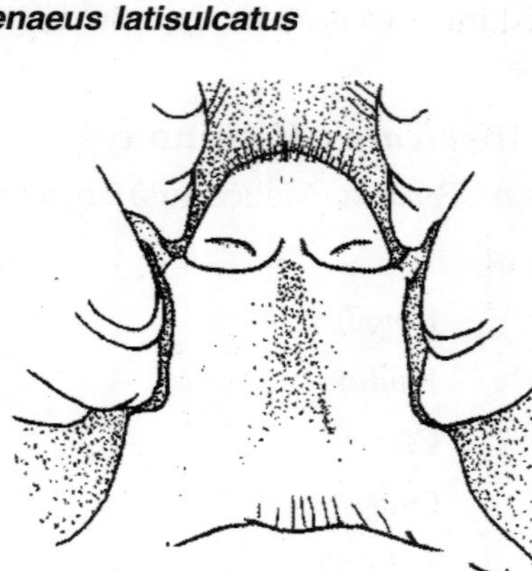

Figure 5.53: Thelycum

Distinctive Characters

Carapace is smooth. Rostrum with 9 to 12 teeth on dorsal, and a single tooth on ventral margin, sometimes with a feeble accessory crest on the blade; adrostral crest and groove long, extending almost to posterior margin of carapace, the groove wide; postrostral crest well developed as far back as adrostral groove, with a deep median groove throughout its length; gastrofrontal crest present; gastrofrontal groove bifurcate posteriorly; hepatic crest almost horizontal to base of antennal crest and from there sloping anteroventrally; telson armed with 3 pairs of movable spines; no ischial spine on first pereopod. Petasma (in males) with distomedian projections reaching to or slightly overhanging distal margins of costae; ventral costae largely broadened apically, unarmed on their free border but incised on their attached border; outer surface of lateral lobes not tuberculate. Thelycum (in females) with lateral plates, their anteromedian angles divergent; anterior process with anterolateral edges raised, forming 2 subtriangular or cylindrical projections; posterior process triangular, its anterior edges raised in lateral ridges delimiting a median depressed area.

Penaeus latisulcatus, were exclusively nocturnal in their active behaviour. Males and females showed no significant difference in the mean time spent in various behaviours. Feeding efficiency increased as food became more dispersed over the tank floor.

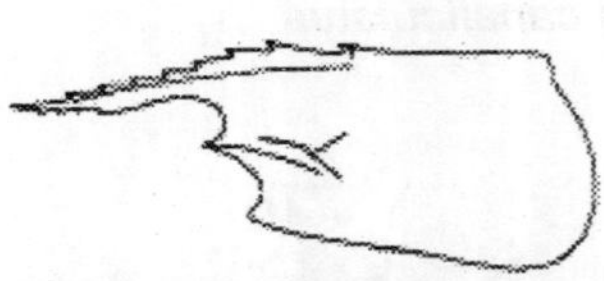

Figure 5.54: Carapace of *P. indicus*

Figure 5.55: Carapace of *P. latisulcatus*

Penaeus canaliculatus

New name: Penaeus (Melicertus) canaliculatus (Olivier, 1811)

Common name: Witch prawn

Distinctive Characters

Carapace is smooth. Rostrum with one ventral teeth; Rostrum armed with 9 to 11 teeth on dorsal, and a single tooth on ventral margin; adrostral crest and groove long, extending almost to posterior margin of carapace, the groove wide; postrostral crest well developed as far back as adrostral groove, with a deep median groove throughout its length; gastrofrontal crest present; gastrofrontal groove bifurcate posteriorly; Rostrum with one ventral teeth; hepatic crest almost horizontal to base of antennal crest and from there sloping anteroventrally; telson lacking lateral spines; no ischial spine on first pereopod. Petasma (in males) with short distomedian projections, reaching or slightly overhanging distal margin of costae; ventral costae broadened apically and bearing minute spinules at tip; outer surface of lateral lobes not tuberculate. Thelycum (in females) with lateral plates, their anteromedian margins diverging, then turning in a broad arc continuous with anterolateral margins; anterior

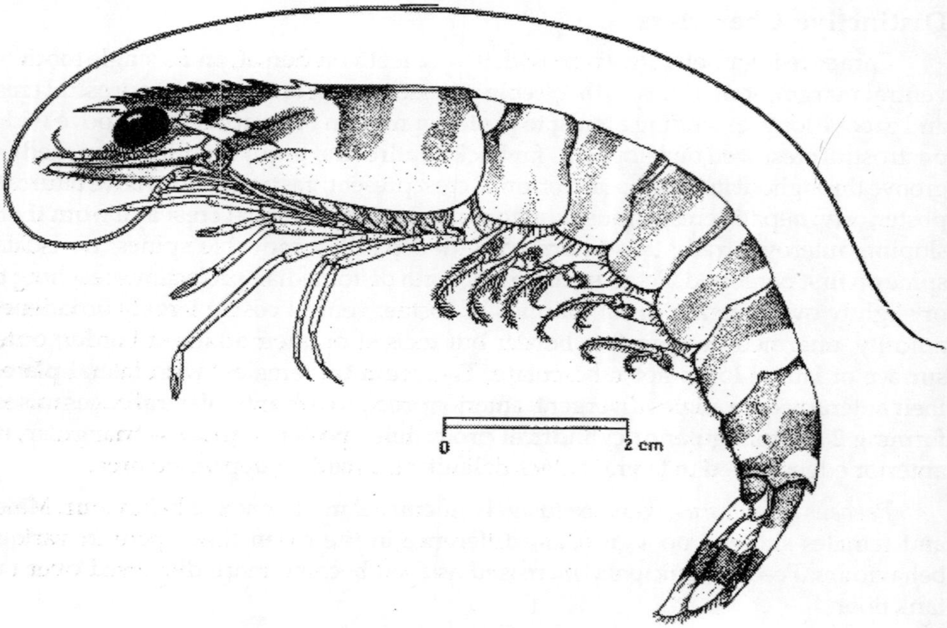

Figure 5.56: *Penaeus canaliculatus*

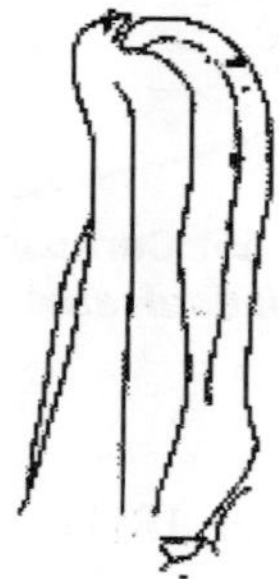

Figure 5.57: Petasma

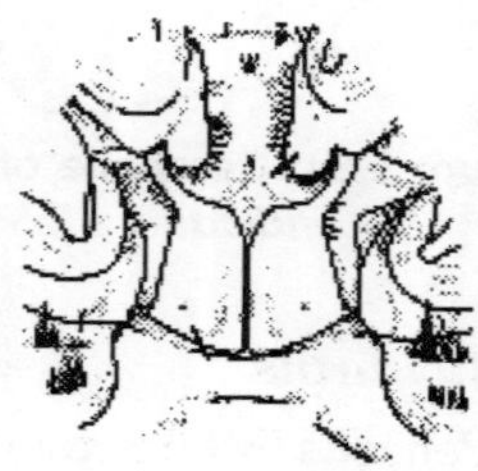

Figure 5.58: Thelycum

process subacuminate or subovate; posterior process triangular, anterior edges raised in lateral ridges delimiting a median depressed area.

Source: FAO Species Identification Sheets Pen 20 1983.

Penaeus carcinus

The giant prawn *P. carcinus* from Kerala is upto 90 cm long. While the dwarf prawn *P. lamarrei,* found almost throughout India, is 25 to 5 cm long. Young stages are translucent and white, but the adults are differently tinted according to the species. Usual colour is dull pale-blue or greenish with brown orange-red patches. Preserved specimens become deep orange-red.

Table 5.1: Distinguishing Characters of Similar Looking Types of *Penaeus* species *i.e. P. indicus, P. merguiensis* and *P. penicillatus*

Sl.No.	Characters	P.indicus	P. merguiensis	P. penicillatus
1.	Dectyle of third maxilliped		In adult males it is half as long as propodus	In adult males is much longer than propodus
2.	Adrostral crest		Not reaching as far as epigastric tooth	Extend just beyond epigastric tooth
3.	Gastro-orbital crest	Reaches posterior as far as hepatic spine	Never reaches in it	Gastro-orbital crest is present
4.	Adrostral crest and groove	Long and extending almost to posterior margin of carapace	Absent	Long and extending almost to posterior margin of carapace

Table 5.2: Character Indicating Difference Between Two Identical Looking Types of Prawn *P. indicus* and *P. merguiensis*

Sl.No.	Character	P. indicus	P. merguiensis
1.	Position of anterior most dorsal tooth	Tooth is *posterior* to the tip of antennular peduncle (15–190 mm size)	Tooth is *anterior* to the tip of antennular peduncle (15–190 mm size)
2.	Mid-ventral prominence	Present on 1st abdominal segment may have short deciduous hair like setae, shorter than the height of the prominence.	Present on 1st abdominal segment hair like setae absent.
3.	Ratio of length of dorsal unarmed portion of rostrum and distance between anterior most dorsal tooth and penultimate tooth	More than 2.0 (15–190 mm size)	Less than 2.0 (15–190 mm size)
4.	Position of anterior most dorsal tooth in relation to third lower tooth	Posterior (15–190 mm size)	Anterior (15–210 mm size)
5.	Dorsal rostral teeth	There are SIX dorsal rosteral teeth behind anterior margin of cornea	There are FIVE dorsal rosteral teeth behind anterior margin of cornea
6.	Colour of antennual flagell	WHITE distally, in live condition	RED throughout in live condition
7.	Spine on abdominal segment	At post larval stage median dorsal spine present, on 5th and 6th abdominal segment	At post larval stage 4th abdominal segment develops a small posterolateral spine

Contd...

Table 5.2–Contd...

Sl.No.	*Character*	*P. indicus*	*P. merguiensis*
8.	Chromatophores on telson tail fan	Present in distal half only	Present from base to distal end
9.	Chromospheres on the inner uropod ramus	May have minute one but usually absent	3 to 4 on median aspects
10.	Uropods	The inner and outer uropods of the earlier post larvae upto 3 rostral spine have chromatophores in the distal half. These chromatophores disappear in the PL with 4 rostral spine	Distal half of uropods translucent their tips are red
11.	Base of pleopods	Without blue spot (15–100 mm size)	With conspicuous blue spot distally on posterior aspect
12.	Colour of antennular flagella	On antennular flagella inner flagellum spotted reddish brown along entire length even in specimens of 15 m in length, outer flagellum with yellowish orange with *reddish brown SPOTS* along entire length in specimen above 40 mm size.	On antennular flagella distal half on inner flagellum remains colourless in juveniles upto 90 mm size; the outer flagellum dull yellow with clear *reddish brown BANDS* distally in specimens above 50–55 mm size.
13.	The exo and endopodites of the pleopods	Reddish in juveniles above 50–60 mm size	Reddish in juveniles above 85–90 mm size.
14.	Posterolateral spines on last three abdominal segment in Post larvae	Absent	Present.
15.	Rostral tip	Length of rostrum anterior to the anterior most dorsal tooth distinct longer than distance between penultimate dorsal tooth and the anterior most dorsal tooth.	Length of rostrum anterior to to the anterior most dorsal tooth as long as or shorter than distance between penultimate dorsal tooth and the anterior most dorsal tooth.
16.	Post rostral carina	Well defined upto posterior 1/5 or 1/6 of carapace	Well defined only for a short distance behind epigastric tooth and then broadens out losting its identity
17.	Antennular flagella	Inner flagellum ¾ or more length of outer	Inner flagellum more or less ½ length of outer

Source: Dholakia, A. D. Ph. D. Thesis 2004, Muthu, M. S., and G. Sudhakar Rao 1973.

Metapenaeus

Classification

Class	Crustacea
Subclass	Malacostraca
Order	Decapoda
Suborder	Natantia
Family	Penaeidae
Subfamily	Penaeinae
Genus	*Metapenaeus* Wood-Mason,1891

Metapenaeus dobsoni (Mitres 1879)

Common Name: Kadal shrimp

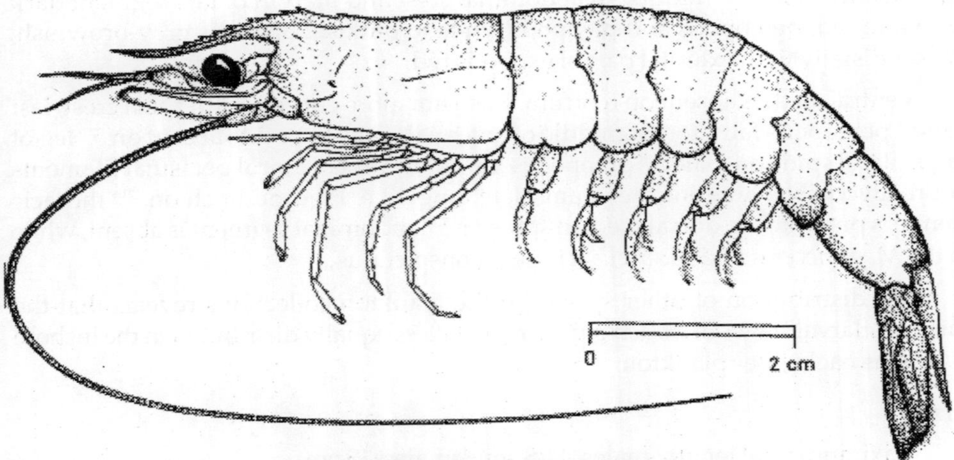

Figure 5.59: *Metapenaeus dobsoni*

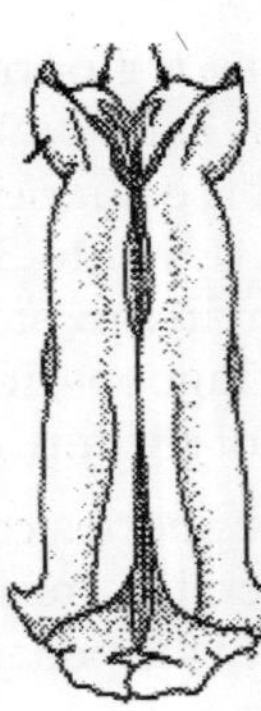

Figure 5.60: Petasma

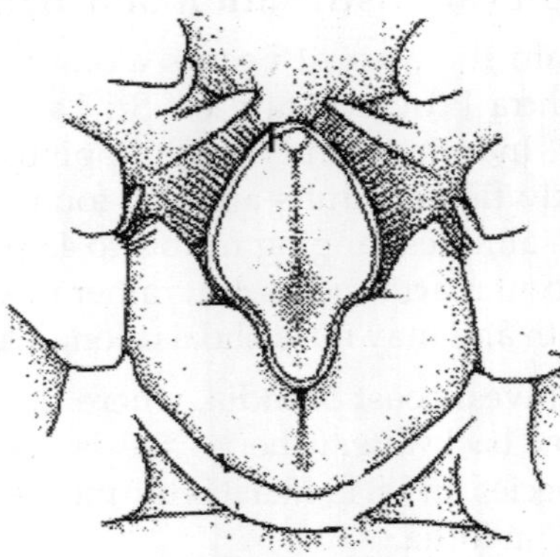

Figure 5.61: Thelycum

Distinctive Characters

Usually almost entire body pubescent, but pubescence can be restricted to a few patches; rostrum long, extending beyond antennular peduncle, slightly sinuous, armed with 7 to 9 dorsal teeth, but toothless on its distal half; postrostral ridge ending near posterior margin of carapace; adrostral crest reaching as far as epigastric tooth, adrostral groove a little beyond; branchiocardiac groove almost reaching to middle of carapace; telson armed only with spinules; no ischial spine on first pereopod. In adult males, basial spine of third pereopod extremely long and barbed, and merus of fifth pereopod with 1 or 2 large, triangular teeth; each distomedian projection of petasma with a short filament on ventral surface and another on dorsal surface; distolateral projections directed forward. In females, fifth pereopod often reduced to coxa and basis; thelycum with a long, grooved tongue-like anterior plate partially ensheathed in a horse-shoe-like process formed by the lateral plates; impregnated (fertilized) specimens with white conjoined pads on thelycum. Colour: body pale yellow to brownish with red, brownish or greenish specks; distal part of rostrum darker; antennae red; middorsal abdominal crest and margin of last segment dark brown to red; pereopods and pleopods white to pinkish; uropods grey-brownish, darker distally with external part of exopods red.

Ventral teeth is absent on rostrum. Rostrum more curved, with basal crest; last pair of pereopods fall short of middle of antennal scale; Expodite absent on 5th leg of torax. Basal spine on male 3rd preopod is long and barbed; apical petasmal filaments not really visible; anterior thelycum plate tong–like. Pleurobranch on 7th thoracic somite is present. The dorsal median spine on 5th abdominal segment is absent, while in the *M. monoceros* and *M. affinis* it is very conspicuous.

The distribution of other species in the plankton collections reveals that the early postlarval stages of *M. dobsoni* are more or less equally distributed in the inshore as well as backwater plankton.

Size

Maximum total length: males, 11.8 cm; females, 13 cm.

Geographical Distribution and Behaviour

Within the area, it occurs along the Indian coast as far north as Gujarat, Maharashtra Province, and off Sri Lanka. Further east it extends to the Gulf of Thailand, Indonesia and the Philippines. Juveniles inhabit estuaries, backwaters and paddy fields; adults are also found at sea on mud, down to 37 m depth; it tolerates salinities ranging from 3 to 43 per cent; abundant in low salinity lagoons and adjacent marine areas, but rather rare in high salinity lagoons; it breeds in 9 to 24 m depth and may form shoals along with other species of shrimp.

Southwest coast of India, where it is one of the major species contributing to inshore and backwater fisheries; the paddy field fishery of Kerala is mostly dependent on this species. In Sri Lanka it is the most abundant shrimp of the penaeid fauna. It is cultivated in India.

Metapenaeus monocersos (Fabricius 1798)

Common name: Speckled shrimp.

Figure 5.62

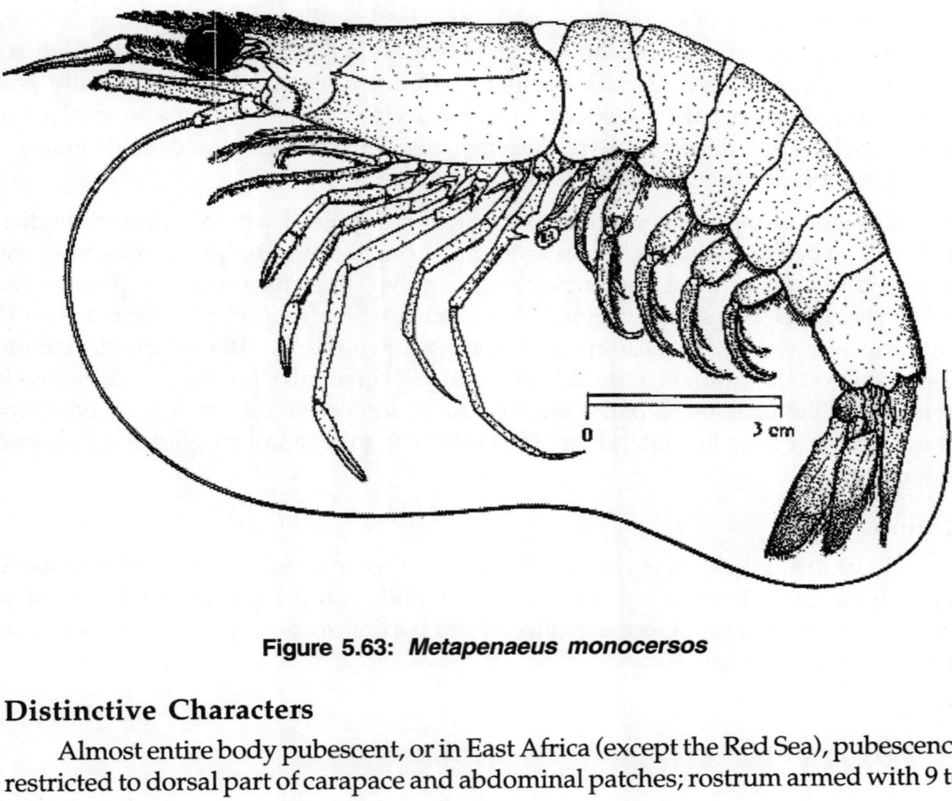

Figure 5.63: ***Metapenaeus monocersos***

Distinctive Characters

Almost entire body pubescent, or in East Africa (except the Red Sea), pubescence restricted to dorsal part of carapace and abdominal patches; rostrum armed with 9 to

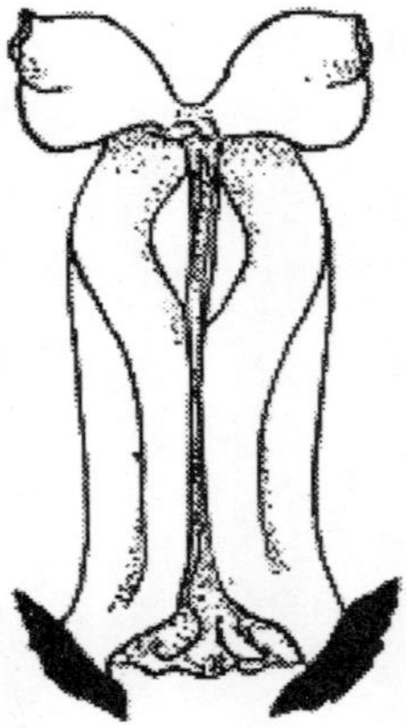

Figure 5.64: Petasma

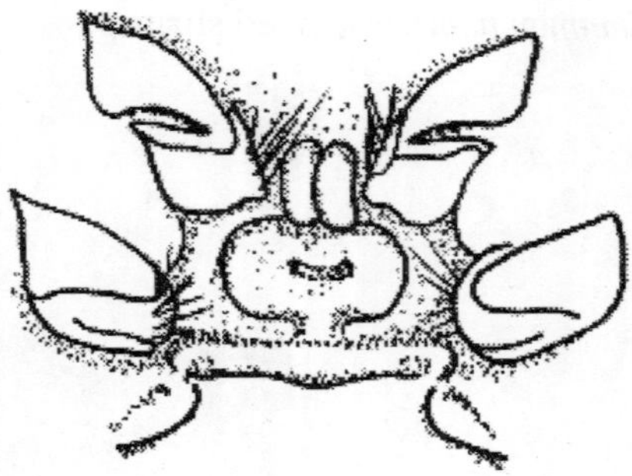

Figure 5.65: Thelycum

12 teeth along entire dorsal margin, straight, reaching as far as, or beyond, tip of antennular peduncle; postrostral crest reaching posterior margin of carapace or nearly so; adrostral crest ending behind second rostral tooth, adrostral groove behind epigastric tooth; branchiocardiac ridge sinuous, reaching posterior extension of hepatic spine; telson armed only with spinules; a small ischial spine on first pereopod. In adult males, merus of fifth pereopod with a proximal notch followed by a long, inwardly curved spiniform process and a row of tubercles; distomedian projections of petasma convoluted, greatly swollen, bulbiform, directed anterolaterally and concealing distolateral projections in ventral view. In females, anterior plate of thelycum long and deeply grooved; lateral plates with strongly raised lateral margins forming 2 longitudinal crests.

Rostrum is without ventral teeth, Post-orbital spine absent. No distinct median tubercle on ocular peduncle; upper antennular flagellum sub-equal to the lower one attached to apex of third antennular segment. No distal fixed pair of spine on the telson; lateral mobile spines present. No exopod on 5th pereopod; pleurobranch on 7th thoracic somite present. Distomedian petasmal projection without apical filament looks as hood–like and directed anterolaterally. Thelycum of impregnated females is without white conjointed pads. Brachocardiac sulcus distinct in at least posterior theylycal plate tong like lateral thelycal plate with salient and paralleled ear shaped lateral ridges.

Colour

Body is pink, green-greyish or whitish with brown specks; rostral and middorsal abdominal crests brown; antennae red; pereopods and pleopods of same colour as body, sometimes more intensely pink; distal part of uropods purple-blue, external margin of exopods red.

Size

Maximum total length: males, 15 cm; females, 20 cm.

The following description is adapted from Aloock (1906).

Body covered with stiff, very short tomentum. Rostrum is nearly straight, uptilted, reaching nearly to, or a little beyond, tip of antennular peduncle; armed dorsally with 9 to 12 teeth. Postrostral crest continued to, or almost to, posterior border of carapace. Anterolateral angles of carapace broadly rounded off, very small postocular (orbital) tooth. Postantennular (antennal) spine strong, produced as ridge to base of small hepatic spine; ridge bounding well marked postantennular groove which meets cervical groove. Gastric region defined anteriorly by short oblique postorbital groove. Branchial region defined (i) anteriorly, by deep and narrow crescentic groove (anterior part of cervical groove) which embraces base of postantennular ridge and meets postantennular groove, (ii) superiorly, by sinuous ridge which is most distinct in posterior half and runs from hepatic spine almost too posterior border of carapace.

Dorsal carina is on 2nd to 6th abdominal terga, usually 1st also, blunt and inconspicuous on (1st) 2nd and 3rd, very sharp on 4th to 6th. Fifth abdominal somite is about two-thirds length of 6th, 6th a little shorter than telson. Telson is shorter than endoped of uropod; without marginal spines. Eyes are very large, slightly surpassed by antennal scale. Outer (upper) antennular flagellum is slightly longer than inner, not much more than half length of peduncle.

Third maxillipeds barely reach middle of antennal scale: dactylus in male not modified, consists of slender, setose, tapering joint, about four-fifths length of propodus. Strong anterior spine is on basis of each cheliped. Last pair of thoracic legs of adult male with proximal end of merus notched on outer side, notch deepened anteriorly by large hook-like spine, and posteriorly by subterminal lobule on posterior border of ischium. Edge of merus is finely denticulate beyond spine. Three terminal joints of 5th legs slender in both sexes, the dactylus rarely reaches much beyond middle of antennal scale. No exopods on the 5th legs.

Petasma

Petasma symmetrical, consists of 2 rigid segments tightly folded longitudinally, interlocked all along anterior margins, in close apposition along most of posterior margins, forming compressed tube; tube ends distally in pair of large gargoyles with posterior lips convoluted like mouth of personate corolla (Figure 5.64).

Thelycum

Thelycum concave, bounded laterally by pair of ear-like lobes with free edge often incurved, bounded anteriorly by median projecting tongue embedded between 2 lobes of sternum corresponding with penultimate pair of legs (Figure 5.65).

Semitransparent, closely covered with small red chromatophores; dorsal carina of carapace, rostrum, bases of eyestalks, dorsal abdominal carinae and carinae of telson and uropods dull red; antennae bright red; first 2 legs colourless; last 3 legs with numerous red chromatophores; setae of uropods golden red; outer uropod bright red along external margin (Kemp, 1915). According to Ahmad (1957), body is flesh-coloured with pigment sparsely distributed, thickly covered with brown dots; spots also present on flagella and thoracio legs. Joubert (1965) described body as white, covered with dark brown speckles.

Metapenaeus kutchensis (George and Rao)

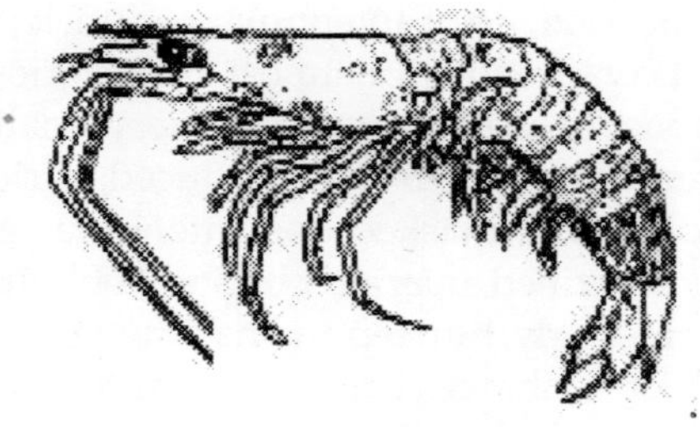

Figure 5.66: ***Metapenaeus kutchensis***

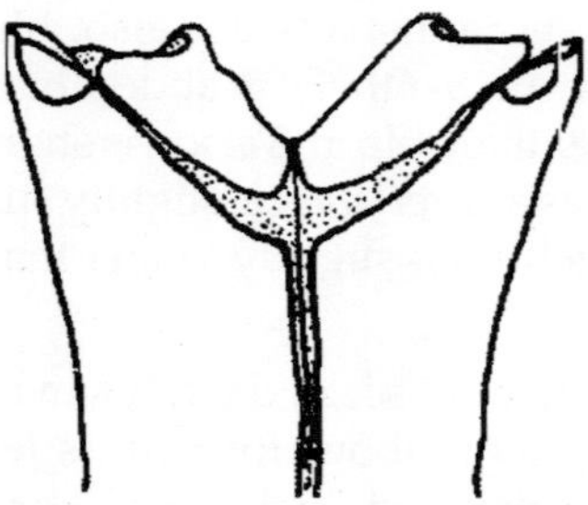

Figure 5.67: Petasma

Figure 5.68: Thelycum

Distinctive Characters

Rostrum is without ventral teeth. Rostrum is with teeth 7–8 + epigastric; straight with small crest. Expodite is absent on 5th leg of thorax. Rostrum straight and extends the 2nd segment of antennular peduncle.Adrostral carina ending near nepogastric sulcus extending to middle of carapace; posterostral carina ending in a glabyous expansion 1/6 length of carapace from its posterior edge.

Postocular sulcus is at an angle of 40° to rostrum; orbito-antennal sulcus meeting the hepatic below hepatic spine; hepatic sulcus descending vertically and the curving towards pterygostomial angle. Cervical sulsus is straight; branchiocardiac sulcus distinct, carina meeting the glabrous posterior extension of the hepatic spine. Antennular spine is strong; epigastric spine at 1/4th length of carapace. Colour of body is light brown in preserved specimen with speckling of minute greenish blue spots on the abdomen. Tips of uropods are greenish blue.

Petasma

Quite symmetrical and has the general form of the species of Metapenaeus group. It resembles that of *M.affinis*, differ much in size and shape of the distal ends. In adult it is comparatively smaller in size and extends only to the base of the fourth periopod. The distomedian lobes are more transversely placed with the proximal end narrow and distal end broad. Distance between tips 1/3 length of patesma.

Thelycum

Anterior plate tongue shaped and wider posteriorly placed at a level with and bounded on either side by expanded coxal projections of the fourth pereopod; the median groove of this plate widens posteriorly. Posterior plates concave, glabrous and transversely cut into two unequal segments; the lateral edges of these segments curve up and are placed one behind the other.

Dorsomedian abdominal carina starts from 4th segment. Rostral tip is straighter than in *M. affinis*; posterior extension of anteromedian thelycal plate not bond by oval plate on either side. Epigastric teeth are 7–8. Adrostral carina ending near epigastric suclus extending to middle of carapace.

Metapenaeus affinis (H. M. Edward)

Figure 5.69

Distinctive Characters

Generally almost entire body pubescent, rarely partly or completely hairless; rostrum armed with 8 to 11 teeth along entire dorsal margin, slightly sinuous and reaching from proximal to distal margin of third antennular article, or exceeding it; postrostral ridge ending near posterior margin of carapace; adrostral crest ending behind second rostral tooth, and adrostral groove a little behind epigastric tooth; branchiocardiac ridge slightly sinuous and reaching posterior extension of hepatic spine; telson armed only with spinules; ischial spine on first pereopod present or absent. In adult males, merus of fifth pereopod with a proximal notch, followed by a twisted, keeled tubercle; distomedian projections of petasma crescent-shaped, leaning

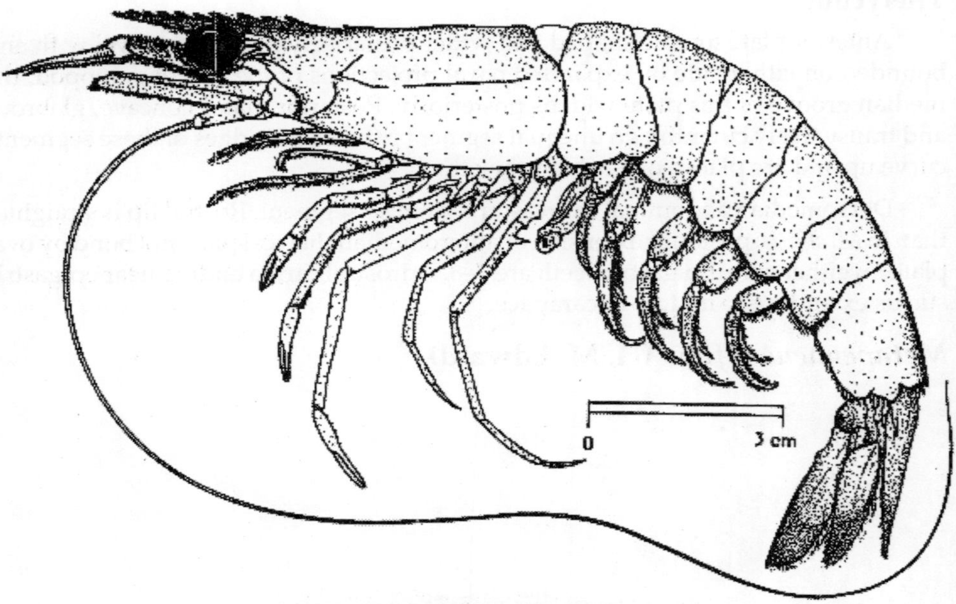

Figure 5.70: *Metapenaeus affinis*

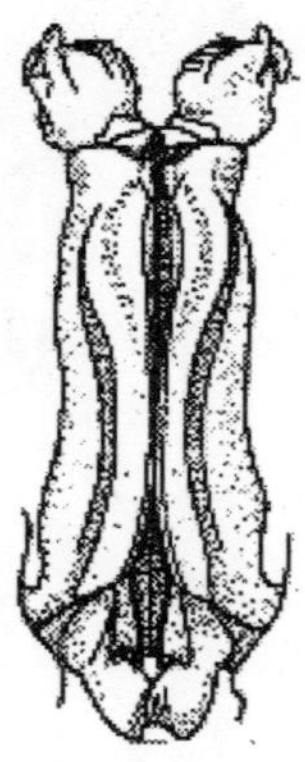

Figure 5.71: Petasma

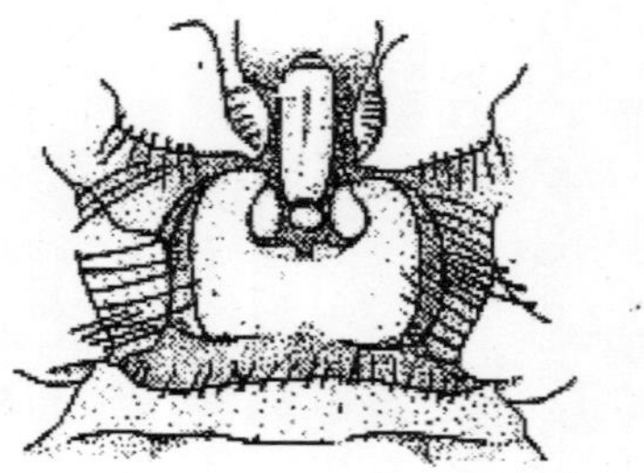

Figure 5.72: Thelycum

on distolateral projections and concealing them partly or completely; distolateral projections directed anterolaterally. In females, anterior plate of thelycum deeply grooved longitudinally and considerably wider posteriorly; posterior transverse ridge with 2 anterolateral rounded projections partly covering lateral plates; impregnated (fertilized) specimens occasionally with white conjoined pads on thelycum.

Ventral teeth are absent on rostrum. Rostrum curved without crest. 5th leg of thorax is surpassing the antennal scale. Rostrum is short and extends to the middle of the eye. The ventral margin is slightly concave and the tip curves down, so that when viewed laterally, the rostrum appears practically hidden in between the eyes. In older individuals, the rostrum, however becomes straight and prominent and reaches almost the tip of eye or slightly beyond it. Expopodite is absent on 5th leg of thorax. Ischial spine on 1st pereopod is small. Branchio cardiac carina distinct from posterior margins of carapace almost no hepatic spine; anterior thelycal plate longitudinally grooved, wider posteriorly than anteriorly; distomedian petasmal projections crescent-shaped. Posterior extension of the anterior median theylycal plate is not bound laterally by oval plate on either side.

Colour

Body is pale greenish to pale pinkish, sometimes green-bluish or pink-brownish, with green or red-brown specks; middorsal abdominal crest brown or brownish-red; antennae red; pereopods white or of same colour as body; pleopods reddish to whitish; distal half of uropods translucent green or rust coloured, tips usually whitish to yellowish.

The ground colour of body varies from blue to bluish brown. The chromatophores of carapace are mostly bluish and distributed almost on the same pattern as that of *M. monoceros*. But those on the mid–lateral region, although they form a short band, are less conspicuous and lighter in colour than those of *M. monoceros*.

In the abdomen, the median dorsal carination is seen from the 3rd segment. The distribution of chromatophores is similar to that of *M. dobsoni*, but more numerous in numbers. The tip of uropod is coloured with brown chromatophores in specimen over 18 mm and in the larger juveniles it has bluish brown border.

Size

Maximum total length: males, 14.6 cm; females, 18.6 (22.2) cm.

Geographical Distribution

Within the area, it occurs in the "Gulf" and the Arabian Sea from the Gulf of Oman to south India; it is also present in Sri Lanka. Further east it extends as far as the Philippines and Taiwan Island. Found from the coastline to depths of about 55 m (occasionally in deeper water to 90 m), mainly on mud or sandymud; juveniles inhabit estuaries and backwaters, but the postlarval migration from the sea to backwaters and estuaries is not as extensive as for other species like *M. dobsoni* or *P. indicus*; adults can form large shoals, along with other species, in mud bank areas.

Present Fishing Grounds

In the "Gulf" (off Kuwait and Bahrain) it is one of the most important species in the artisanal fisheries; important catches are also taken in Pakistan and Sri Lanka; along the Indian coast it is the most important commercial species of Metapenaeus, particularly near Bombay.

Table 5.3: A Comparison of the Diagnostic Feature of *M. kutchensis, M. monoceros* and *M. affinis*

Sl.No.	*Character*	*M. kutchensis*	*M. monoceros*	*M. affinis*
1.	Pubescence	Body only partly covered with harsh and very short tomenium	Body fully covered with harsh and very short tomentum	Carapace finely setose; abdomen may have some glabrous area
2.	Rostrum	Straight and with a small crest	Nearly straight and uptilted	More curved and less uptilted
3.	Mid-dorsal carination of the abdominal segments	Carination commences from the fourth segment	Carination commences from the second segment	Carination commences from the second segment
4.	Ischial spine on 1st pereopod	Present	Present	Generally absent; if present small denticle only
5.	Length of 5th pereopod	Reaches a little beyond the middle of antennal scale	Reaching a little beyond the middle of antennal scale	Surpasses the tip of the antennal scale by dectylus
6.	5th pereopod of adult male	With a shallow notch and feeble tooth at the base of merus	With a notch and hook-like spine at the base of merus	With a notch and tooth at the base of the merus
7.	Petasma	Distomedian lobes more transversely placed with proximal end narrow and distal end broad	Distomedian lobes hood-like	Distomedian lobes ending in a pair of two-lipped spouts resembling a pair of short horns
8.	Thelycum	Posterior plate concave without ear like lobes, cut transversely into two unequal segments with no apparent clusters of setae between them	Posterior plate concave, bounded laterally by elevated ear like lobes	Posterior plate laterally flat, cut transversely into two unequal segments, with conspicuous clusters of setae between them

Source: George, P.C. *et al.*, 1963.

Metapenaeus brevicornis (H. Milne Edward 1837)

Rostrum is without ventral teeth. Pleurae are of 2nd abdominal somite overlapping that 1st segment, and 3rd leg with a chela. Rostrum curved and rarely reaching middle of 2nd joint of antennular peduncle, sometimes only just surpassing eyes, bearing dorsal crest of 7 teeth. Post-orbital spine is absent. No distinct median tubercle on ocular peduncle; upper antennular flagellum sub-equal to the lower one attached to apex of third antennular segment. No exopod on 5th pereopod; Pleurobranch on 7th thoracic somite present. No distal fixed pair of spines on the telson. Posterior part of rostrum with distinctly elevated crest; basial spine on male 3rd pereopod simple;

Colour Plates

1

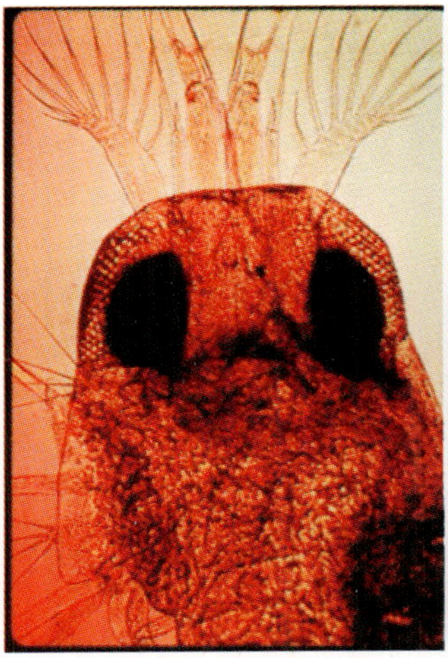

Figure 2.1 (Page 65)

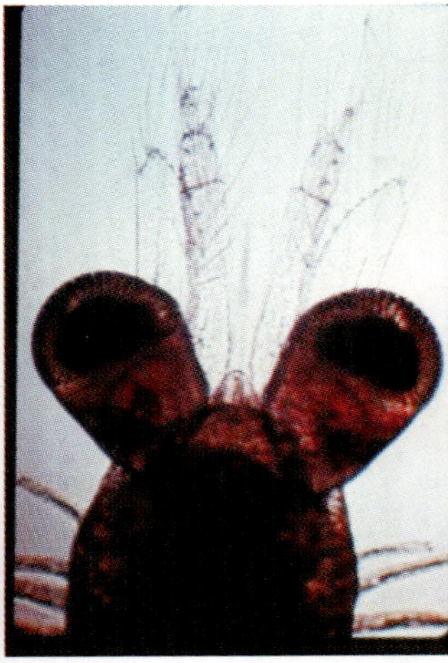

Figure 2.2 (Page 65)

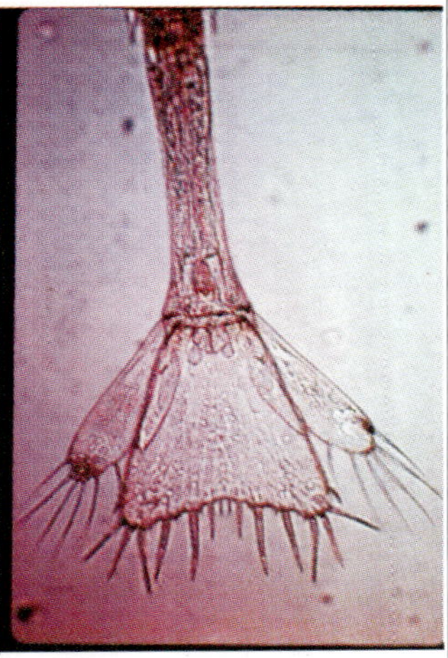

Figure 2.3 (Page 65)

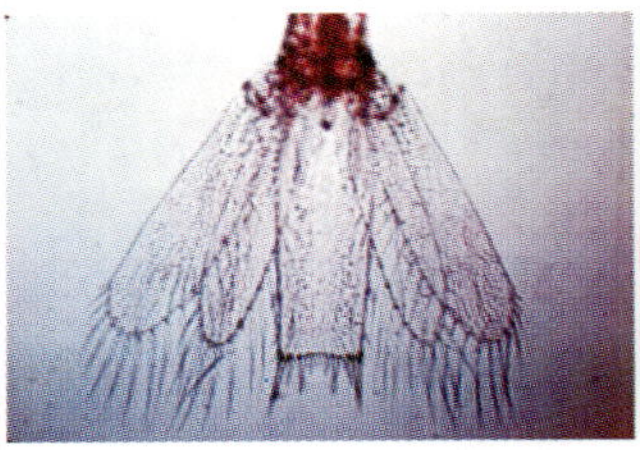

Figure 2.4 (Page 65)

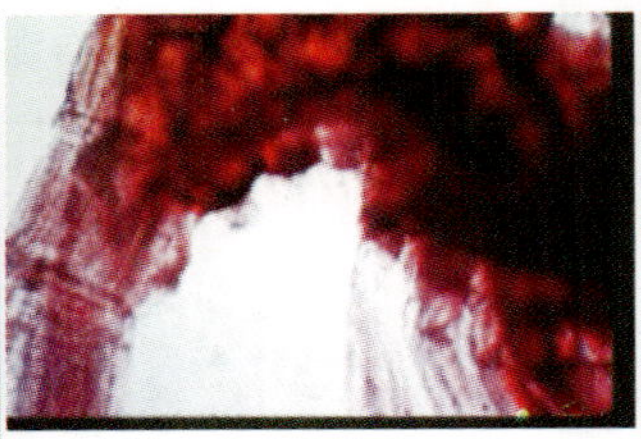

Figure 2.5 (Page 65)

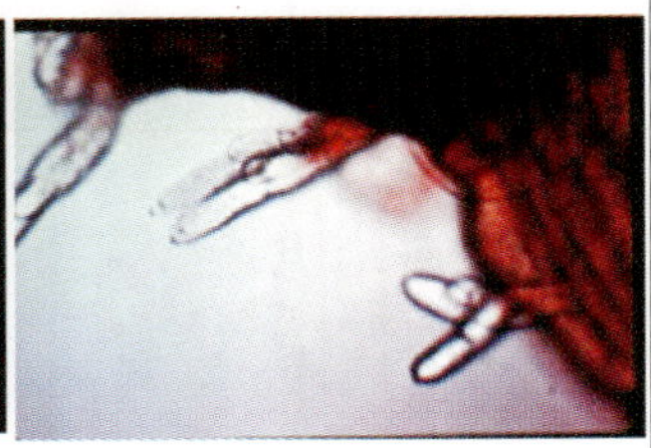

Figure 2.6 (Page 65)

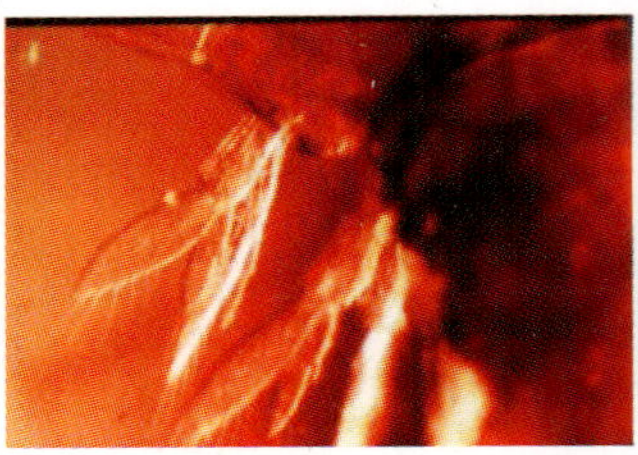

Figure 2.7 (Page 65)

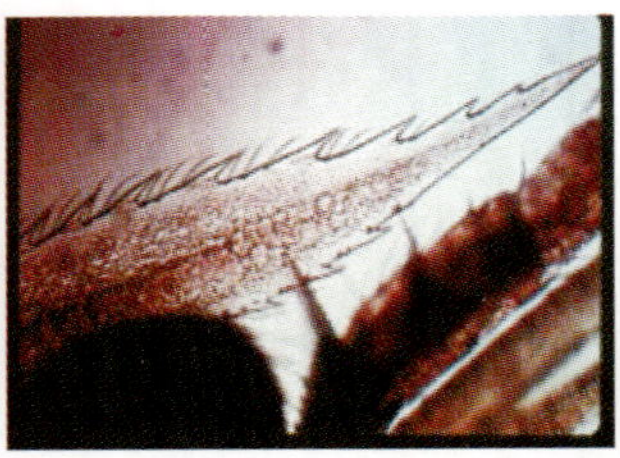

Figure 2.8 (Page 65)

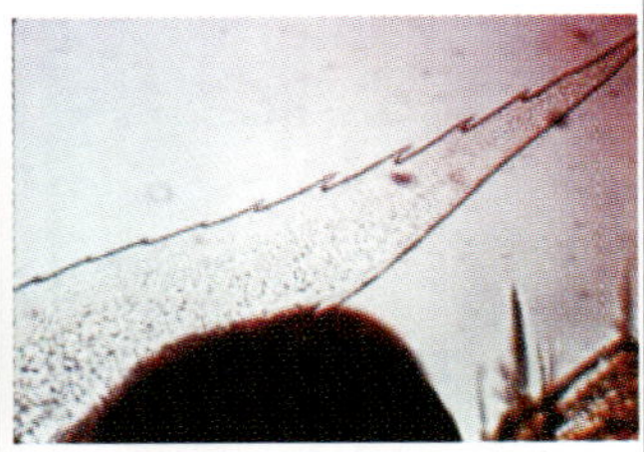

Figure 2.9 (Page 65)

Figure 3.5: Kuruma Prawn (*Penaeus japonicus*) Post Larvae As "PL10" (10 Days old postlarvae) (Page 82)

2

Figure 4.3: Marine Water Prawn (A = Two Antinnuals) (Page 102)

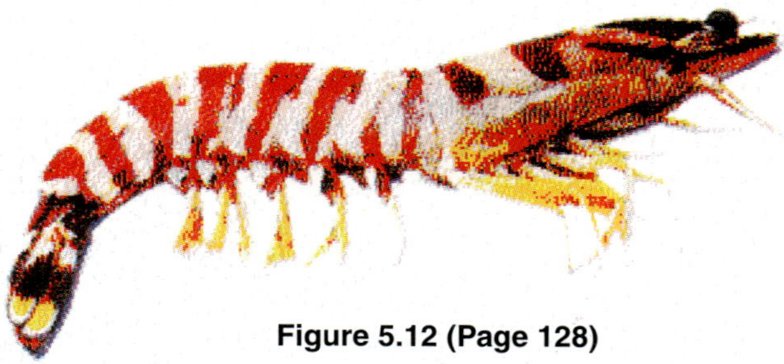

Figure 5.12 (Page 128)

Figure 5.17 (Page 129)

Figure 5.21 (Page 132)

Figure 5.34 (Page 136)

Figure 5.44: *Penaeus esculentus* (Page 141)

Figure 5.43 (Page 139)

Figure 5.45: *Penaeus plebejus* (Page 141)

Figure 5.46 (Page 142)

(Page 195)

Figure 5.62 (Page 151)

Figure 5.69 (Page 155)

Figure 5.151: *Macrobrachium rosenbergi* (Page 196)

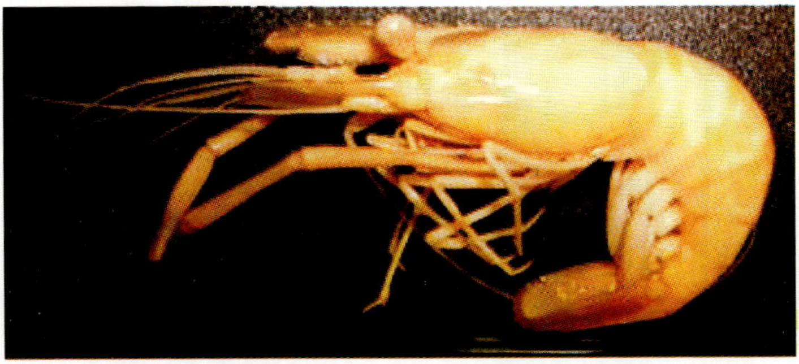

Figure 5.158: Photograph of *Macrobrachium nipponense* (Page 205)

6

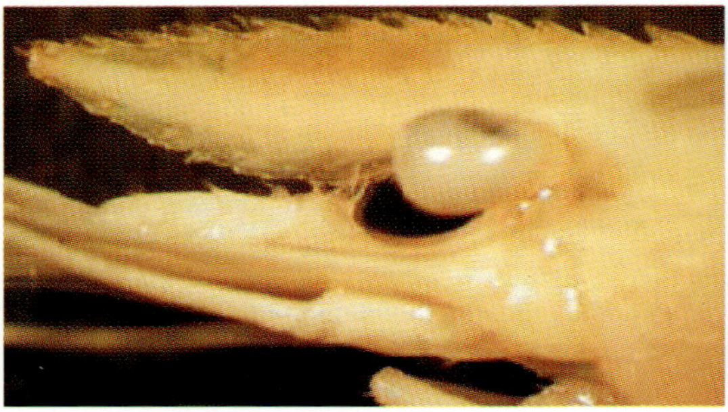

Figure 5.159: Rostrum (Page 205)

Figure 5.160: Uropods (Page 205)

Figure 5.161: Telson (Page 205)

Figure 5.167: *Palaemon scabriculum* (Page 212)

Figure 5.174: *Palaemon longirostris* (Top View) (Page 215)

Figure 5.175: *Palaemon longirostris* (Side View) (Page 215)

7

Figure 5.168: *Palaemon serenus* (Page 213)

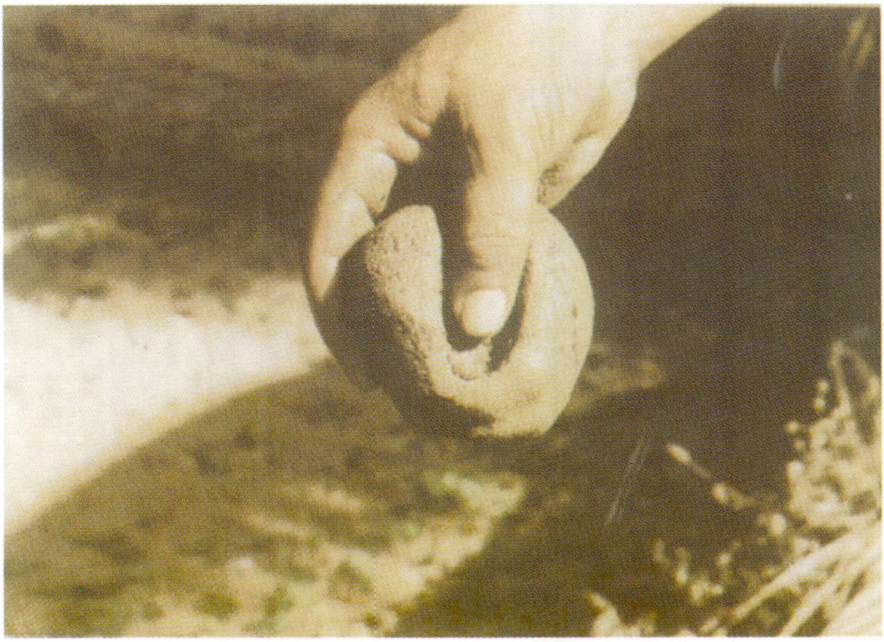

Figure 8.1: Soil Test for Water Retention (Page 240)

Figure 8.6: Inlet with Filter Bags (Page 246)

Figure 8.7: Drainage/Outlet (Page 247)

Figure 10.1: Holding Tank: Site for Outdoor Nursery (Page 261)

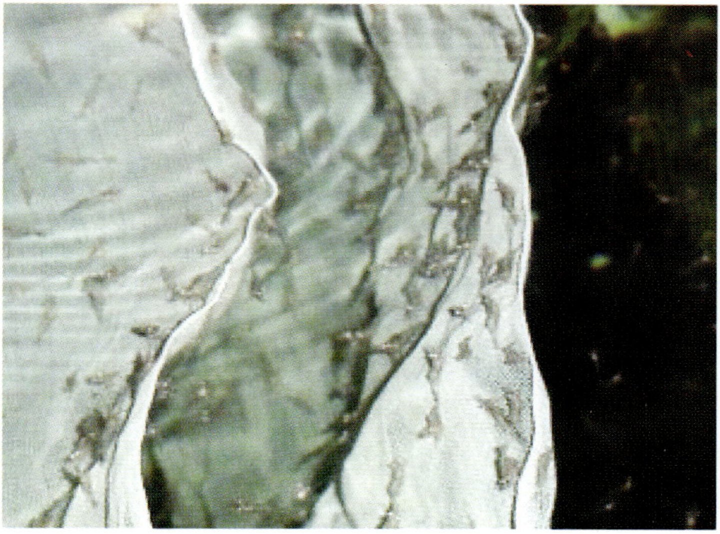

Figure 10.2: Postlarval Freshwater Prawns can Use a Nylon Screen as an Additional Surface Srea in Holding Tanks (Page 262)

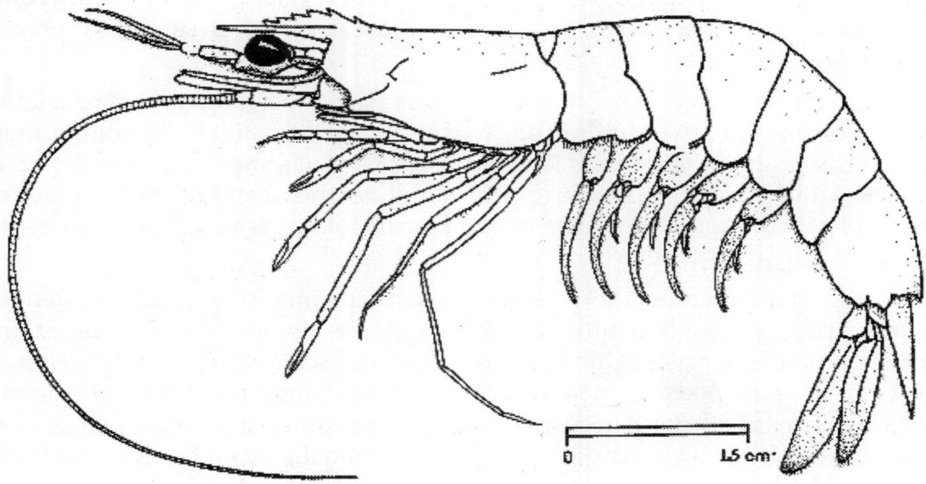

Figure 5.73: *Metapenaeus brevicornis*

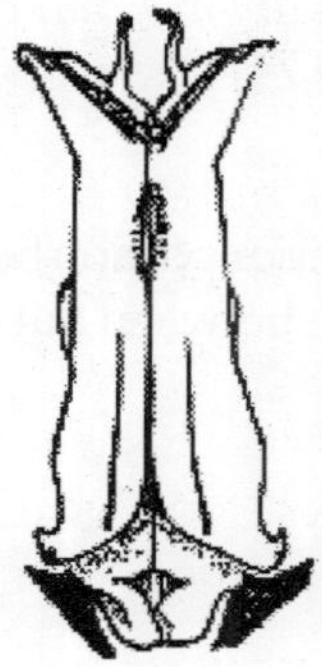

Figure 5.74: Petasma

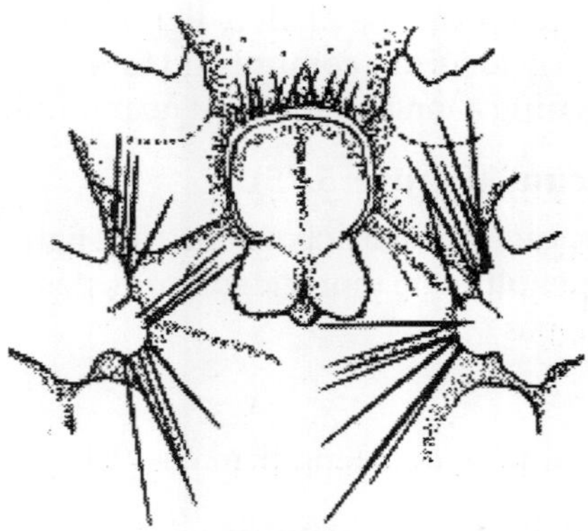

Figure 5.75: Thelycum

apical petasmal filament slender slightly converting; thelycum with a large anterior and small lateral plates.

Source: FAO Species Identification Sheets Pen Metap 4 1983.

According to Rajyalakshmi (1961), males and females attain lengths of 45.8 mm and 47.4 mm, respectively, at the end of the 1st year of life, and 80.5 mm and 89.0 mm, respectively, by the end of the 2nd year of life. Thus the females show a faster growth rate than males, particularly during the 2nd year. No information is available on the larval development of the species. An account of the biology of juveniles in the fishery of the Hooghly estuarine system is given by Rajyalakshmi (1961).

Postrostral crest very indistinct, only just reaches posterior third of carapace. Postantennular (antennal) spine weak; hepatic spine is very small. Postantennular groove shallow; subhepatic groove (anterior part of cervical groove) shallow, does not meet hepatic spine. Indistinct ridge defining branchial region superiorly, present only on posterior part of carapace.

Median carination is on abdominal terga hardly perceptible on 3rd somite, distinct on posterior two-thirds of 4th and on 5th and 6th somites; 5th somite about two-thirds length of 6th, 6th as long as telson. Telson is shorter than endopod of uropod; without marginal spines or with pair of clearly perceptible distal spine and series of minute spinules. Outer antennular flagellum is nearly as long as the peduncle. (George, M. J. 2000)

Third maxilliped barely reaches middle of antennal scale; dactylus slender, setose, tapering about four-fifths length of propodus, not modified in male. Strong antrorse spine is on basis of all 3 pairs of chelipeds; ischial spine on 1st pereiopod. Last pair of pereiopods reach more than a dactylus length beyond tip of antennal scale; adult male with notch in posterior border of proximal end of merus, small tooth at end of notch but no other denticles; no subterminal lobe on border of ischium. (George, M. J. 2000)

Petasma

Symmetrical, the 2 halves lightly folded, interlocked anteriorly, closely apposed posteriorly to form a compressed tube. Tube ends distally in a pair of simple spouts, each bearing a longish filament near middle (Figure 5.74)

Thelycum (Figure 5.75)

Concave; median lobe shaped like figure eight, anterior portion between processes of antepenultimate thoracic sternum, posterior portion between flat crescent-shaped lateral lobes.

Size

Maximum total length: males, 9.8 cm; females, 13.2 cm.

Present Fishing Grounds

In eastern Pakistan and on the north-west coast of India, the species is common and is considered to be of moderate to great commercial importance. The species is cultivated in India and several other countries outside the area.

Metapenaeus elegans De Man, 1907

Common Name: Fine shrimp

Distinctive Characters

Almost entire body pubescent, occasionally pubescence restricted to dorsal part of carapace and posterior abdominal segments (juveniles hairless); rostrum armed with 8 to 11 teeth along entire dorsal margin, nearly straight, and reaching to, or nearly to, tip of antennular peduncle; postrostral crest reaching to, or almost to posterior margin of carapace; adrostral crest ending behind second rostral tooth, adrostral

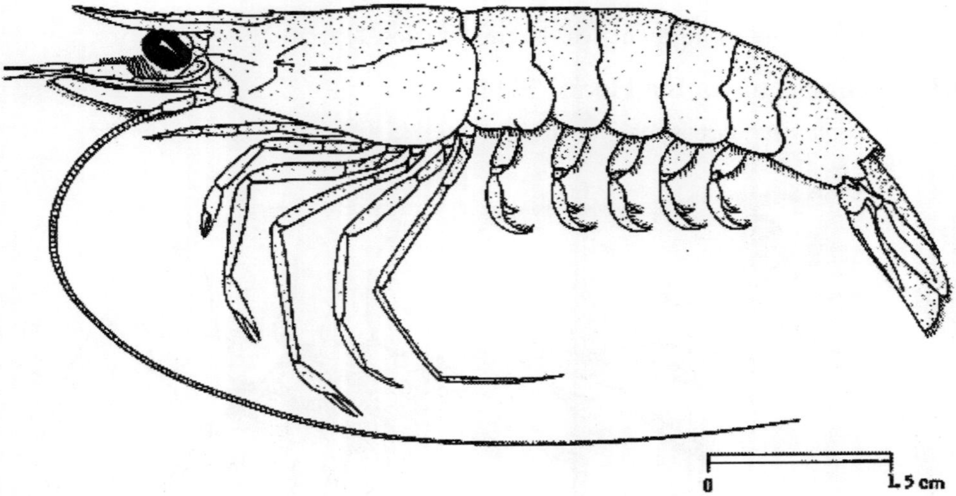

Figure 5.76: *Metapenaeus elegans*

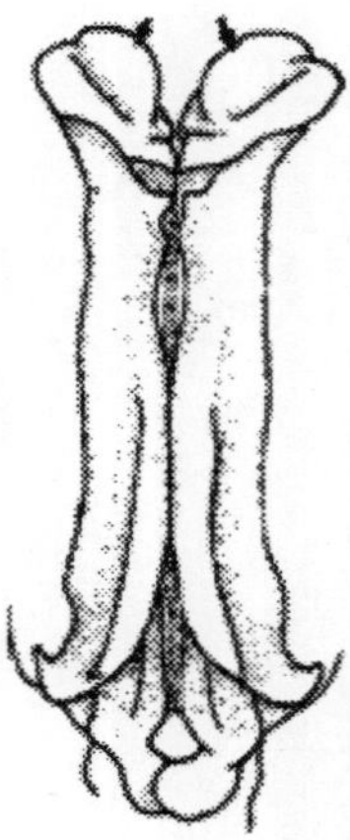

Figure 5.77a: Petasma

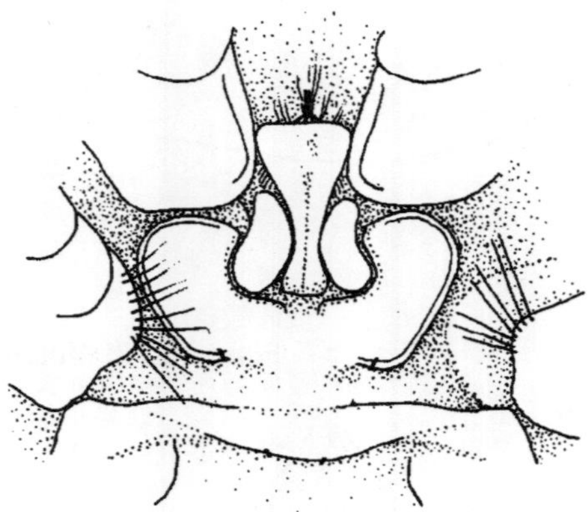

Figure 5.77b: Thelycum

groove behind epigastric tooth; branchiocardiac ridge usually sinuous, reaching posterior extension of hepatic spine; telson armed only with spinules; a small ischial spine on first pereopod. In adult males, merus of fifth pereopod with a proximal notch followed bye a long, inwardly curved spiniform process and a row of tubercles; distomedian projections of petasma convoluted, greatly swollen and directed forward, triangular in shape and concealing almost entirely distolateral projections in ventral view. In females, anterior plate of thelycum lone and deeply grooved; lateral plates with strongly raised lateral margins farming posteriorly 2 inwardly curved triangular.

Size

Maximum total length: males, 8.4 cm; females, 11.8 cm

Metapenaeus ensis De Haan, 1844

Common name: Greasyback shrimp.

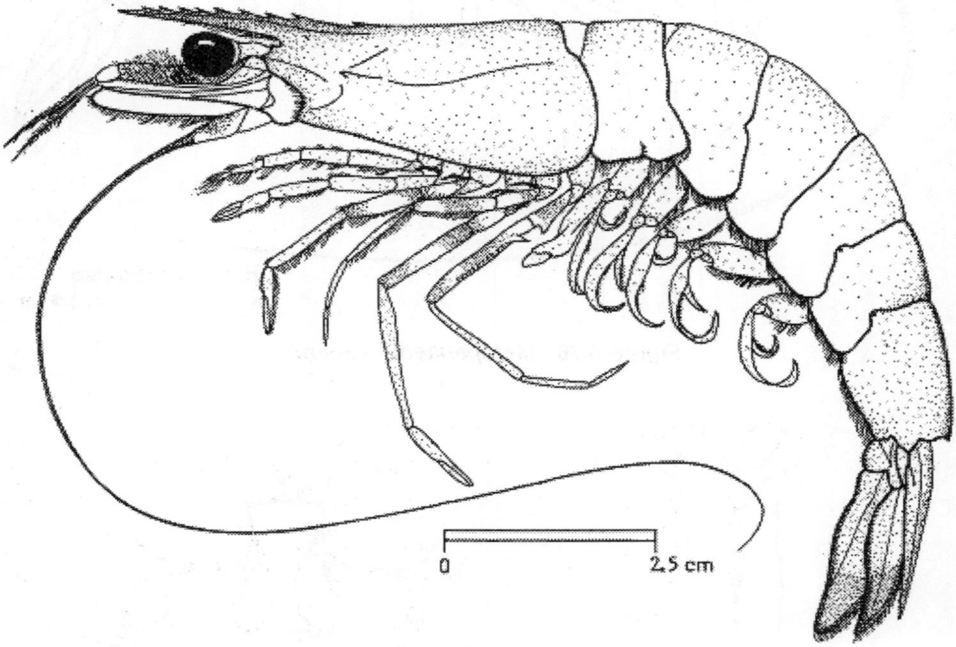

Figure 5.78: *Metapenaeus ensis*

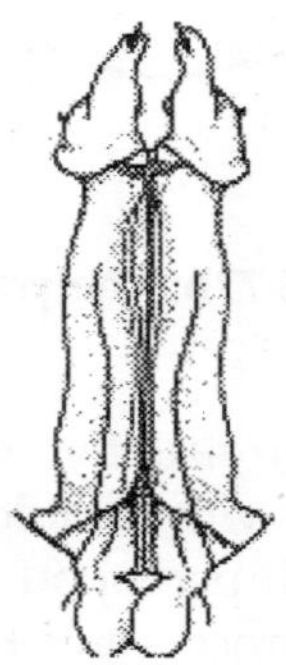

Figure 5.79: Petasma

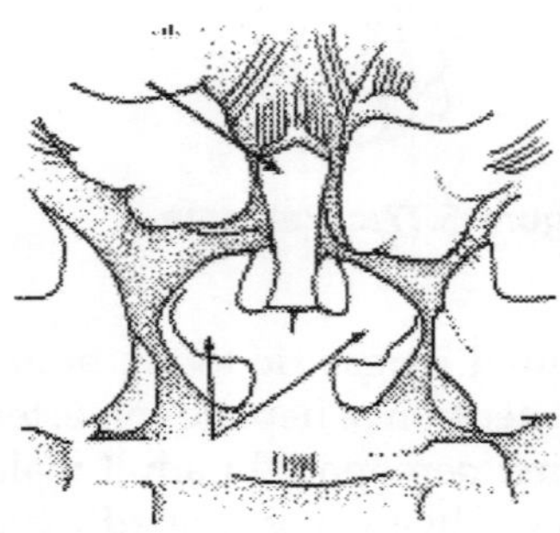

Figure 5.80: Thelycum

Distinctive Characters

Almost entire body pubescent, occasionally pubescence restricted to dorsal part of carapace and posterior abdominal segments (juveniles hairless); rostrum armed with 8 to 11 teeth along entire dorsal margin, nearly straight, and reaching to, or nearly to, tip of antennular peduncle; postrostral crest reaching to, or almost to posterior margin of carapace; adrostral crest ending behind second rostral tooth, adrostral groove behind epigastric tooth; branchiocardiac ridge usually sinuous, reaching posterior extension of hepatic spine; telson armed only with spinules; a small ischial spine on first pereopod. In adult males, merus of fifth pereopod with a proximal notch followed by a long, inwardly curved spiniform process and a row of tubercles; distomedian projections of petasma convoluted, greatly swollen and directed forward, triangular in shape and concealing almost entirely distolateral projections in ventral view. In females, anterior plate of thelycum lone and deeply grooved; lateral plates with strongly raised lateral margins farming posteriorly 2 inwardly curved triangular projections.

Source: FAO Species Identification Sheets Pen Metap 12 1983.

Size

Maximum total length: males, 8.4 cm; females, 11.8 cm.

Distribution

Juveniles are found in estuaries, backwaters and near shore areas; adults in deeper, and often turbid waters down to 70 m depth on mud, sandy-mud or silt.

Metapenaeus lysianassa (De Man, 1888)

Common Name: Bird shrimp

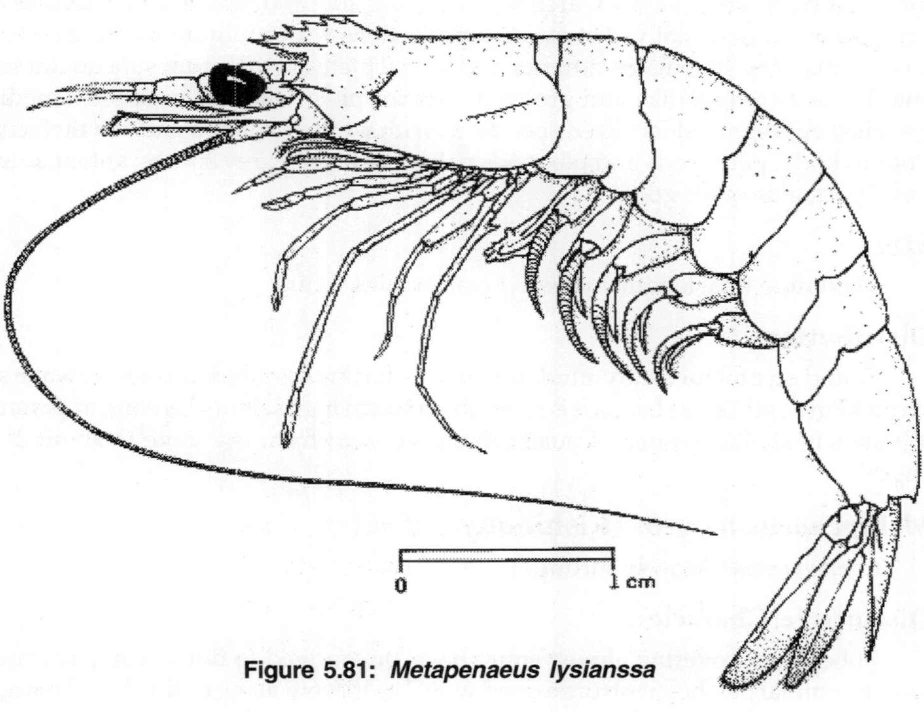

Figure 5.81: *Metapenaeus lysianssa*

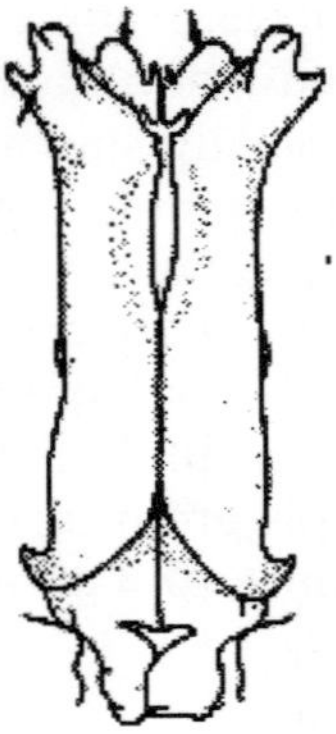

Figure 5.82: Petasma

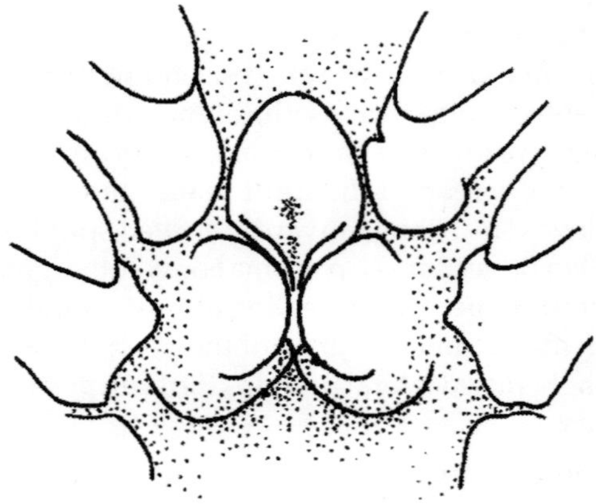

Figure 5.83: Thelycum

Distinctive Characters

Almost entire body pubescent; rostrum very short, reaching as far as middle of first antennular article and armed with 6 or 7 teeth along entire dorsal margin, rostral crest high; postrostral ridge ending near posterior margin of carapace; adrostral crest and groove reaching as far as third rostral tooth; branchiocardiac groove not reaching to middle of carapace; telson armed only with spinules; ischial spine on first pereopod absent or minute. In adult males, merus of fifth pereopod with a proximal notch followed by a large, slightly arcuate triangular tooth; distolateral projections of petasma bifurcate distally; distomedian projections with a minute filament on their median margins. In females, anterior and lateral plates of thelycum subequalin size, anterior plate tongue-like and grooved, osterior plates suboval with thick median margins; impregnated fertilized specimens with white conjoined pads on thelycum. Colour: body, pereapods and pleopods pale yellow with grey specks; antennae and distal part of uropods grey.

Size

Maximum total length: males, 6.1 cm; females, 9 cm.

Distribution

Found on mud or sandy-mud in estuaries, backwaters and near-shore waters to about 45 m depth; in Sri Lanka it is more abundant in high salinity lagoons; apparently it breeds in shallow waters. Usually found on mud from the shore to about 28 m depth.

Metapenaeus moyebi (Kishinouye, 1896)

Common name: Moyebi shrimp

Distinctive Characters

Pubescence covering almost entire body or confined to dorsal carapace and a few abdominal patches; rostrum armed with 7 to 10 teeth along entire dorsal margin,

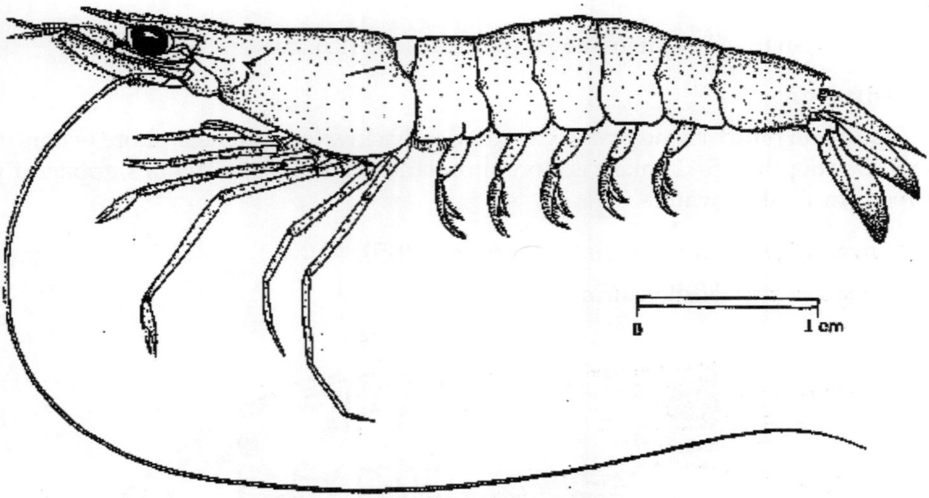

Figure 5.84: ***Metapenaeus moyebi***

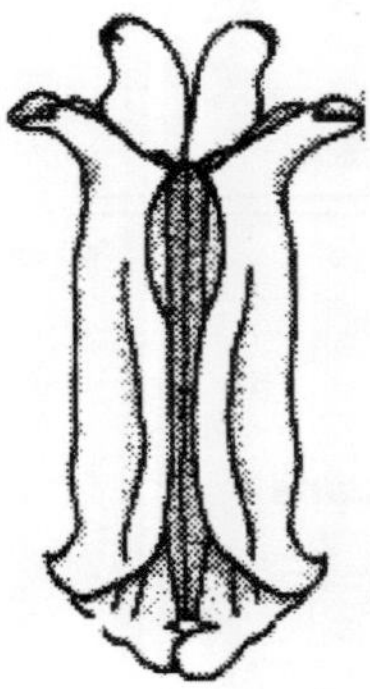

Figure 5.85: Petasma

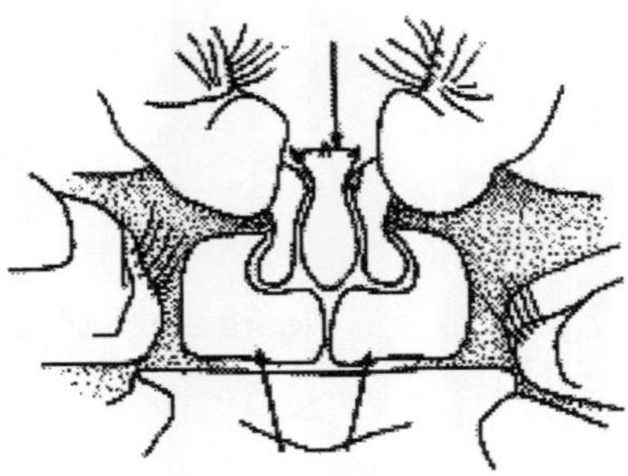

Figure 5.86: Thelycum

nearly straight, slightly uptilted and reaching distal half of third antennular article or just beyond; postrostral ridge low, generally ending near posterior margin of carapace; adrostral crest ending behind second rostral tooth, adrostral groove behind epigastric tooth; branchiocardiac ridge feeble, ending near posterior third of carapace; telson armed only with spinules; ischial spine on first pereopod minute or absent. In adult males, merus of fifth pereopod with a proximal notch followed by a twisted keeled tubercle; distomedian projections of petasma laminose and diverging; distolateral projections directed anterolaterally. In females, anterior plate of thelycum is flask-shaped, its anterior margin slightly convex and bearing 3 tubercles of subequal size; lateral plates kidney-shaped, often with angular contours. Colour: body semi-translucent pale green with brownish-green specks; pereopods and pleopods of same colour as body; distal part of uropods green.

Size

Maximum total length: males, 8.3 cm; females, 12.6 cm.

Distribution

Found on mud or sandy-mud in estuaries, backwaters and near-shore waters to about 45 m depth; in Sri Lanka it is more abundant in high salinity lagoons; apparently it breeds in shallow waters.

Metapenaeopsis stridulans (Alcock, 1905)

Common name: Fiddler shrimp

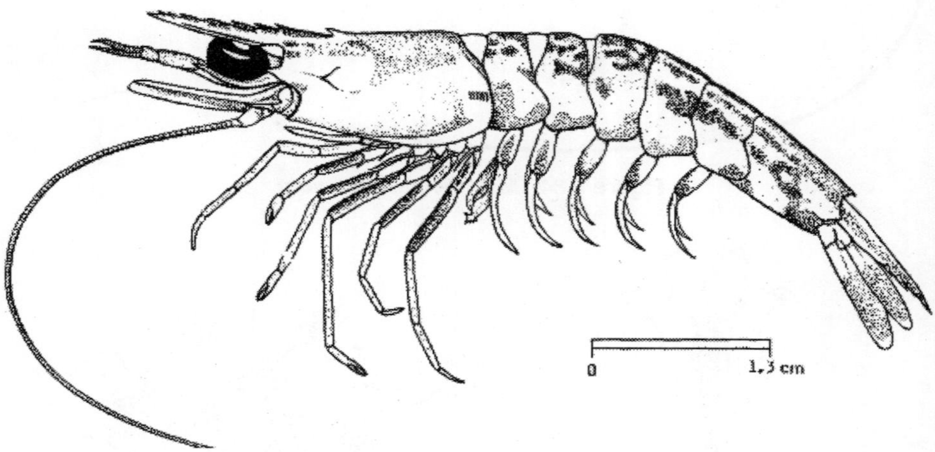

Figure 8.87: ***Metapenaeus stridulans***

Figure 5.88: Petasma

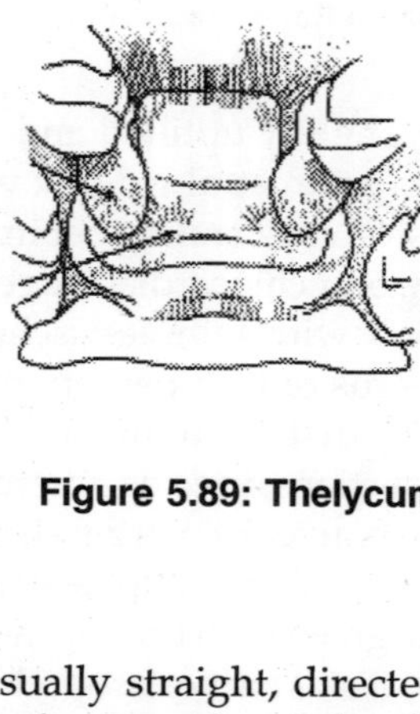

Figure 5.89: Thelycum

Distinctive Characters

Body densely pubescent. Rostrum low, usually straight, directed forward or slightly upward, reaching to, or almost to, tip of antennular peduncle and armed with 7 or 8 dorsal teeth, the penultimate tooth generally anterior to orbital margin of carapace; stridulating organ (on posterior part of carapace) consisting of 5 to 7 very strong ridges in a wide, straight band at 4/10 of carapace depth; middorsal crest on third abdominal segment with a usually broad groove. Petasma (in males) asymmetrical, right distoventral projection shorter and bearing a few small apical processes, left distoventral projection with 5 to 12 larger apical processes; in dorsal view: inner intermediate strip (i.i.s.) broadly quadrangular, distomedian lobule (d.l.) slightly shorter, but much broader distally than i.i.s. In females, thelycal plate subquadrate with rounded corners and slightly wider than long; intermediate plate broadly trapezoidal, much wider than long, flat or with a shallow median groove; coxal plates of fourth pereopods smaller than thelycal plate. Colour: white to reddish-brown, with red to dark brown mottlings; ereopods pinkish to dark red except on their proximal parts; uropods red to brown except for their proximal third and often their tips.

Size

Maximum total length: males, 7.2 cm; females, 10.5 cm.

Geographical Distribution and Behaviour

Their distribution is within the area, the "Gulf" and the Gulf of Oman to south India and Sri Lanka. Further east it extends as far as the Gulf of Thailand, China and Indonesia. A marine species found from 9 to 90 m depth on sandy or muddy bottom.

Present Fishing Grounds

In the "Gulf" and the northern Arabian Sea it is caught rather frequently but the fishery for this species is of minor or no importance; near Bombay it is caught in fairly large numbers; also fished in Sri Lanka where it is apparently the most abundant penaeid on the mud banks of the northeast coast.

Source: FAO Species Identification Sheets Pen Meta 16 1983.

Metapenaeus stebbingi Nobili, 1904

Common name: Peregrine shrimp

Distinctive Characters

Almost entire body hairless; rostrum armed with 7 to 10 teeth along entire dorsal margin, reaching or exceeding distal margin of antennular peduncle except for specimens from the Gulf in which it only reaches distal margin of second antennular article; postrostral ridge broad and low, ending near middle of carapace; branchiocardiac ridge indistinct; telson armed on each side with a row of small, movable spines; ischial spine on first pereopod absent. In adult males, merus of fifth pereopod with a proximal notch followed by a keel-shaped tubercle; each distomedian projection of petasma with a stiff, styliform appendix directed forward and ventrally serrated; distolateral projections directed laterally and separate in ventral and dorsal processes. In females, thelycum with posterior transverse ridge protruding forward between lateral plates and forming an inverted T-shape plate; lateral plates triangular.

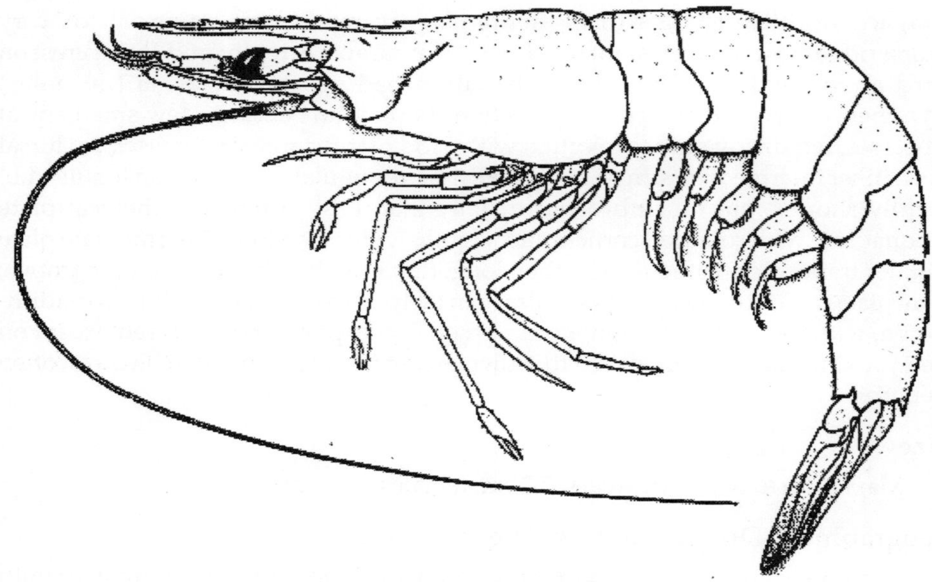

Figure 5.90: *Metapenaeus stebbingi*

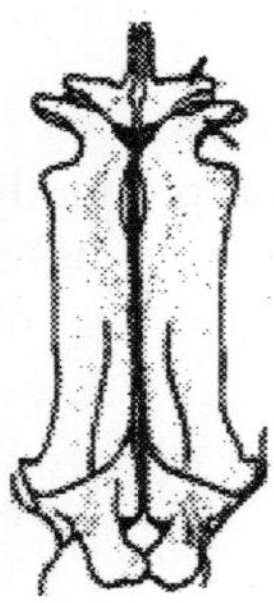

Figure 5.91: Petasma

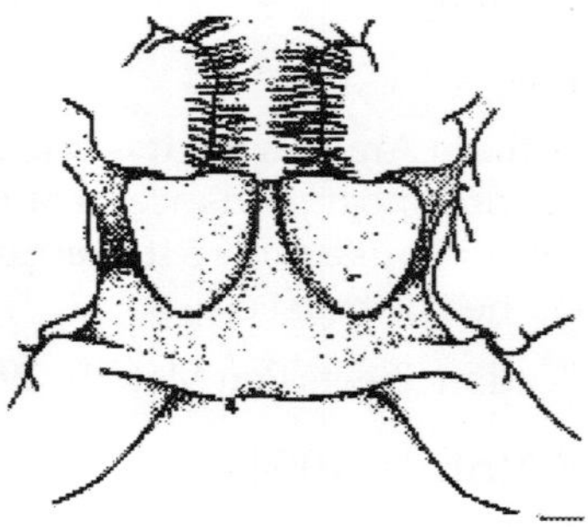

Figure 5.92: Thelycum

Figure 5.93: Telson

Colour: body, perepods and pleopods white to creamy-yellow with grey and rust coloured specks; antennae and distal part of uropods rusty colour to grey-purplish.

Size

Maximum total length: males, 11.1 cm; females, 13.9 cm.

Distribution

All along the coast from South Africa to the Gulf of Kutch in north-west India, including the Red Sea and the" Gulf," also present along the west coast of Madagascar; the species has entered the eastern Mediterranean through the Suez Canal. Found from the shore to about 90 m depth on soft bottom, mud or sand; juveniles inhabit near-shore, and adults in deeper waters, but usually in less than 45 to 50 m depth.

Source: FAO Species Identification Sheets Pen Metap 22 1983.

PARAPENOPSIS

Parapeneaopsis stylifera (Milne, Edward)

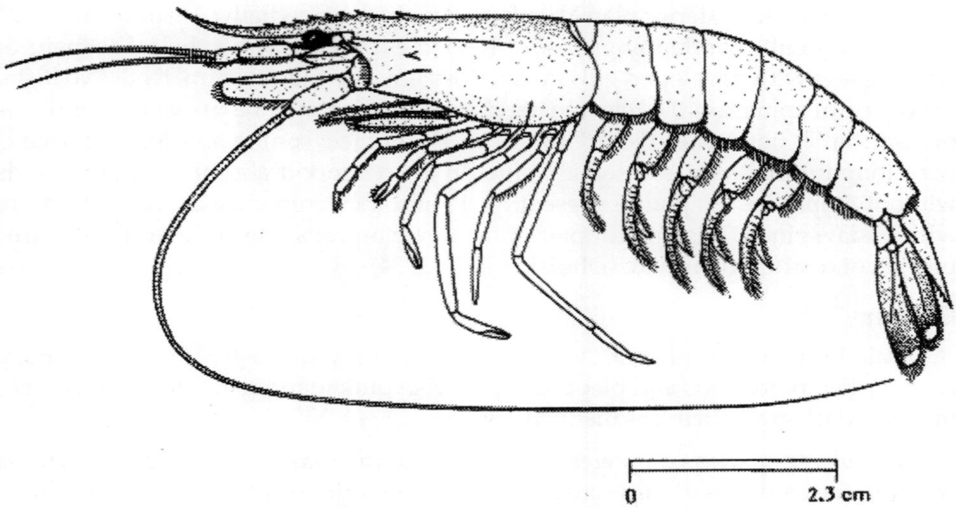

Figure 5.94: *Parapeneaopsis stylifera*

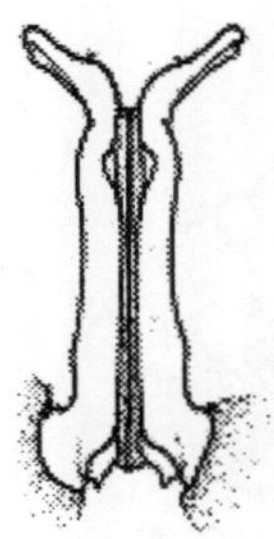

Figure 5.95: Petasma

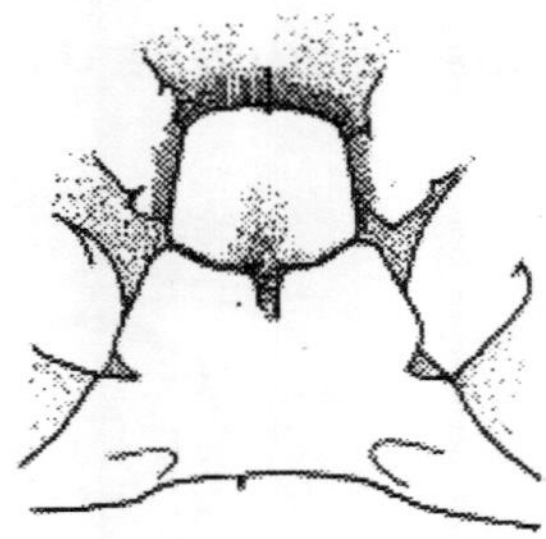

Figure 5.96: Thelycum

Distinctive Characters

Rostrum is with dorsal teeth only. Rostrum sigmoid-shaped, strongly upcurved and by far overreaching tip of antennular peduncle (in males somewhat shorter), armed with 7 to 9 dorsal teeth; but toothless in distal half or more; epigastric tooth present; Carapace is without longitudinal or transverse sutures or lateral keels. Dorsal keel on 4th–6th abdominal segments; lateral keels on 6th segment discontinuous and inconspicuous. Telson grooved, not trifid. No exopod on 3rd maxilliped or 5th pereiopod. postrostral crest almost reaching posterior margin of carapace; longitudinal suture long, reaching 2/3 of carapace length; telson armed with 4 pairs of lateral fixed spines; antennular flagellae as long as carapace; epipod and basial spine present on first and second pereopods, basis of third pereopod unarmed. In males, distolateral projections of petasma slender, horn-like and straight, directed antero-laterally and with ventro-external openings; distomedian projections small and curved ventrally. In females, anterior plate: of thelycum square, concave, with a slender stem-like posterior process; posterior plate deeply notched arteromedially. Pleurae are of 2nd abdominal somite overlapping those of 1st segment, 3rd leg with a chela. Last pairs of walking legs with developed gills many. Rostrum is without ventral teeth. No distal fixed pair of spines on the telson; lateral mobile spine may be present. Exopode is present on 5th pereopod; Pleurobranch on 7th thoracic somite absent. Carapace is with longitudinal sutures; ischial spine on 2nd pereopod absent. 3rd pereopod is without epipodite. Epipodite present on 1st and 2nd pereopods. 2nd pereopods are with basial spines. Telson with pair of fixed sub apical spine; at least distal ½ free portion of rostrum unarmed. (Dholakia A. D. 2004)

Colour

Pale brownish or pinkish white, sometimes greyish; rostrum and abdominal crest darker; pereopods and pleopods yellowish pink to reddish pink; distal part of uropods dark grey, their tips distinctly white.

P. stylifera species completes its life cycle in the marine environment without entering the estuaries at any stage of its life. The rarity of its larvae in the inshore plankton and the complete absence of it in the backwater collections support this view and suggest the possibility of their preference for offshore waters.

Parapeneaopsis sculptilis (Hellar)

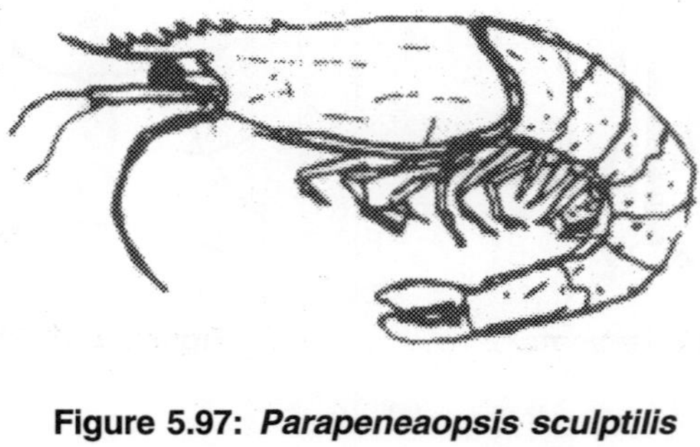

Figure 5.97: ***Parapeneaopsis sculptilis***

Figure 5.98: Petasma

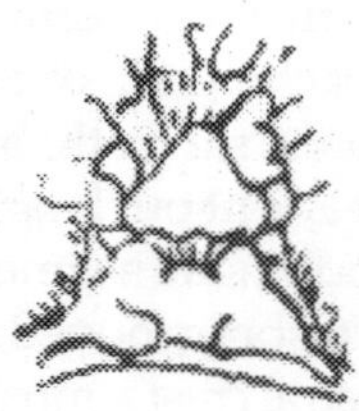

Figure 5.99: Thelycum

Pleurae are of second abdominal somite overlapping those of first and third segment; no chela of third pereopods. Carpus of 2nd pair of pereopods divided into two or more articles; if not, 1st pair of pereopods not chelate. Chelate of 2nd pair of pereopods small and slender. Rostrum is without ventral teeth. No distal fixed pair of spines on the telson; No exopod on 5th pereopod; Pleubranch on 7th thoracic absent. Carapace with longitudinal sutures; ischial spine on 2nd pereopod absent, but basal spine is present, 3rd pereopod without epipodite. Epipodites present on 1st and 2nd pereopods. Telson without fixed subapical spine, with or without lateral movable spine; 1/3 or less free portion of rostrum unarmed. Petasma with a pair of disto-lateral projections usually short and spout like and pair of cap like distal projections; Post-rostral carina reaching almost to posterior border of carapace; Antennular flagellum 0.5–0.6 length of carapace; thelycum with median tuft of setae on posterior plate. (Dholakia A. D. 2004)

Parapenaeopsis acclivirostris Alcock, 1905

Common name: Hawknose shrimp

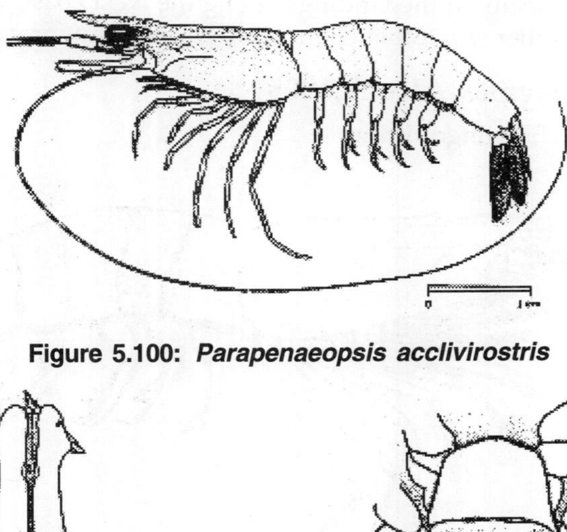

Figure 5.100: *Parapenaeopsis acclivirostris*

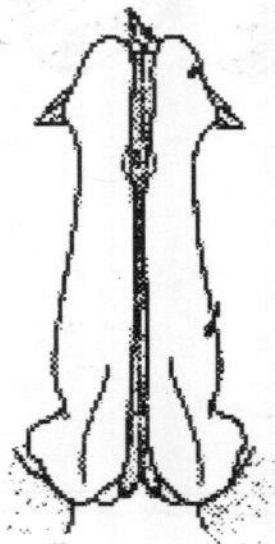

Figure 5.101: Petasma

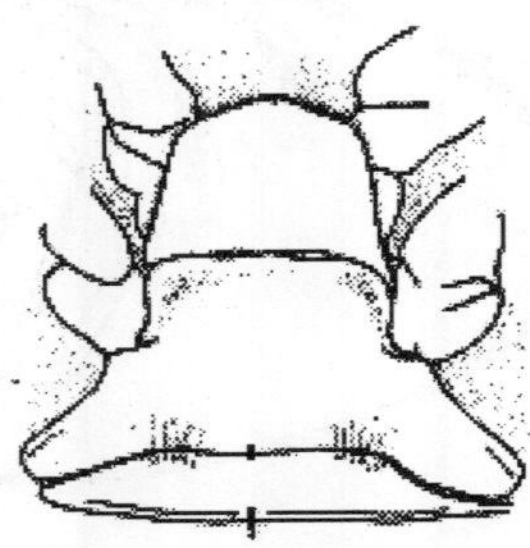

Figure 5.102: Thelycum

Distinctive Characters

Rostrum nearly straight, uptilted and armed with 6 to 9 teeth along entire dorsal margin, reaching or exceeding third antennular article in females, not extending beyond distal margin of second article in males; epigastric tooth absent; postrostral crest broad ending near middle of carapace; longitudinal suture reaching 2/3 of carapace length; antennular flagella 0.3 to 0.4 times the length of carapace; epipod absent on first and second pereopods; basial spine present on first and second pereopods; basis of third pereopod and telson unarmed. In males, distolateral projections of petasma with distal part slender, in dorsal position and directed proximolaterally distance between tip of projections about 2/5 of length of petasma; distomedian projections very small; lateral lobes of petasma without wing-like projections. In females, anterior plate of thelycum is semicircular and concave; posterior plate broad and trapezoidal.

Colour

Colour is pale with dull red specks. on carapace; antennal flagella and pereopods banded with dark pink; pleopods orangish; uropods and telsci, dark red.

Size

Maximum total length: males, 4.7 cm; females, 7.3 cm.

Distribution

Present sporadically in the landings; along the west coast of India it occurs in small numbers together with other commercially important prawns.

Parapenaeus longipes Alcock,1905

Common name: Flamingo shrimp

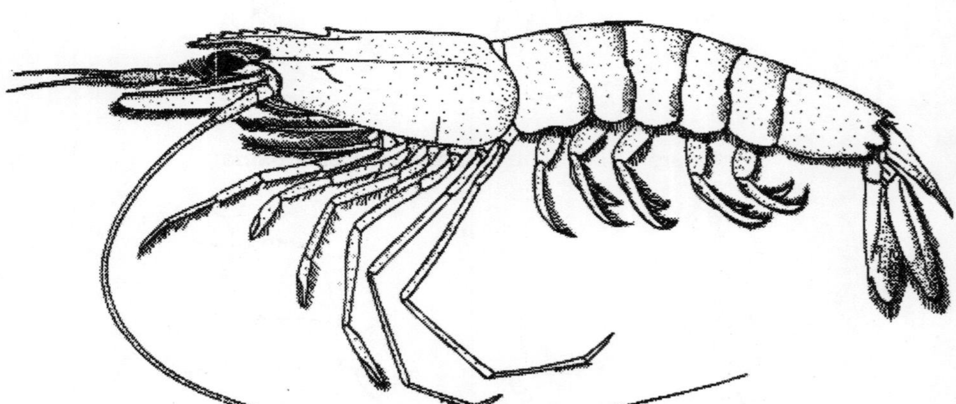

Figure 5.103: *Parapenaeus longipes*

Distinctive Characters

Body is hairless. Rostrum reaching, or exceeding distal margin of basal antennular segment, slightly curved downward and armed with 6 or 7 dorsal teeth; postrostral crest distinct; branchiostegal spine absent; no trace of hepatic crest; longitudinal and

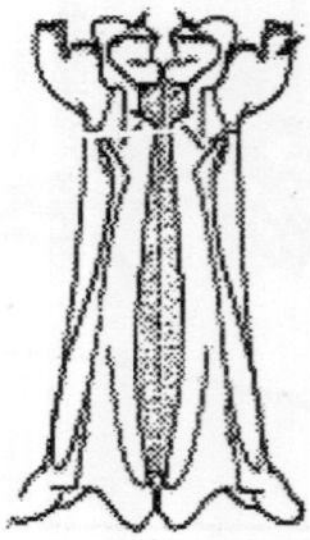

Figure 5.104: Petasma

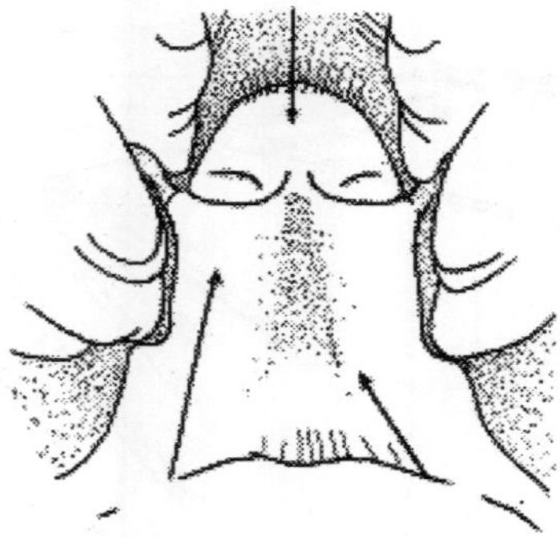

Figure 5.105: Thelycum

transverse sutures present, the former extending almost to posterior border of carapace; antennular flagella shorter than carapace; fifth pereopod exceeding antennal scale by length of dactyl often also by distal part of propodus. Petasma (in males) is with distolateral lobes sprout-like and as long as distomedian ones. In females, anterior plate of thelycum semicircular, articulating to intermediate plate and with a median groove which continues on intermediate plate; the latter is broad, quadrate and continuous to posterior sternal plate.

Colour

Pink to whitish, speckled with pink chromatophores; a red spot on distomedian part of outer uropods.

Size

Maximum total length: males, 7.6 cm; females, 7.9 cm.

Geographical Distribution and Behaviour

The species is found within the eastern Arabian Sea, from Pakistan to south India and off Sri Lanka. A marine species found between 10 and 90 m depth.

Parapenaeopsis cornuta (Kishinouye,1900)

Common name: Coral shrimp

Distinctive Characters

Rostrum slightly sigmoidal or straight, tip uptilted, sometimes toothless on distal 1/4, reaching from middle of second to distal margin of third antennular article and armed with 7 or 8 (rarely 9) dorsal teeth; epigastric tooth present; postrostral crest ending near posterior margin of carapace; longitudinal suture reaching about 0.4 of carapace length; antennular flagella 0.4 to 0.6 times the length of carapace; epipod and basial spine present on first and second pereopods; basis of third pereopod unarmed in both sexes; telson armed with 2 to 4 pairs of distolateral spinules. Petasma (in males) with long and slender, horn-like distolateral projections, diverging

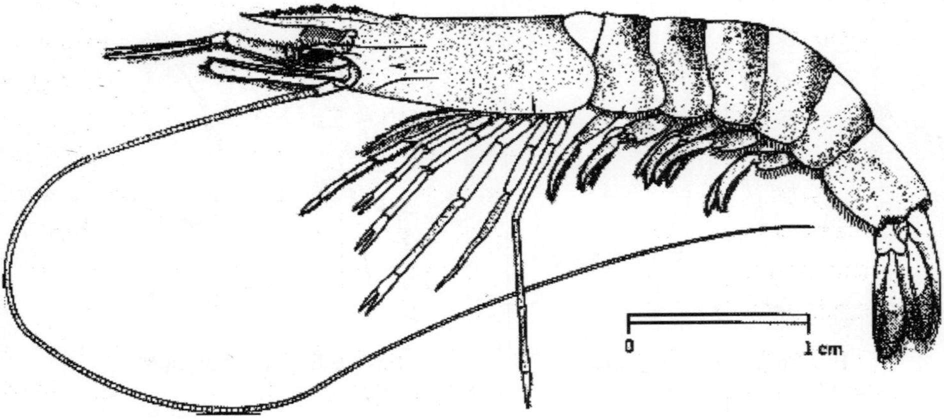

Figure 5.106: ***Parapeopsis cornuta***

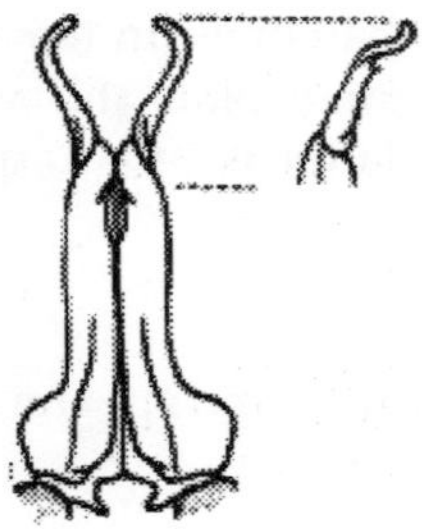

Figure 5.107: Petasma

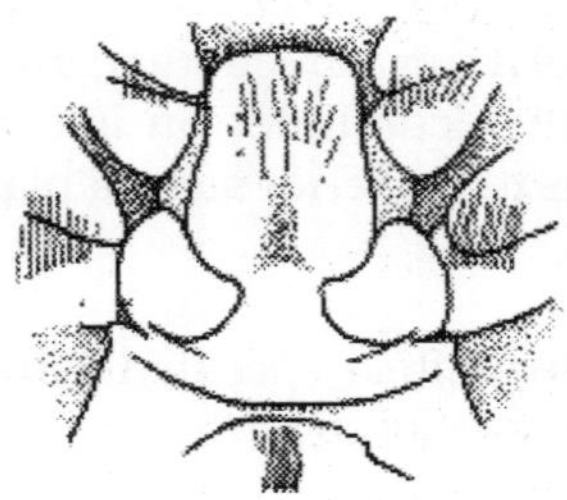

Figure 5.108: Thelycum

proximally and curving inward distally, each with a small dorsal spiniform process; distomedian projections extremely small. In females, anterior plate of thelycum oblong and concave, fused posteromedially with posterior late the latter with a air of lateral depressions; a median tuft of long setae (hairs) behind the thelycum.

Colour

Colour is pale brown or faintly pinkish with brown-grey specks and dorsal transverse brawn bands on abdomen; pereopods whitish, pleopods brownish, both speckled; uropods dark grey distally.

Size

Maximum total length: males, 7.6 cm; females, 7.9 cm.

Geographical Distribution and Behaviour

They are found from Pakistan to south India and off Sri Lanka. Further east, it extends as far as China, Japan and Indonesia.

A marine species found between 10 and 90 m depth.

Source: FAO Species Identification Sheets Pen Para 4 1983.

Parapeneaopsis hardwickii **(Morse 1878)**

Common name: Spear shrimp

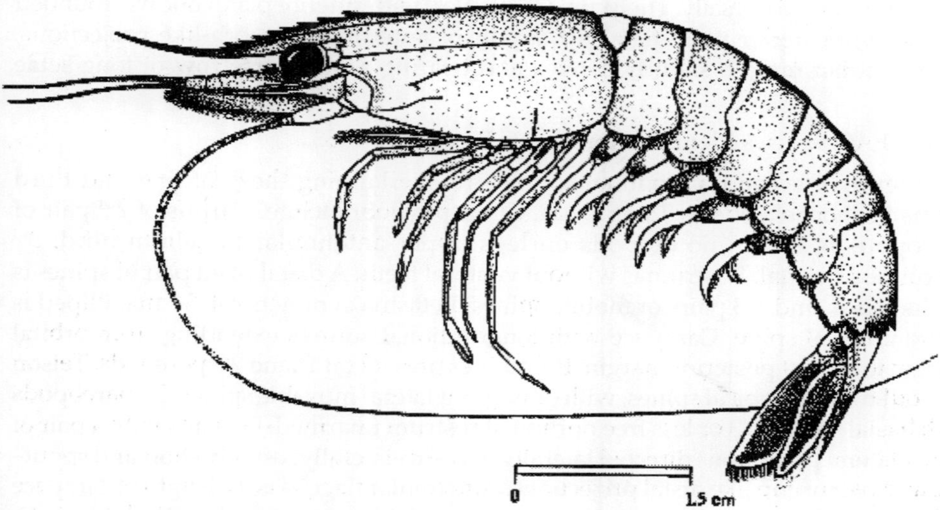

Figure 5.109: *Parapeneaopsis hardwickii*

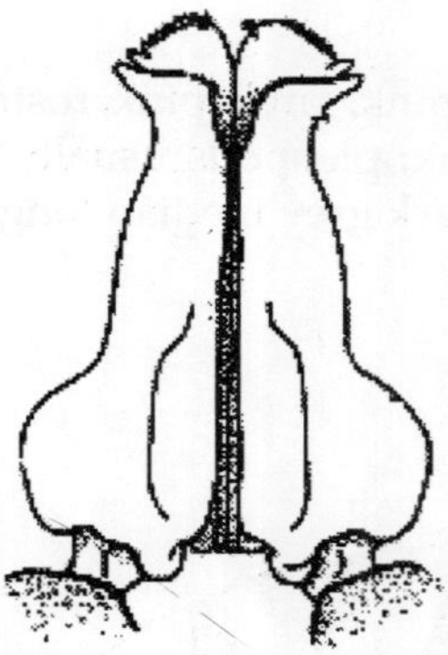

Figure 5.110: Petasma

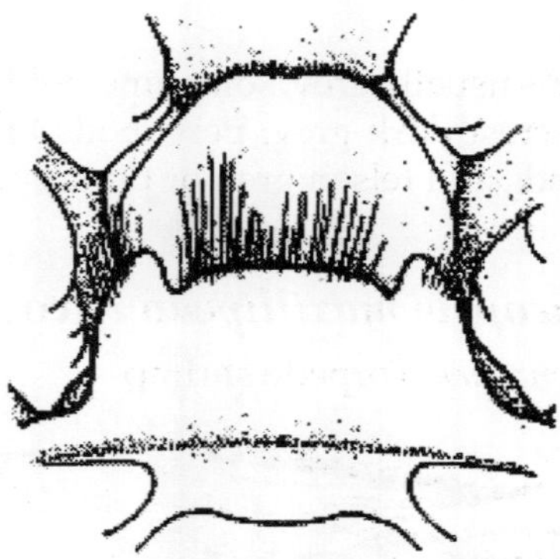

Figure 5.111: Thelycum

Distinctive Characters

Rostrum armed with 8 to 10 dorsal teeth; in females, sigmoidal, toothless on distal 1/3 to 1/2, upcurved and exceeding antennular peduncle; in large males curving downward, unarmed portion absent and reaching middle of second antennular article; both shapes found in young males. Epigastric tooth small; postrostral crest almost reaching posterior margin of carapace, grooved in females; longitudinal suture reaching 3/4 or more of carapace length; telson armed with 3 to 5 (usually 4) pairs of small mobile spines; antennular flagella at least 0.6 times the length of carapace, usually longer; epipod and basial spine present on first and second pereopods; basis of third pereopod unarmed. Petasma (in males) is with distomedian projections wing-like, wider than long, their anterior margin often crenulate; distolateral projections short and directed laterally; proximolateral lobes very large, curved dorsally. Thelycum (in females) with anterior plate concave, rounded anteriorly; posterior plate flat, with a pair of anterolateral tooth-like projections, anteromedian margin slightly convex and bearing a transverse row of long setae hairs.

Source: FAO Species Identification Sheets Pen Para 6 1983.

Pleure of second abdominal somite are overlapping those of first and third segments; bochela on third peropods. Gill is phyllobranchite. Carpus of 2nd pair of pereopods is entire; no epipods on legs upper antennular flagellum bifid; 3rd maxilliped normal. Rostrum is without venteral teeth. A distal fixed pair of spines is on the telson and 1-3 pairs of mobile spines. Petasma symmetrical; 3rd maxilliped is without basial spine. Carapace with longitudional sutures extending from orbital margin to almost posterior margin. Epipodites present on 1st and 2nd parepods. Telson without fixed subapical spines, with or without lateral movable spines; 2nd pareopods with basial spine; 1/3 or less free portion of rostrum unarmed. Petasma with a pair of distro-lateral projections directed laterally or distro-laterally, usually short and spout–like and pair of cup like distal projections. Antennular flagella is 0.7 length of carapace or larger; thelycum without median tuft of setae on posterior plate. (Dholakia A. D. 2004)

Colour

Colour is usually grey, sometimes with a touch of pink, rarely pink; rostrum and postrostral crest dark grey; pereopods brownish pink; pleopods usually reddish pink; uropods and telson grey or pink, each with a dark grey median longitudinal stripe.

Parapenaeopsis maxillipedo Alcock, 1905

Common name: Torpedo shrimp

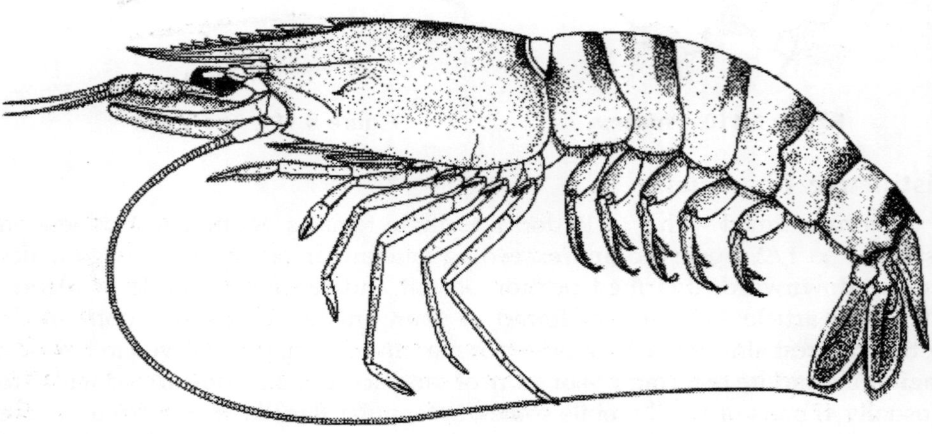

Figure 5.112: *Parapenopsis maxillipedo* 0 2 cm

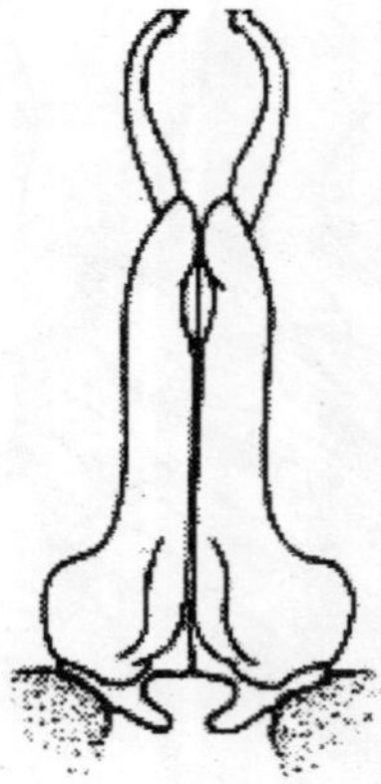

Figure 5.113: Petasma

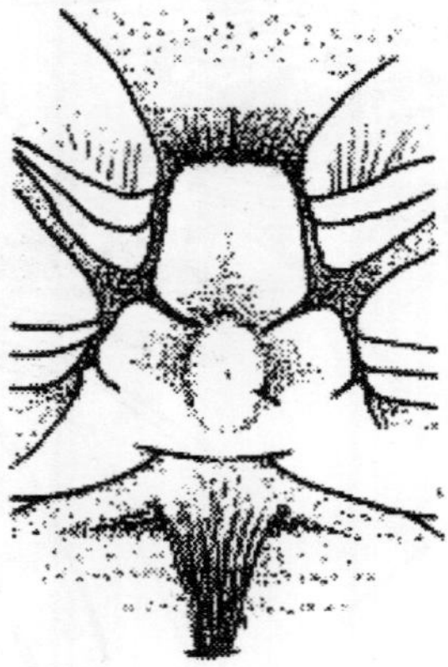

Figure 5.114: Thelycum

Distinctive Characters

Rostrum straight, tip uptilted, generally toothless on distal 1/4, reaching from middle of second to distal margin of third antennular article and armed with 8 to 11 (usually 9 or 10) dorsal teeth; epigastric tooth present; postrostral crest reaching, or almost, posterior margin of carapace; longitudinal suture reaching 1/3 to 1/2 of carapace length; antennular flagella 0.35 to 0.55 times the length of carapace; epipod present on first and second pereopods; basial sine resent on first 3 pairs of pereopods; telson unarmed. Petasma (in males) is with long and slender, horn-like distolateral projections, diverging proximally and curving inward distally, without small dorsal spiniform processes; distomedian projections extremely small. In females, anterior plate of thelycum subquadrate, posteriorly depressed and medially fused to posterior plate, the latter with a pair of lateral depressions and a median boss; a median tuft of long setae (hairs) behind the thelycum.

Colour

Colour is usually grey, sometimes pale brown; rostrum and postrostral crest dark brown; abdomen with dorsal transverse dark bands; pereopods brownish; pleopods red to brown; uropods greenish to red-brown with a pale stripe along margins.

Source: FAO Species Identification Sheets Pen Para 8 1983.

Penaeopsis jerryi Pérez Farfante, 1979

Common name: Gondwana shrimp

Distinctive Characters

Body is hairless. Rostrum almost horizontal, straight or slightly sinuous (occasionally convex basally, straight anteriorly) and long, almost reaching to, or exceeding, distal margin of antennular peduncle, armed with 12 to 16 dorsal teeth;

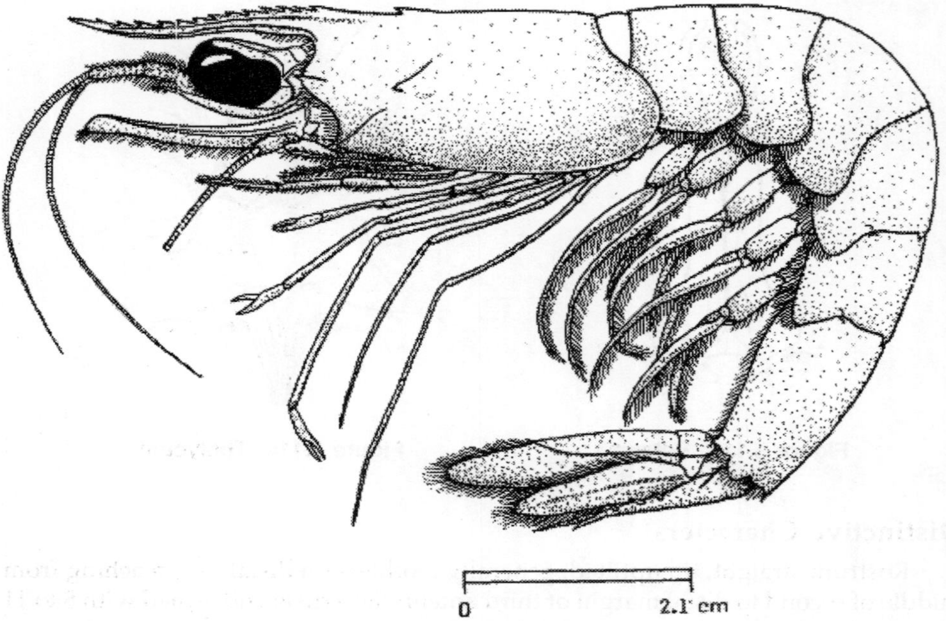

Figure 5.115: *Parapenopsis jerryi*

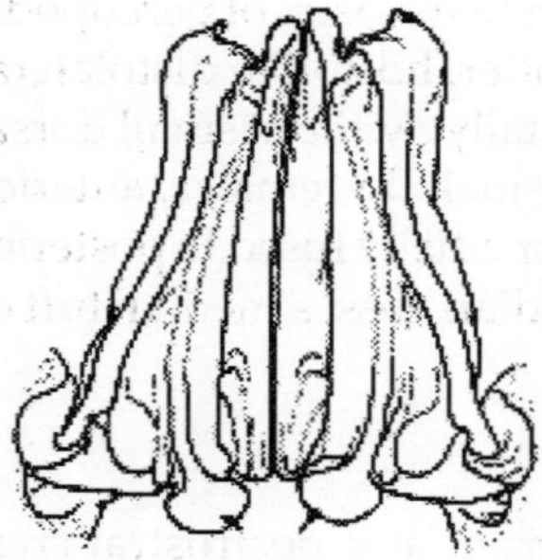

Figure 5.116: Petasma

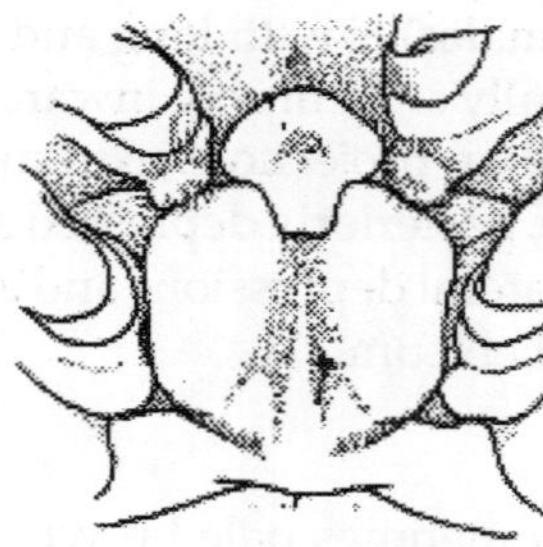

Figure 5.117: Thelycum

Figure 5.118: Telson

anteroventral angle of carapace almost rectangular; cervical and hepatic crests and grooves well defined; hepatic spine situated at about same level as antennal spine; telson armed with a pair of fixed lateral spines and 3 pairs of small, movable spines. Petasma (in males with distal part of ventral costae curving abruptly dorsomedially and ending in short, relatively narrow processes; ribs of dorsolateral lobules terminating proximally in semicircular or subcircular processes. In females, anterior plate of thelycum subsemicircular to trilobed; posterior plate with anterior border broadly arched on each side of posteromedian projection of anterior plate, and strongly inclined posterolaterally, its median ridge sometimes reduced to a posterior tubercle.

Colour

Red or dark brown with a reddish tint.

Source: FAO Species Identification Sheets Pen Para 4 1983.

Penaeopsis balssi Ivanov and Hassan, 1976

Common name: Scythe shrimp

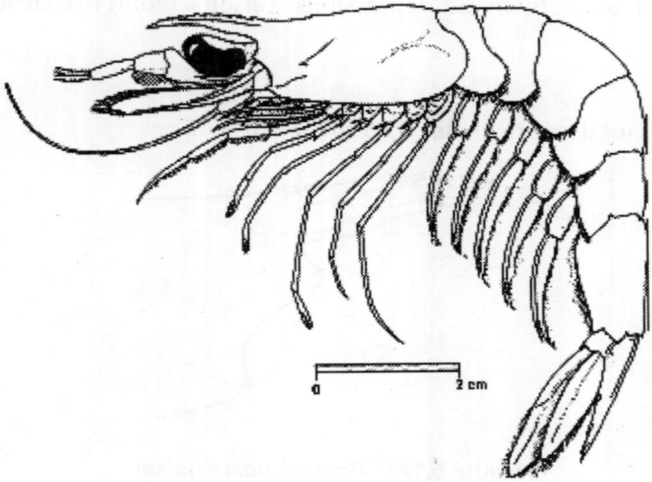

Figure 5.119: *Parapenopsis balssi*

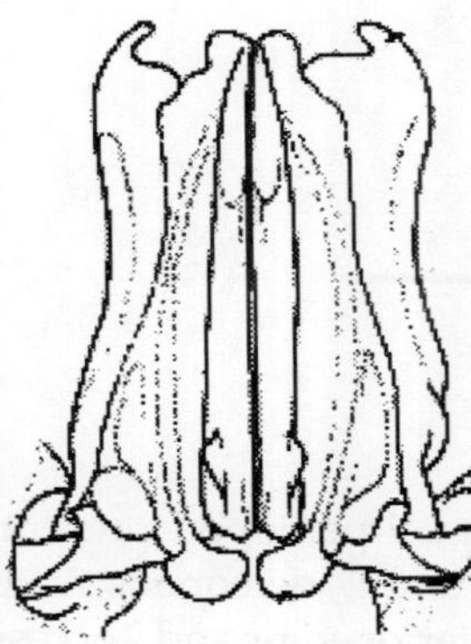

Figure 5.120: Petasma

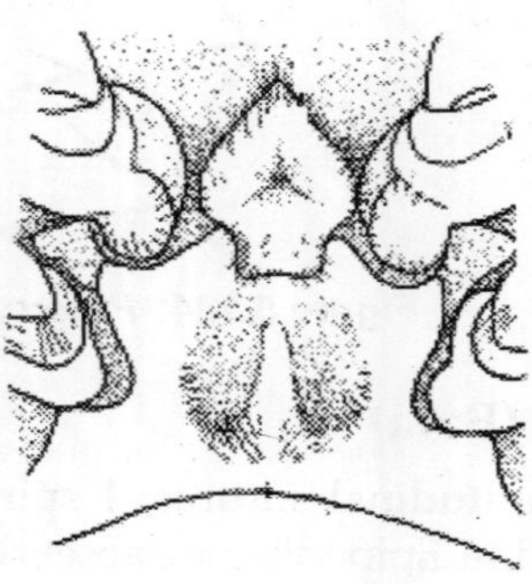

Figure 5.121: Thelycum

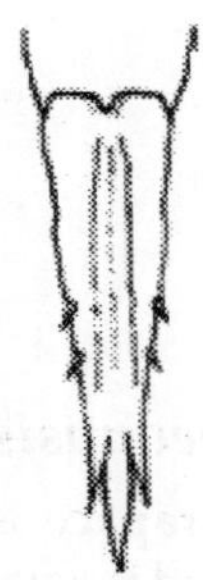

Figure 5.122: Telson

Distinctive Characters

Body is hairless. Rostrum usually markedly curved (always strongly so in young) and short, reaching at most to midlength of second antennular segment, armed with 9 to 13 dorsal teeth; anteroventral angle of carapace obtuse; cervical and hepatic crests and grooves well defined; hepatic spine situated ventral to level of antennal spine; telson armed with a pair of long, fixed, lateral spines and 2 pairs of small, movable spines. Petasma (in males) with distal part of ventral costae curving dorsomedially and ending in conspicuous subsemicircular processes; ribs of dorsolateral lobules ending proximally in flattened, mesially directed, suboval processes. In females, anterior plate of thelycum subtriangular to orbicular; posterior plate with anterior border straight or, usually, concave on each side of posteromedian projection of anterior plate; its median ridge (sometimes reduced to posterior protuberance) flanked by broad depressions. Telson Colour: frozen specimens are red to pinkish.

Size

Maximum total length: males, 12.8 cm; females, 15 cm.

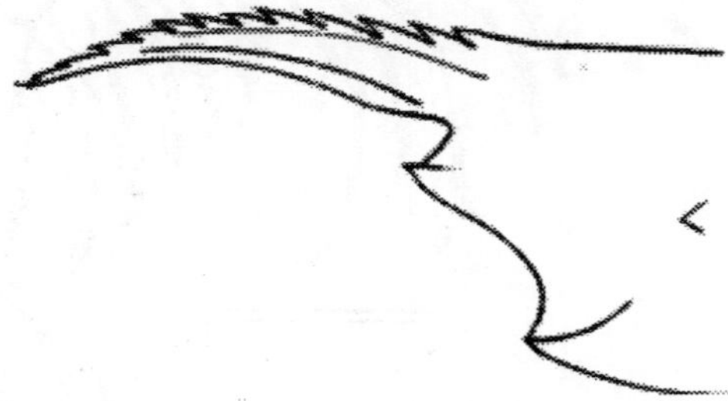

Figure 5.123: ***Parapenopsis balssi***

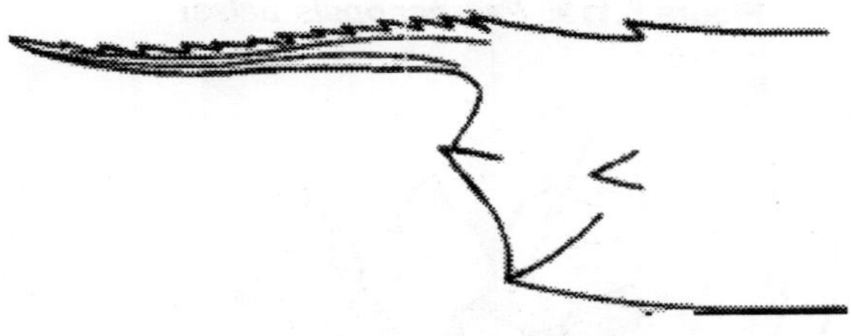

Figure 5.124: ***P. jerry***

Parapeonpsis tenella (Bata)

Carapace is with longitudinal sutures; 1 spine on 2nd pereopod present. 3rd pereopod is without epipodite. Epipodites are absent on 1st and 2nd pereopods. Anterior plate of thelycum is with V-shaped posterior edge of posterior edge; and 2 accessory ridges on anterior edge of posterior plate; rostrum with proximal 1/3 rising from carapace; reminder more or less horizontal.

Parapenaeopsis uncta Alcock, 1905

Synonyms: *Parapenaeopsis probata* Hall, 1961

Common name: Uncta shrimp

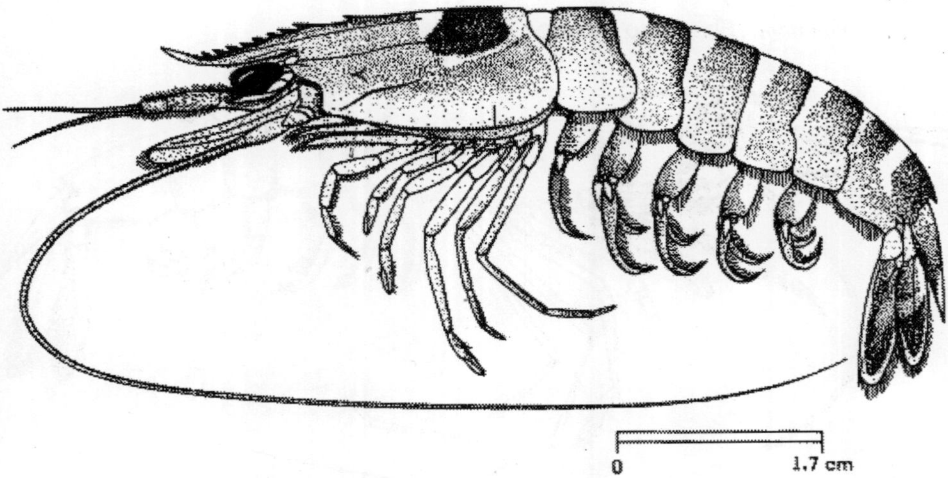

Figure 5.125: *Parapenopsis uncta*

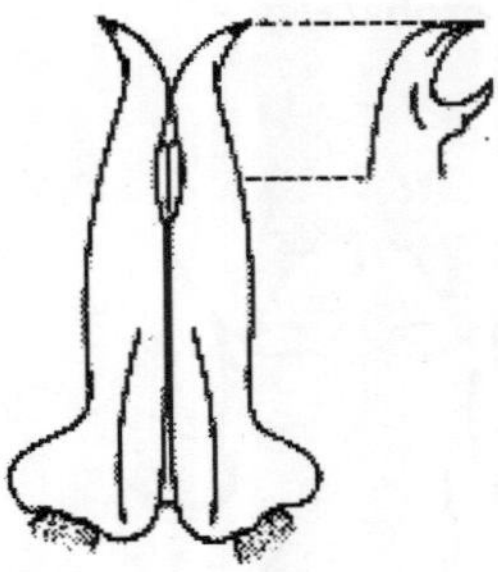

Figure 5.126: Petasma

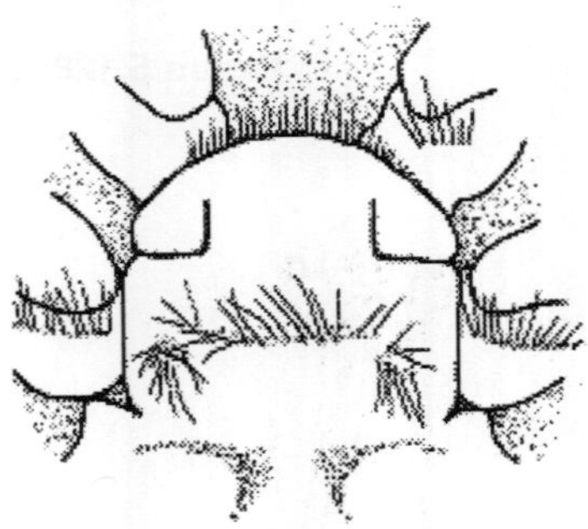

Figure 5.127: Thelycum

Distinctive Characters

Rostrum armed with 9 to 11 dorsal teeth; in females and young males sigmoid-shaped, toothless on distal 1/4 to 1/3 and styliform, usually reaching third antennular article or beyond; in large males, straight, not reaching beyond second antennular article; epigastric tooth present, but small in males; postrostral crest grooved, almost reaching posterior margin of carapace; longitudinal suture reaching 3/4 of carapace length; telson unarmed; antennular flagella 0.45 to 0.60 times the length of arapace; epipod and basial spine present on first pereopod, both also present on second pereopod of females, but in males the basial spine is absent or very small. In males, distolateral projections of petasma is simply tapering to ends each with a long dorsomedian spine-like process; distomedian projections very small. In females, anterior plate of thelycum is wide and short, with curved anterior margin and with 2 longitudinal ridges, medially fused with the quadrate posterior plate.

Parapenaeopsis probata Hall, 1961

Common name: Flamingo shrimp

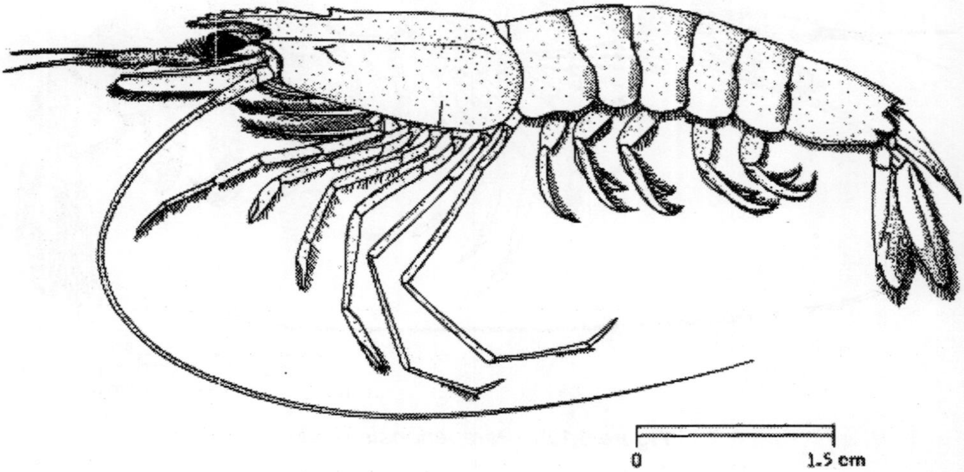

Figure 5.128: *Parapenopsis probata*

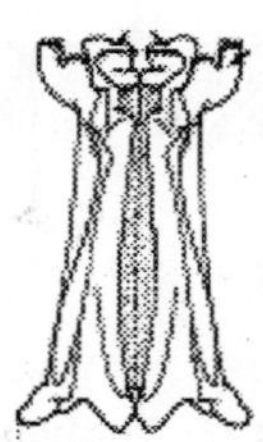

Figure 5.129: Petasma

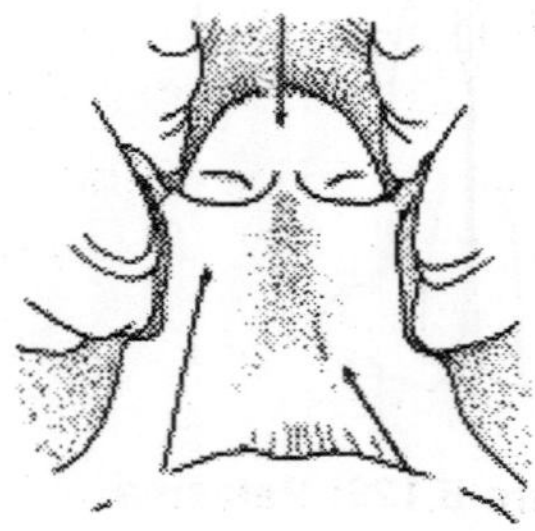

Figure 5.130: Thelycum

Distinctive Characters

Body is hairless. Rostrum reaching, or exceeding distal margin of basal antennular segment, slightly curved downward and armed with 6 or 7 dorsal teeth; postrostral crest distinct; branchiostegal spine absent; no trace of hepatic crest; longitudinal and transverse sutures present, the former extending almost to posterior border of carapace; antennular flagella shorter than carapace; fifth pereopod exceeding antennal scale by length of dactyl often also by distal part of propodus. Petasma (in males) is with distolateral lobes sprout-like and as long as distomedian ones. In females, anterior plate of thelycum semicircular, articulating to intermediate plate and with a median groove which continues on intermediate plate; the latter is broad, quadrate and continuous to posterior sternal plate.

Colour

Colour is pink to whitish, speckled with pink chromatophores; a red spot on distomedian part of outer uropods.

FAMILY: PENAEIDAE

SOLENOCERIDAE

Solenocerid Shrimps

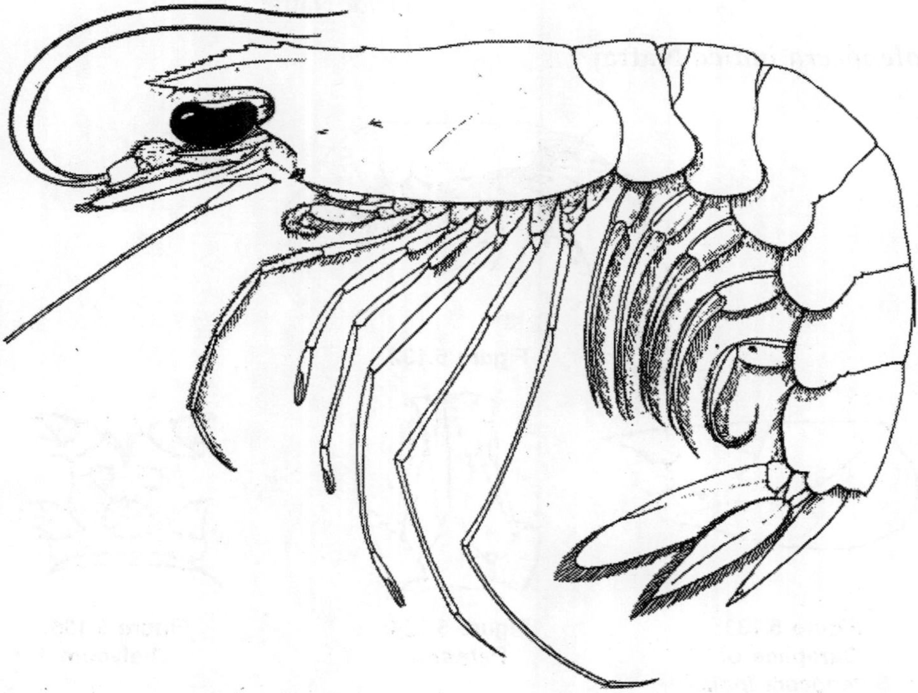

Figure 5.131: *Solenocerid*

Distinctive Characters

Shrimps with a well developed and toothed rostrum which extends at least to centre of eye diameter; no styliform projection at base of eyestalk, but a tubercle present on its mesial (inner) border. Carapace is with postorbital spine and long cervical groove which end at, or close to, dorsal midline. Last 2 pairs of pereopods well developed; endopods of second pair of pleopods in males bearing appendix masculina appendix interna and lateral projection; third and fourth pairs of pleopods biramous. Telson tridentate in most species (with a fixed spine on each side of tip). Two well eveloped arthrobranchs on the penultimate thoracic segment (hidden beneath the carapace).

This family includes only marine representatives. All except four of the species occurring in the W. Indian Ocean are too small or not abundant enough to be of present or potential economic interest.

Classification

Family	Penaeidae
Sub family	Solenocerinae
Genus	*Solenocera*
Species	*indica* Natraj.

Solenocera indica Natraj

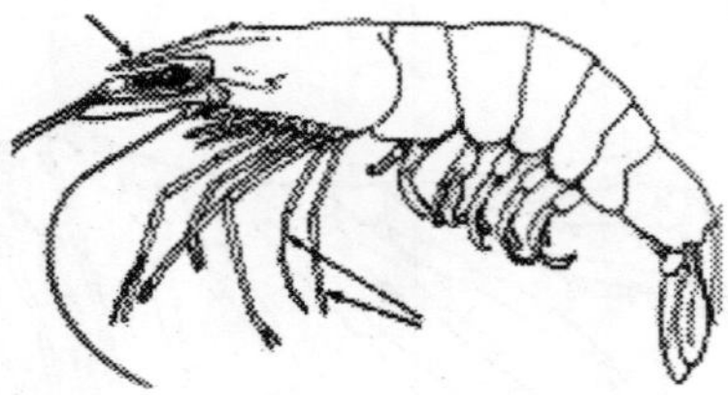

Figure 5.132

Figure 5.133: Carapace of *Solenocera indica*

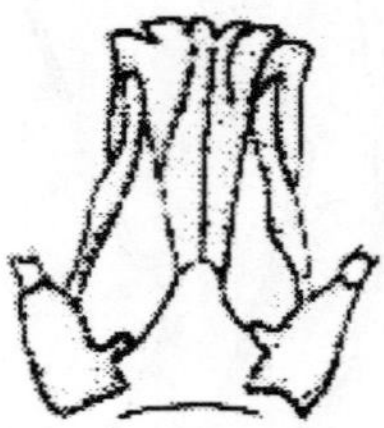

Figure 5.134: Petasma

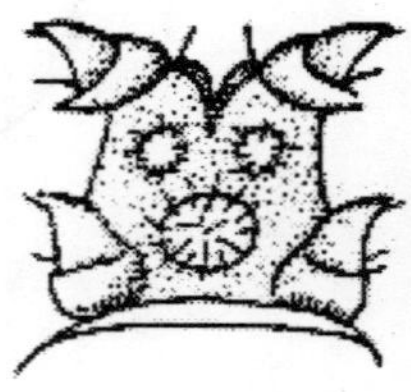

Figure 5.135: Thelycum

Chelae of 1st pair of pereopods is microscopically small or absent; chelae of 2nd pair of pereopods small and slender. Post orbital spine present; Antennular flagella foliaceous; telson simple and devoid of any spine on lateral margin.

Carapace is with median dentate crest extending nearly or quite to the posterior margin. Antennular flagella are foliaceous. Telson is simple and devoid of any spine on lateral margin. Last two pairs of walking legs well developed; gills are many. Carapace measurements were grouped in 1.0 mm classes. Average growth rates of adult females and males, as estimated from length frequency studies, are 6.96 mm and 6.49 mm per month respectively. Juveniles grow faster. Estimated life spans of female and male are 14 to 15 months and 9 to 10 months, respectively. Females always dominate the population.

Relationship

The relationship was found to be (L = 6.91 + 3.20 C) where L is the total length (mm) and C the carapace length (mm). The largest female and male encountered during this study measured 114 mm and 80 mm respectively, and taking into consideration the monthly average growth and the approximate age of the juvenile recruits discussed above.

Availability

S. indica is found all along the coast of India. Nataraj (1945) recorded the species from Gulf of Cambay, Bombay, Madras, Visakhapatnam, Orissa and Ganjam coasts, and from the sand-heads and the mouth of River Hooghly. All these records were from within 40 m depth. Two types of migratory movements were discernible in *S. indica*, one in connection with spawning and the other in relation to salinity.

Spawning Season

In the peak spawning months of December and April the temperature is relatively low (25 to 26°C) in the former and high (29 to 31°C) in the latter month, Jayaraman indicating that spawning does not probably depend on temperature conditions. But both these peak periods coincide with the period of highest salinity of 35 to 37‰ in the surface waters.

Sex Ratio

The disparity between the sexes was least in the month of October and high in the months of January, April and May when more than half the population was constituted by females. A breakdown of the data for the entire period of investigation showed that the number of females was twice that of males.

Breeding Period

The breeding period is protracted, with two peaks in December and April. Main spawning grounds are outside, and probably contiguous with, the fishing area. Males do not leave the fishing grounds for purposes of breeding. (Source: Kunju, M.M., 2005).

Solenocera pectinata (Bate)

Externo-distal margin of the exopod of the uropod is without spine; Post rostral carina is not extending beyond cervical groove. Telson is trifurcate.

Solenocera hextii Word Mason

Solenocera hextii Wood-Mason and Alcock, 1891

Common name: Deep-sea mud shrimp.

Distinctive Characters

Body hairless, except, the rostrum; rostrum high, reaching to about distal margin of eye, armed with 6 to 8 dorsal teeth, its ventral margin nearly straight; postrostral crest elevated and laminose, reaching posterior margin of carapace and interrupted by a notch just in front of cervical groove (in small specimens, the crest is less developed

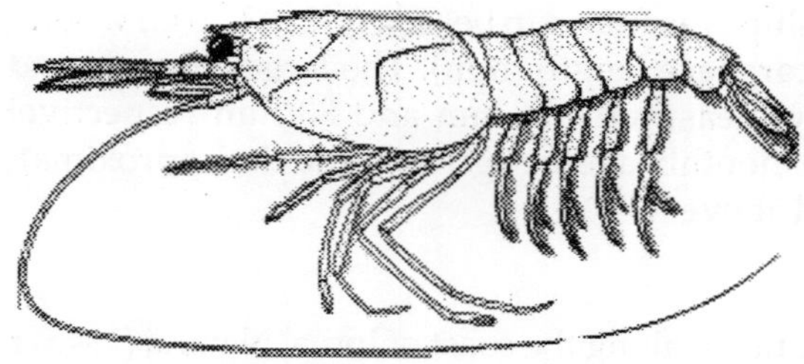

Figure 5.136: ***Solenocera hextii***

and the notch may be absent); cervical groove deep, reaching to, or almost, to dorsal midline; postorbital and suprahepatic spines present; ranchiostegal and pterygostomian spines absent; hepatic crest curved ventrally on its anterior part, with a sharp bending near its anterior end; branchiocardiac crest very distinct and L-shaped (its posterior half nearly horizontal its anterior half turning ventrally at right angle); telson with a fair of fixed distal lateral spines (trifurcate); fifth pereopod with a coxal spine. Telson is trifurcate. Externo-distal margin of the exopod of the uropod is with spine. Spine on cervical groove ventral to the posterior-most spine of the rostral series is present; "L" shaped groove on either branchiostergal region is present.

Colour

Bright pink.

Size

Maximum total length: males, 12.7 cm; females, 13.8 cm.

Geographical Distribution and Behaviour

Within the area, it occurs in the Arabian Sea from the Gulf of Aden to the south coast of India. Also present in the Bay of Bengal.

Solenocera choprai Nataraj, 1945

Common name: Ridgeback shrimp

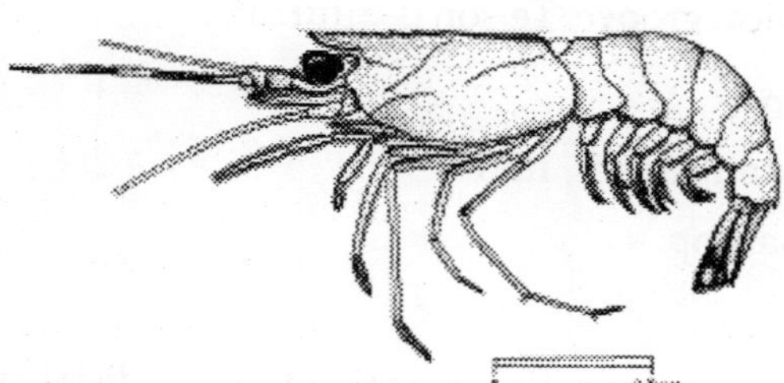

Figure 5.137: ***Solenocera choprai***

Figure 5.138: Telson

Distinctive Characters

Body hairless except at base of rostrum where it is distinctly pubescent; rostrum reaching middle to 3/4 of eye, convex on ventral margin and with 6 to 9 dorsal teeth; postrostral crest markedly elevated and laminose, reaching posterior margin of carapace and interrupted by a notch just ahead of cervical groove; the latter is deep and reaches or almost dorsal midline; postorbital spine present; suprahepatic and branchiostegal spines absent; pterygostomian angle broadly rounded and unarmed; hepatic crest curved downward anteriorly, with a sharp bending near its anterior end delimiting a round loop just behind frontal margin of carapace; branchiodardic crest oblique, its anterior part curving ventrally; telson trifurcate, with a pair of fixed distolateral spines; fifth pereopod with a coxal spine. Telson is trifurcate. Telson solenocera crassicornis: postrostral crest low and rounded (markedly elevated and laminose in *S. choprai*); telson unarmed (with lateral spines in *S. choprai*); anterior part of hepatic crest differently shaped.

Spine on cervical groove ventral to the posterior-most spine of the rostral series is absent; "L" shaped groove on either branchiostergal region is absent. Externo-distal margin of the exopod of the uropod is with spine.

Colour

Body, pereopods and pleopods are red; antennae banded dark red and white; uropods dark red, except for some white areas.

Size

Maximum total length: males, 9.5 cm; females, 13 cm.

Family SERGESTIDAE

Acetes indicus (H. Milne)

Pleurae of 2nd abdominal somite are overlapping those of 1st segment, 3rd leg with a chela. Last one or two pairs of walking legs reduced or absent; gills few. Procurved spine is present between 1st pair of pleopods. Basis of the third pereopod is with tooth on inner free margin; petasma without membranous coupling folds.

Figure 5.139: Petasma

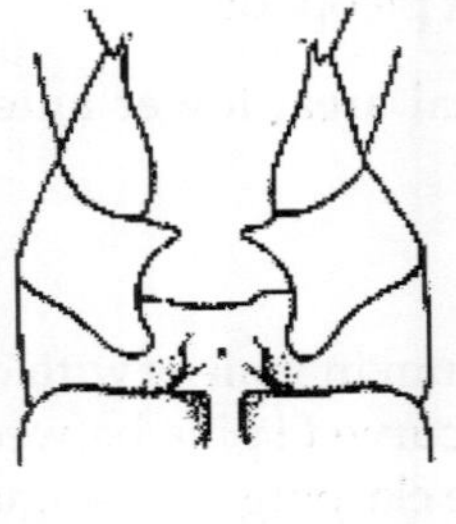

Figure 5.140: Thelycum

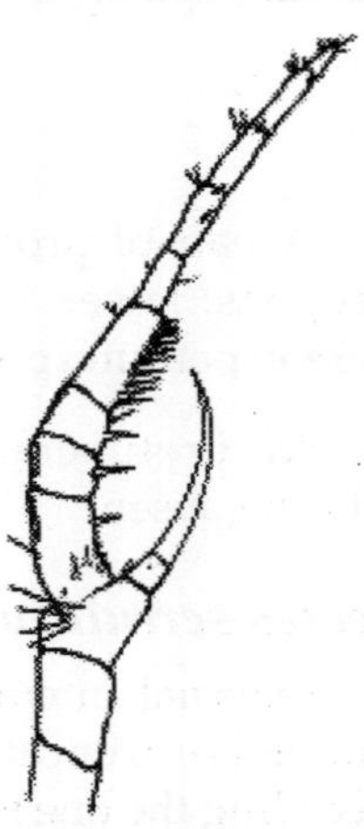

Figure 5.141: Lower Antennular Flagellum

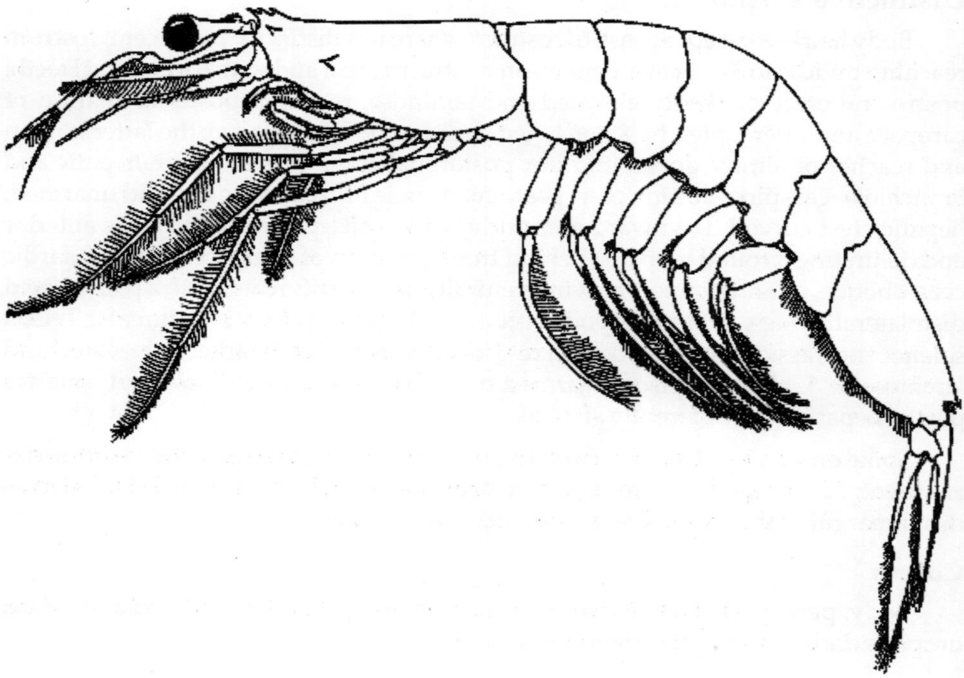

Figure 5.142: *Acetes indicus*

Inner margin of basis of third pereopods with a sharply pointed projection; third and fourth thoracic sternites deeply channelled longitudinally. Procurved tooth present between bases of first pair of pleopods; lower antennular flagella with 1 clasping spine.

KEY TO SEXES OF ACETES

A pair of protuberances (genital coxae) between third pereopods and first pleopods; lower antennular flagella with 1 or 2 clasping spines or modification thereof; petasma present on first pleopods .... Male

No protuberance in genital area; lower antennular flagella without spine; petasma absent .... Female

Acetes serrulatus Kroyer

External antennular flagellum in male is with clasping spines; apex of telson is rounded or truncated. No pre-curved spine between 1st pair of peopods. Segment preceding the one is bearing the clasping with angular process pointing backwards apex of telson truncated and with a tooth at each corner.

Acetes japonicus Kishinouye 1905

Common name: Akiami paste shrimp.

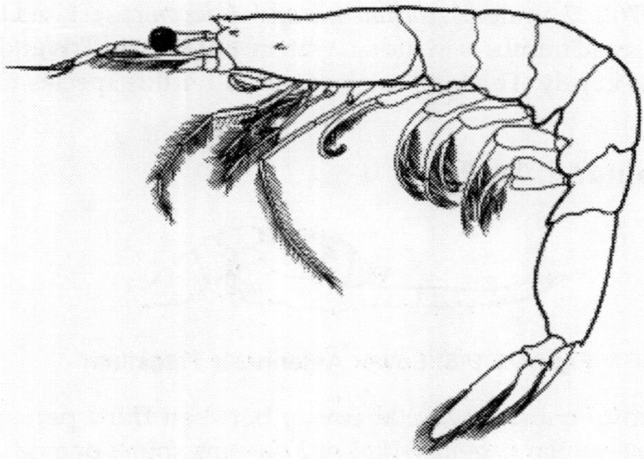

Figure 5.143: *Acetes japonicus*

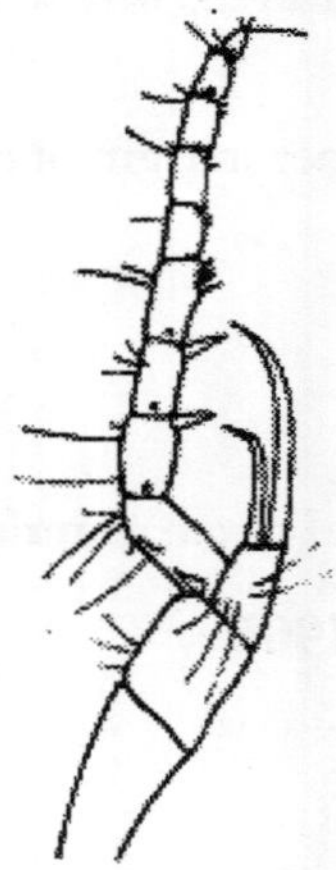

Figure 5.144: Lower antennular flagellum

Ciliated and non-ciliated portions of external border of exopod of uropod is not separated by a tooth; distal portions of pars external with tubercles. Segment proceeding the one bearing the clasping spines without any process; apex of telson round and third thoracic sternite produced posteriorly as large plate as in female. External antennular flagellum in male is with clasping spines; apex of telson is rounded or truncated. First segment of main branch of lower antennnular flagella is without triangular projection. Third thoracic sternite produced posteriorly.

Size

Total length 11 to 24mm (male), 15 to 30 mm (female)

Availability

According to Nataraj (1949:139) this species together with other species of the genus is "of considerable commercial importance" in Travancore, S.W. India. Kurian and Sebastian (1976:102, under *A. cochinensis* and *A. japonicus*) stated that the species usually occurs in small numbers in Indian waters, but that off Trivandrum it is taken in great quantities in July. The total catch reported for this species to FAO for 1999 was 598 602 t.

Acetes johni Nataraj, 1947

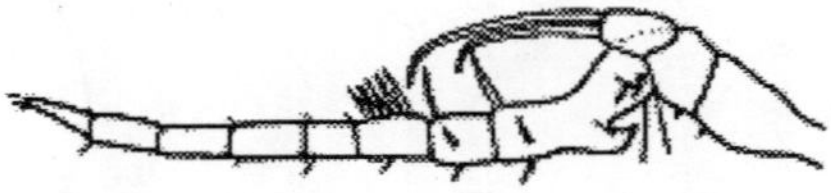

Figure 5.145: Lower Antennular Flagellum

A pair of protuberances (genital coxae) between third pereopods and first pleopods lower antennular flagella with 1 or 2 clasping spines or modification thereof petasma present on first pleopods .... Male

No protuberance in genital area; lower antennular flagella without spine; petasma absent .... Female

Tooth is present on distal inner margin of coxa of third pereopods.

A. sibogae

Procurved tooth absent.

Males

1a. Anterior margin of genital coxae rounded; petasma without pars astringens

Acetes erythraeus Nobili, 1905

Figure 5.146: Petasma

Figure 5.147: Thelycum

Inner margin of basis of third pereopods is without a sharply pointed projection; third and fourth thoracic sternites not channelled longitudinally petasma with pars stringens. Procurved tooth between bases of first pair of pleopods.

Family SERGESTIDAE

Family	Hippolytidae
Genus	*Hippolysmata*
Species	*ensirostris* Kemp

New name: *Hippolysmata* (Exhippolysmate) *ensirostris* Kemp

Hippolysmata ensirostris (Kemp)

Figure 5.148: *Hippolysmata ensirostris*

Chelae of 1st pair of pereopods distinct on both sides; ends of fingers of this chela usually dark coloured; eyes are free, never extremely elongated. Pleaurae of second abdominal somite is overlapping those of first and third segments; no chela on third pereopods. Rostrum is longer than carapace, with an elevated dentate basal crest; lateral margins of telson concave, apex acute and unarmed.

Aristeomorpha woodmasoni Calman 1925

Figure 5.149: Carapace of *Aristeomorpha woodmasoni* Cakman

Chelae of 1st pair of pereopods is microscopically small or absent; chelae of 2nd pair of pereopods small and slender. Post orbital spine present. Carapace is without a median dentate crest except occasionally over the eyes. Rostrum is with many teeth on upper border; hepatic spine present; length of pterygostominian region more than 2.5 times its greatest breadth.

Habitat

Depth 330 to 500 m. Marine.

Size

Maximum total length 153 mm.

As per Kurian and Sebastian (1976) they are commercially important in India

Aristeus alcocki Ramdan

Size

Maximum total length 150 mm, average length 110 mm, 140 mm

Interest to Fishery: Potential. Important prawns in S.W. India, where the species was obtained in small numbers.

Aristeus semidentatus Bate, 1881

Habitat

Depth 180 to 1 100 m. Bottom mud. Marine.

Size

Maximum total length 90 mm, 178 mm.

Kurian and Sabastian (1976-95) pointed its commercial importance in view of the fact that it was obtained in fair numbers at exporatory crusis off Cochin in S. W. India.

Atypopenaeus stenodactylus (Stimpson, 1860)

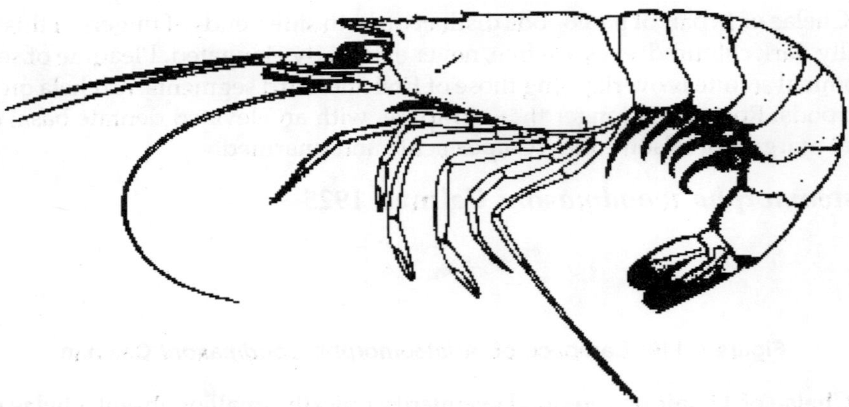

Figure 5.150: *Atypopenaeus stenodactylus*

Distinctive Characters

Rostrum is short and straight, with 7 to 9 dorsal teeth. Post rostral carina runs almost too posterior border of carapace. Hepatic spine is present, epigastric tooth placed usually for back in rostrum. Cervical groove well defined. Petasma with distolateral projections directed forward, narrowing from base to tip with inner surface deeply concave and with an indentation at tip in thelycum, anterior plate tongue-like, with rounded anterior and posterior end.

Carapace is without longitudinal sutures; ischinal spine on 2nd pereopod present. Rostrum short and straight, not exceeding distal margin of first antennular segment and armed with 6 to 9 dorsal teeth. Postrostral crest is ending near posterior margin of carapace. Hepatic spine present; Epigastric teeth far back in rostrum;

PETASMA with distolateral projections directed forward, narrowing from base to tip, with inner surface deeply concave and with an indentation at tip.

THELYCUM anterior plate tongue like, with rounded anterior and posterior ends.

Colour

Pink to Reddish Pink.

Family PALAEMONIDAE

The *Palaemonidae* contains freshwater prawns Tiwari has described the distribution of 34 species of the Gens *Palaemon* in India. Many of these Palaemon are now described as *Macrobranchium.* According to Tiwari the Genus Palaemon has a marine origin, and has acquired freshwater habitat by immigration from sea to the interior of land through rivers. The process of adaptation to freshwater is not yet complete, because many species are found in esturies and still depend on brackish water for breeding. Several have become completely acclimatized to freshwater and are found in inland river regions and hill streams. Table 5.4 is a key for the identification of species occurring along the coastal areas living in fresh and brackish water areas, and which are important commercially.

Table 5.4: Key to Commercially Important Coastal Species of the Family *Palaemonidae*

1. Branchiostergal spine present; mandible with a palp; eyes distinctly pigmented; first pleopod of male with or without a rudimental appendix interna on the endopod; branchiostergal spines present as a sharp line; propodus of 5th pereopod with transverse rows of setae on the distal part of the posterior margin; the two median hairs of the posterior margin of telson are slender .... *Palaemon*

— Branchiostegal spine absent; hepatic spine present; dectylus of last 3 legs simple *Macrobranchium* 3

2(1) Dectylus of last three pereopods very long and slender; 4th and 5th pairs excessively long, flagelliform with dactylus much longer than carapace; pleopods very long; 1st pair much longer than carapace; carpus of 2nd pereopod much more than half as long as palm; basal crest of rostrum with at most 7 teeth .... *Palaemon (Nematopalaemon) tenuipes* (Henderson)

— Dactylus of last three pereopods not abnormal in length, that of 3rd scarcely ½ length of propods and that of 5th at most 1/3 length propodus; pleopods normal in length, one or more subapical dorsal teeth on rostrum. Last four abdominal somites bluntly carinate dorsally.................................... *Palaemon (Exopalaemon) styliferus* H. Milne Edw.

3(1) Carpus of 2nd pereopod longer than merus 4

— Carpus of 2nd pereopod about as long as or shorter than merus. 10

4(3) Rostrum with a distinct elevated basal crest, generally very long or with a distinct naked portion in the distal half of the upper margin............5

— Rostrum without a distinct elevated basal crest 8

5(4) Tips of telson reaching beyond the tip of the longer posterior spines 6

— Tips of telson over reached by the tip of the longer posterior spines; rostrum generally straight distal past of the rostrum without dorsal teeth 7

6(5) Carpus of the 2nd pereopod in adult male slightly longer than half as long as chela; fingers of that leg of the same length as the palm. .... *Macrobrachium rosenbergii* (de Man)

— Carpus of the 2nd pereopod in adult male slightly longer or slightly shorter than chela; fingers of that leg a little less half as long as the palm........... .... *Macrobranchium villosimanus* Tiwari

7(5) Basal crest not much elevated, provided with 5–9 teeth; palm of 2nd leg not swollen, fingers shorter than palm .... *Macrobranchium lamarreri* (H. Milne Edw)

— Basal crest distinctly elevated, provided with 5–9 teeth; in younger specimen the 2nd leg has the palm swollen and the fingers longer than the palm; carpus of 2nd leg in adult male shorter than chela .... *Macrobhanchium malcomsonii* (H. Milne Edw.)

8(4) Larger chela of 2nd of adult male with tubercles at both side of the cutting edges; carpus of 2nd leg in adult male shorter than chela; all joints of 2nd legs in adult male pubescent .... *Macrobranchium rude* (Heller)

— Larger chela of 2nd of adult male without tubercles at both side of the cutting edges 9

9(8) Rostrum with 9–11 teeth dorsally, 3 of which generally placed behind the orbit; carpus of 2nd leg in adult male larger than chela .... *Macrobranchium idella* (Heller)

— Rostrum curved upwards and lower margin with 5–7 (seidon 4) teeth; fingers covered with stiff or velvety hairs on the entire surface or in the proximal part ....*Macrobranchium equidens* (Dana)

10(3) Fifth legs conspicuously (about 1 1/3) longer than the 4th; rostrum short and high with many dorsal teeth; second legs of adult male smooth........ .... *Macrobranchium mirabile* Kemp

— Fifth legs of about the same length as the fourth. 11

11(10) Fingers of 2nd legs of adults male with 1 or 2 fairly large teeth; smaller teeth may be oresent between the first tooth and the base of the fingers; anterior tooth of the dactylus placed in or slightly before middle of the finger .... *Macrobranchium javanicum* (Heller)

— Fingers of 2nd legs of adults male with more than 4 teeth placed at regular intervals, sometimes restricted to the proximal part: teeth are generally of equal size, but one of the proximal may sometimes be larger; fingers with a velvety pubescence in their basal portion; dorsal teeth of the rostrum being in the distal third of the carapace ...... *Macrobranchium scabriculum* (Heller)

FRESHWATER PRAWNS

World production of prawns is rising. The genus *Macrobrachium*, includes about 200 species. Almost all of which live in freshwater for at least part of their life cycle, is circumtropical and native to all continents except Europe. The favored species for farming has always been *M. rosenbergii*, sometimes called the "giant river prawn" or the "Malaysian prawn", but recently, China began culturing *M. nipponense*, a species native to Japan, Taiwan and Vietnam, which has also been introduced into Russia, the Philippines and Singapore. In India, some *M. malcolsmonni* are farmed. In the United States, there are several hundred small freshwater prawn farms that grow *M. rosenbergii.* Beginning in 2000, freshwater prawns (defrosted shell-on tails) began showing up in USA grocery stores. They look a lot like giant tiger shrimp, but they're bigger, chunkier, lighter in color, and their shells are always on. In fact, if you look carefully at the second tail segment, you can easily distinguish prawns from shrimp. If the bottom part of the shell on the second tail segment overlaps the shell on the first and third segments, it's a freshwater prawn.

Freshwater Prawn *Macrobrachium rosenbergii*

Classification

Phylum	Arthropoda–insects, spider, crustaceans etc.
Sub-phylum	Crustacea
Class	Malacostraca
Order	Decapoda
Sub-order	Pleocyemata
Super family	Palaemonoidea
Family	Palaemonoidea

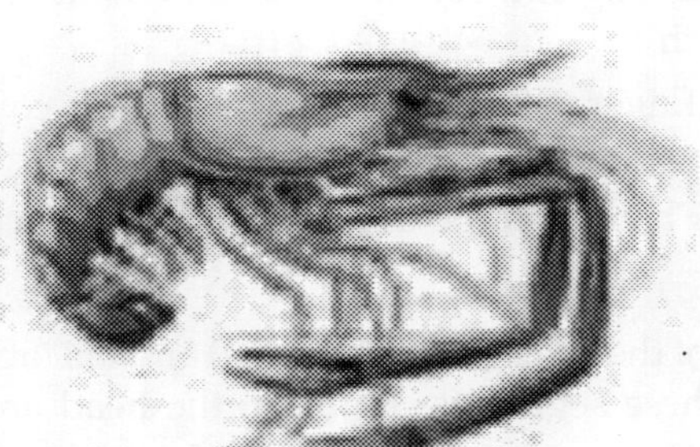

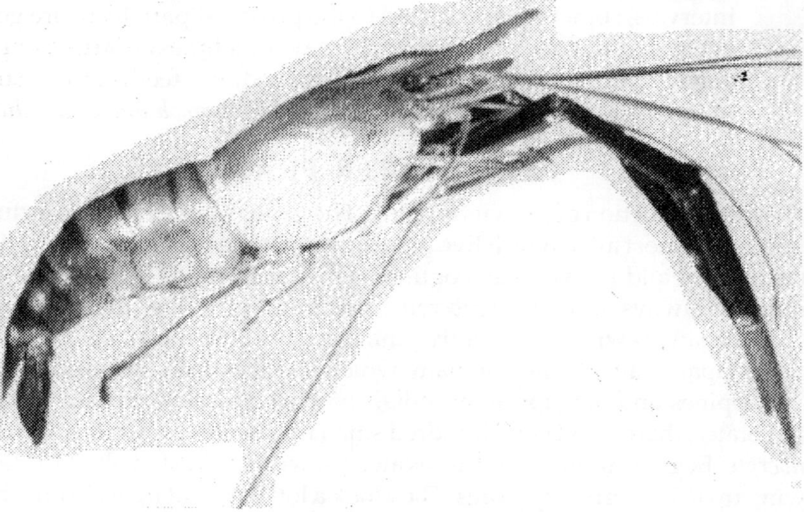

Figure 5.151: *Macrobrachium rosenbergii*

Genus	Macrobrachium
Species	rosenbergii
Common name	Giant River prawn

M. rosenbergii has become the main freshwater prawn species for small-scale and large-scale farming because of its fast growth, large size, better meat quality, omnivorous feeding habit, and established domestic and export markets in Asia. Its farming is well developed in China, India, Thailand, Viet Nam, Bangladesh, Malaysia and Taiwan, as well as Ecuador in South America. There are also capture fisheries for *M. rosenbergii* in India, Viet Nam, Bangladesh, Papua New Guinea and several other countries in Asia.

M. rosenbergii (Figure 5.151) is the largest natantian (swimming) prawn in the world and belongs to the family Palaemonidae. The adult prawn can easily be identified from other species in the genus by the following characteristics:

- ☆ Adult male has a pair of very long legs (chelipeds)
- ☆ The rostrum is long and bent in the middle with 11–13 dorsal teeth and 8–10 ventral teeth
- ☆ The movable finger of the leg of the adult male is covered by a dense mat of spongy fur
- ☆ Distinct black bands on the dorsal side at the junctions of the abdominal segments

The body consists of the head (cephalothorax) and tail (abdomen) and is divided into 20 segments. Of these segments, 14 are in the head and covered by a shield known as the carapace.

Identity and morphology

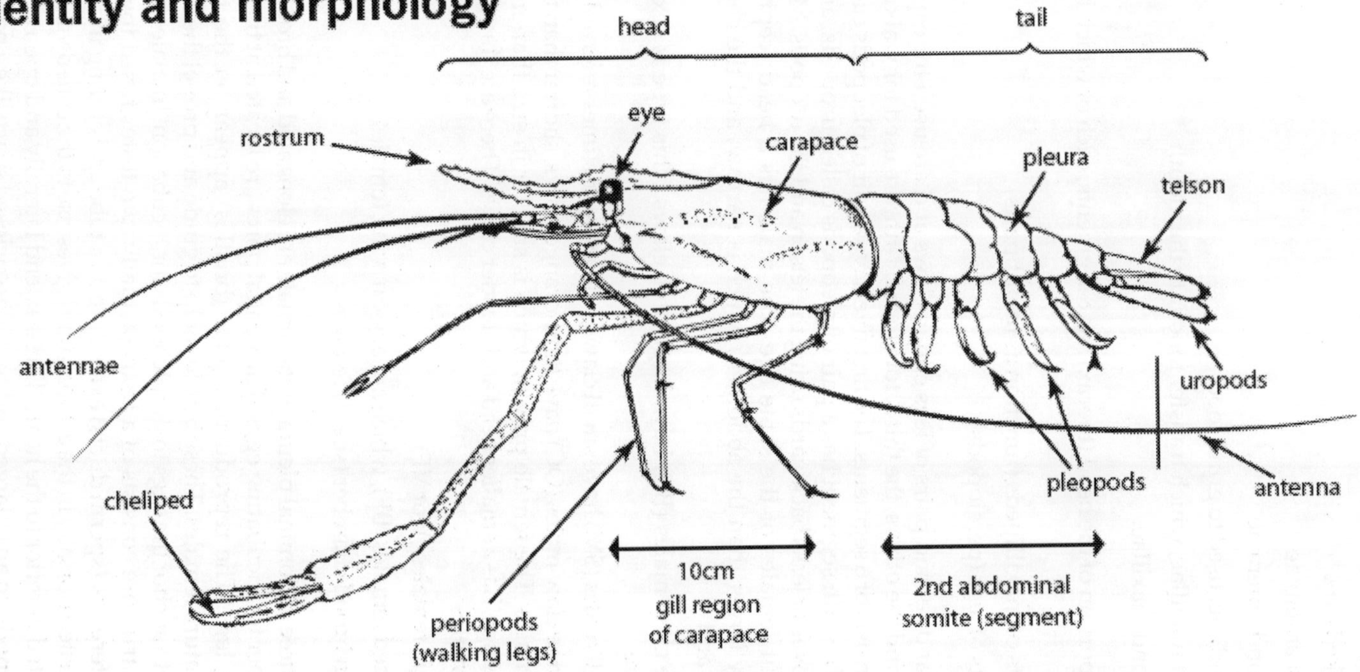

Figure 5.152: External Anatomy of Freshwater Prawn *Macrobrachium rosenbergii* (Male)

The front portion of the head has 6 segments, and features:

- ☆ Stalked eyes
- ☆ First antennae
- ☆ Second antennae
- ☆ Mandible, used to grind food
- ☆ First maxillae, which transfer food into the mouth
- ☆ Second maxillae.

The rear portion of the head (thorax) has 8 segments, each of which has a pair of appendages:

- ☆ 3 sets of maxillipeds (function as mouthparts)
- ☆ 5 sets of legs (pereiopods)

The first and second pairs of legs end in claws and are used for capturing and holding food and the others, the third, fourth and fifth, are used for walking. The tail (abdomen) consists of 6 segments. The first five have a pair of pleopods each, used for swimming. The sixth segment has a pair of pleopods called uropods, and a telson. The prawn moves or jerks backwards using the telson and the uropods. Male prawns are larger than females of the same age. The male has a head (cephalothorax) proportionally larger than the abdomen, which is narrow, and the chelipeds are long, massive and

- ☆ Blue claw males (BC) have large blue claws. This type is sexually most active.
- ☆ Small males (SM) have small claws, and are sometimes referred to as runts.
- ☆ Orange claw males (OC) have light-orange claws, shorter than the claws of BC males. The female prawn, which is smaller in size than males of the same age, has a smaller head and slender claws. There are three main types of females:
- ☆ Virgin females (V or VF)
- ☆ Berried females (BF), which are egg carrying females
- ☆ Open brood chamber (spent) females

The first three abdominal pleura of the female are elongated and broad, and form a brood chamber for incubating eggs. The genital pores are located at the base of the third walking legs. The reproductive setae (bristles) appear on the thorax and pleopods of mature females. These bristles help to guide and propel the eggs during spawning and to anchor the eggs to the pleopods for incubation. In males, the internal reproductive structure consists of a testis (actually two testes fused together), the coiled vasa deferentia (sperm ducts) extending as a tube and ending in an ampulla. The testis is situated dorsally in the carapace and gives rise to the coiled vasa deferentia which are located anterior to the heart. These extend laterally and open at the base of the fifth pereiopods, in an enlarged and partially enclosed space like a box (ampulla).

During ejaculation, muscles surrounding the ampulla contract, extruding the sperm into a spermatophore.

There are three different morphotypes of males. The first stage is called "small male" (SM); this smallest stage has short, nearly translucent claws. If conditions allow, small males grow and metamorphose into "orange claws" (OC), which have large orange claws on their second chelipeds, which may have a length of 0.8 to 1.4 their body size. OC males later may transform into the third and final stage, the "blue claw" (BC) males. These have blue claws, and their second chelipeds may become twice as long as their body.(Wynne, F.2005)

Male *M. rosenbergii* have a strict hierarchy: the territorial BC males dominate the OCs, which in turn dominate the SMs. The presence of BC males inhibts the growth of SMs and delays the metamorphosis of OCs into BCs; an OC will keep growing until it is larger than the largest BC male in its neighbourhood before transforming. All three male stages are sexually active though, and females who have undergone their pre-mating molt will cooperate with any male to reproduce. BC males protect the female until their shell has hardened, OCs and SMs show no such behavior.

Like all crustaceans, freshwater prawns regularly cast off their exoskeleton in order to grow, a process known as moulting. There are four distinct phases in the prawn life cycle: egg, larva (zoea), postlarva (PL) and adult. The time spent in each phase and its growth rate is affected by the environment, especially water temperature and food. The male and females reach first maturity at about 15–35 g within 4 to 6 months.

Mating

Ripe females undergo a pre-mating moult and are soft-shelled. Males do not undergo any change before mating and remain hard-shelled. Fertilization is external and takes place soon after the eggs are extruded. During mating, the male deposits the sperm as a gelatinous mass on the ventral thoracic region between the pereiopods (walking legs) of the female. The female starts to lay eggs about 5–6 hours after mating. As the eggs extrude, they are fertilized by the sperm attached to the exterior of the female body. The fertilized eggs adhere to the setae of the first four pairs of pleopods.

Embryonic Development

The egg development begins with the successful mating between ripe females and mature males. Incubation of the fertilized eggs takes 18–21 days, depending on the temperature (best results are obtained when the water temperature is 28°–30°C). During this time the berried female aerates the eggs by movement of the pleopods. A 'berried' female is an adult female carrying egg under its tail. The eggs are slightly elliptical in shape and initially yellowish to bright orange in colour, then gradually change to grayish a few days before hatching. This colour change occurs as the eyes get larger and embryos utilize their food reserves and grow in size. The number of eggs carried by a female depends on her size, and varies from 3000 to 80,000.

Calculation of Post Larvae

The following is an example of the calculation for a postlarval stock estimation:

Let us assume that the small container had a volume of 25 L. You counted 80, 86, 92 and 98 post larvae in the four beakers, each of 100 ml. The total number of post larvae in the 25 L container can be calculated as follows:

Average number of PL in 100 ml = (80 + 86 + 92 + 98) ÷ 4 = 89

Number of PL in the 25 L container = 89 x 25 x 1 000 ÷ 100 = 22 250

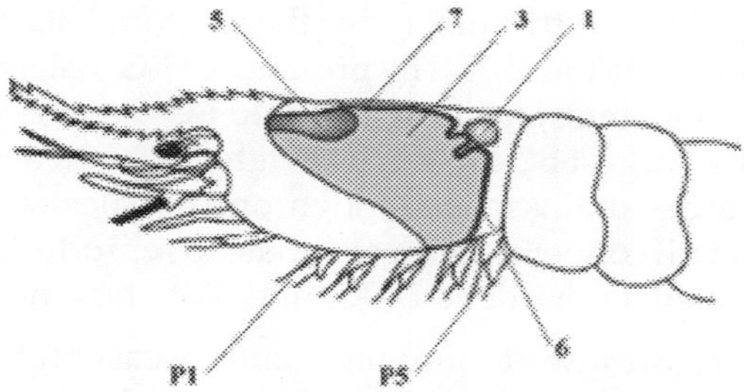

Figure 5.153: A. Male

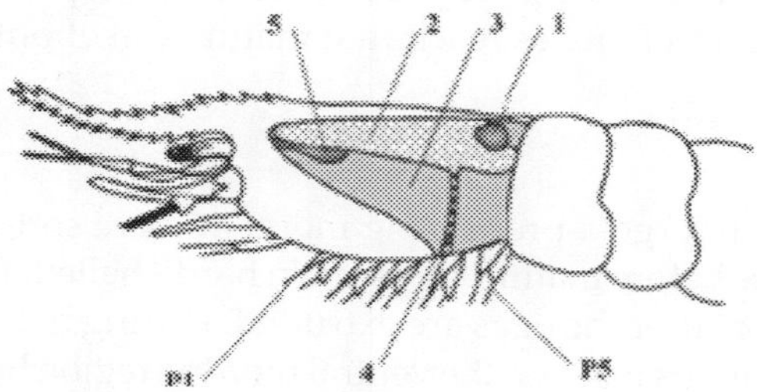

Figure 5.154: B. Female
Internal reproductive structures of Macrobrachium rosenbergii. A: male. B: female. P1 to P5 are pereiopods

In females, the ovaries are located dorsal to the stomach and hepatopancreas in the carapace cavity (Figure 5.154, B. Female). When the female is in ripe condition the orange coloured ovaries are visible through the carapace, extending from just behind the eyes to the first abdominal segment. An oviduct extends from each ovary (anterior to the heart) backwards to the gonopore of the third pereiopod. control of growth in *Macrobrachium rosenbergii* following unilateral eyestalk ablation on jumpus and lazzards, it was observed that, eyestalk ablation had no effect on growth of male jumpus and ablation had a highly significant effect on growth of male and female lazzards, manifested in both, increase in size increment per moult and shortening of moult interval (V. Venkitraman *et al.*)

Table 5.4: Selected Characteristics of *Macrobrachium rosenbergii* Larvae and Post Larvae

Stages	*Characteristics*						
	Eyes	*Rostrum*	*Antennal Flagellum*	*Uropod*	*Telson*	*Pleopods*	*Pereiopods*
I	Sessile						
II	Stalked						
III		1 dorsal tooth		First appearance			
IV		2 dorsal teeth		Biramous with setae			
V			2 or 3 segments		More elongated and narrower		
VI			4 segments		More narrow	First appearance of buds	
VII			5 segments			Biramous and bare	
VIII		About 7 segments				Biramous with setae	
IX			About 9 segments			Endopods with appendices internae	
X		3 or 4 more dorsal teeth	About 12 segments				1st and 2nd fully chelated
XI		Many dorsal teeth	About 15 segments				
PL			Rostrum has dorsal and ventral teeth; behaviour predominantly benthic, as in adults				

Source: Derived from Ismael and New, 2000.

Macrobrachium malcomsonii

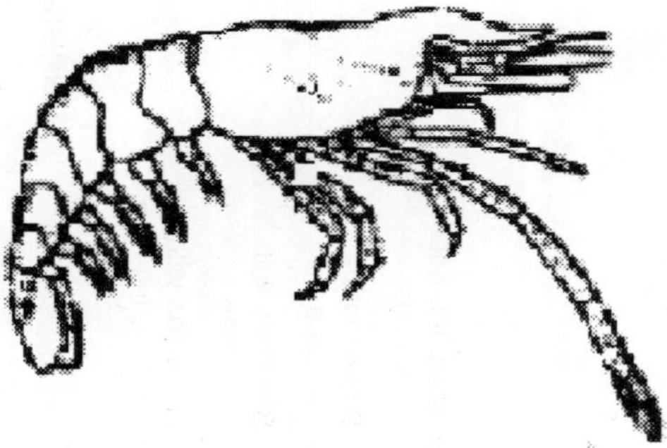

Figure 5.155: *M. molcolmsonii*

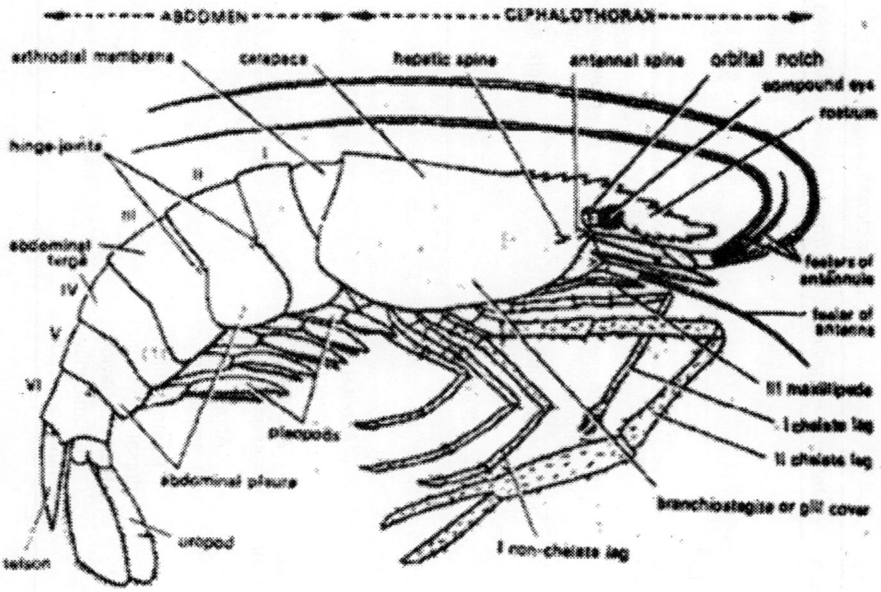

Figure 5.156: External Features of Prawn *Macrobrachium malcomsonii*

Body is elongated, more or less spindle-shaped and bilaterally symmetrical. It offers least resistance in swimming. Size of adult varies from species to species. *Macrobrachium malcomsonii,* found in Central India and Tamil Nadu, measures 25 to 40 cm in length. The giant prawn *P. carcinus* from Kerala is upto 90 cm long. While the dwarf prawn *P. lamarrei,* found almost throughout India, is 25 to 5 cm long. Young stages are translucent and white, but the adults are differently tinted according to the species. Usual colour is dull pale-blue or greenish with brown orange-red patches. Preserved specimens become deep orange-red.

Segmentation and Body Divisions

Body of adult prawn is distinctly divided into 1 9 segments or somites, all bearing jointed appendages. The segments are arranged into two main regions: an anterior cephalothorax (fused head-thorax) and a posterior abdomen.

Cephalothorax

Cephalothorax is large, rigid, unjointed and more or less cylindrical in shape. It consists of 13 segments. The joints between segments are obliterated. Cephalothorax is formed by the union of two regions: (i) head and (ii) abdomen. Head consists of 5 segments, while thorax includes 8 segments, all bearing jointed appendages.

Abdomen

Well-developed abdomen is jointed, unlike cephalothorax. It is composed of 6 distinct movable segments, and a terminal conical piece. The tail-plate or telson. which is not considered a segment because of postsegmental origin. Abdominal segments are dorsally rounded, laterally compressed and normally bent under the cephalothorax, so that the animal looks like a comma (,) in shape. The abdomen looks almost circular in a cross section. Each abdominal segment carries a pair of jointed appendages, called pleopods or swimmerets.

External Apertures

The slit–like mouth opens mid-ventrally at the anterior end of cephalothorax. Anus is a longitudinal aperture lying ventrally at the base of telson. Paired renal apertures open on raised papillae on the inner surface of coxa of antenne. Paired female genital apertures in females are on the inner surface of coxae of the third pair of walking legs. Paired male genital apertures in the male are situated on the inner surface of coxae of the fifth pair of walking legs. There are two minute openings of statocysts, one lying in a deep depression dorsally on the basal segment (precoxa) of each antennule (L. R. Richardson and J. C. Yaldwyn)

Macrobrachium nipponense (De Haan, 1849)

Identification

Rostrum is strait, broad in the middle (width about ¼ length, from orbit to tip of rostrum), extending well beyond antennal peduncle, and beyond tip of scaphocerite (excluding setae). About 2 dorsal rostral teeth are found behind orbit. Hepatic spine is at a level lower than antennal spine. Second pair of pereiopods of adult male equal in size, long and all segments covered with a short and dense pubescence; carpus shorter than propodus and longer than merus; chela with stiff or velvety hairs on entire surface; cutting edge of finger of propodus with 1 proximal tooth; cutting edge of dactylus without tubercules. The maximum size of *M. nipponense* (de Haan 1849) recorded in the literature coincides with the measurements reported here. Holthuis (1950) reported a size limit of 61-99 mm from 12 Japanese specimens, four of which were ovigerous females. For three specimens from Takao, South Formosa, he reported lengths of 44-61 mm, and 92 mm for a males and 75 mm for the females.

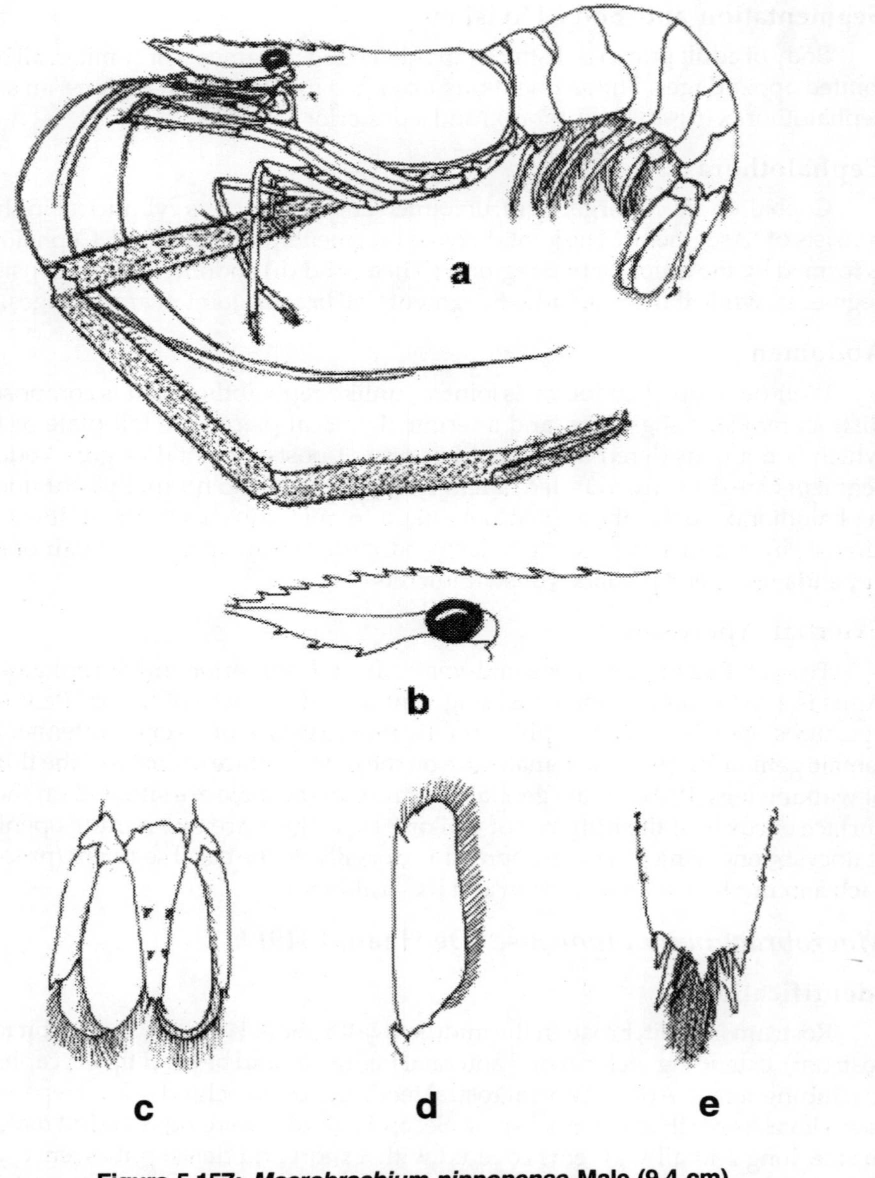

Figure 5.157: *Macrobrachium nipponense* Male (9.4 cm)
a: Lateral view; b: Rostrum of another specimen (9.0 cm); c: Scaphocerite; d: Uropods; e: Telson

Most *Macrobrachium* species are subtropical or tropical, *M. nipponense* survives well in colder temperate climates, thus making it a useful alternative to other popular aquaculture species such as *M. rosenbergii* (de Man, 1879) (New 2005). The wide ecological tolerance of *Macrobrachium nipponense* encompasses temperature and

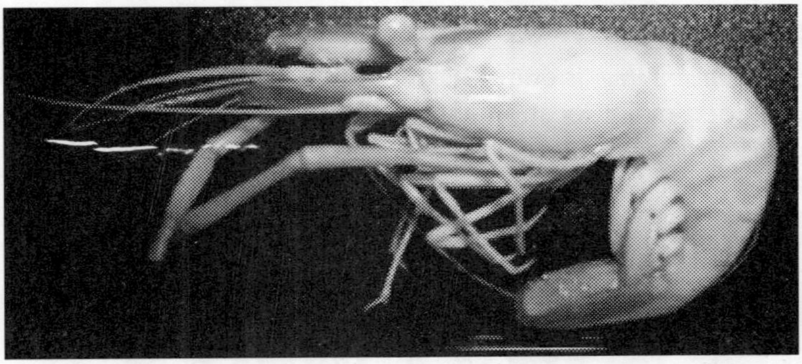

Figure 5.158: Photograph of *Macrobrachium nipponense*

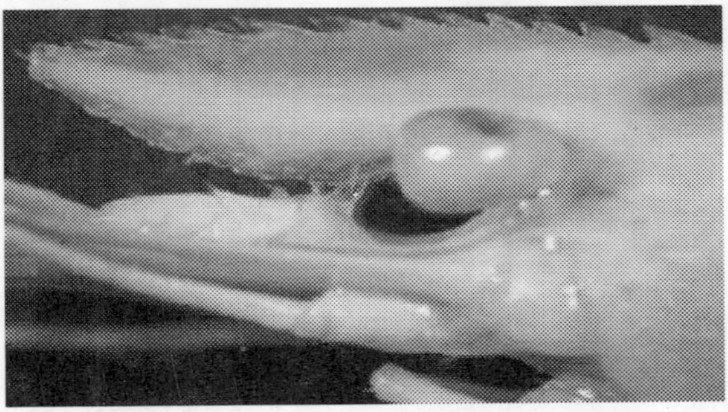

Figure 5.159: Rostrum

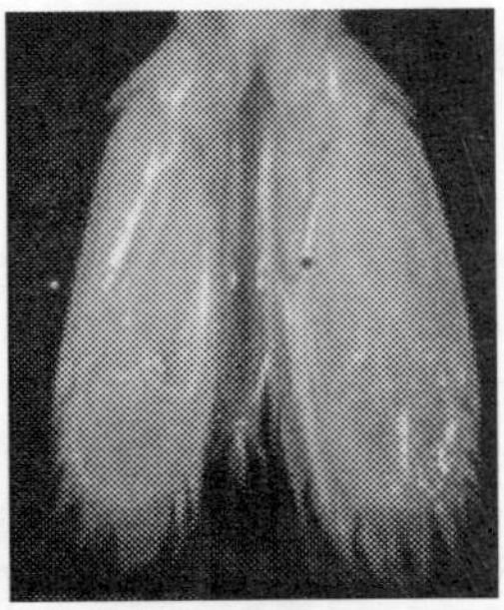

Figure 5.160: Uropods

Figure 5.161: Telson

salinity. It can live at a range of salinities from brackish to fully freshwater, and can quickly adapt to a change to fully freshwater in three generations.The biological basis for this species being a competent coloniser of new areas, and therefore an invasive species threat, is completed by a significant intra-population and intra-individual variation in egg size (Mashiko and Numachi 2000) and larval characters (Alekhnovich and Kulesh 2001). All of these factors make this species effective in dispersing to and surviving in new environments, as it has naturally throughout its native range in East Asia.

Macrobrachium lamarrei (H. Milne-Edward 1937)

Macrobrachium lampreys is one of the medium-sized freshwater prawn. Holthius (1980), has provided major location and maximum size of the species and its FAO English name *M. lamarrei* is known as "Kuncho river prawn" which is 69 mm in its maximum size.

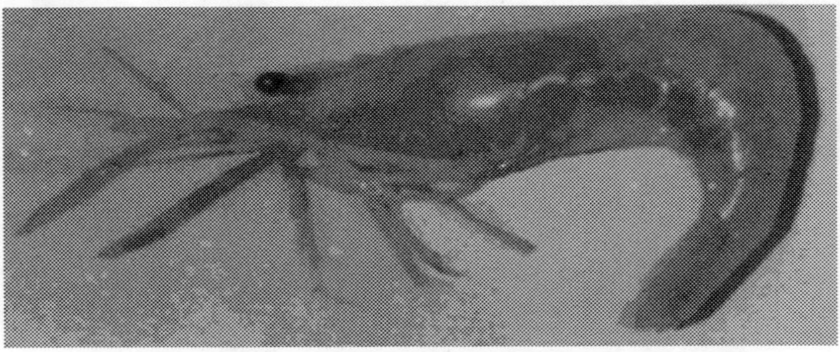

Figure 5.162: Male of *Macrobrachium lamarrei*

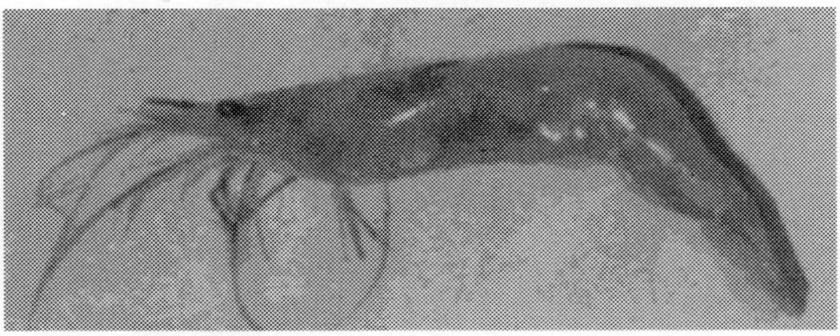

Figure 5.163: Female of *Macrobrachium lamarrei*

General Morphology

Macrobrachium lamarrei (H. Milne-Edward 1937) is a medium-sized freshwater prawn. The males can reach a total length of 80 mm (from the tip of rostrum to tip of telson) whereas, the females up to 75 mm. The body is usually creamy white to light brownish white with greenish brown pigmentation all over the cephalothorax. Like all species of decapod, prawns body consists of two distinct parts cephalothorax and abdomen. The carapace covers over the cephalothorax which is smooth and hard. The rostrum of this species is long reaching beyond the antennal scale. The rostrum is slightly slender. Dorsally rostrum bears 7-9 teeth and 5-8 below. The hepatic spine present on the anterior carapace margin. Five pairs of peraeopods, or true legs present in cephalothorax region. In contrast to the cephalothorax the segmentation of the abdomen is very distinct. This part possesses six segments where each segment bears a pair of ventral appendages called pleopods or swimmerets. The swimmerets of the sixth abdominal somites are stiff and hard, and the telson serves as tail fan (Ismael and New, 2000).

The female, when berried carries very numerous eggs under the abdomen attached with swimmerets. The antennules and antennae of number of crustaceans are considered the most important sites of sensory reception (McLaughlin, 1980). Smooth rounded dorsal body surface present in *Macrobrachium* spp., while penaeids have a simple or complex ridge at the dorsal apex of the abdomen.

Sexual Dimorphism

Mature males can be easily recognized by their longer and stronger chelipeds with larger spines than in case of females. Male possesses appendix masculina, a spinous process adjacent to the appendix interna on the endopod of the second pleopod, (Sandifer and Smith, 1985). Mature females have proportionally small body size as well as head and claw (Sandifer and Smith, 1985). They exhibit a typical brood chamber formed by the first, second and third abdominal pleurae. There is presence of central lump on the ventral side of the first abdominal somite in male whereas this feature is absent in female. Sexual differentiation is controlled by the presence of the androgenic hormone, which induce the male characteristics of the genital tract Females *M. lamarrei* experience a moult known as pre-spawning or pre-mating moult which usually occurs at night (Sharma and Subba, 2005).

Fecundity

The fecundity can be expressed in a number of ways. It is the total number of weight of eggs produced by a female during the average life-span. It can also be known by the number of ripening eggs in the ovaries of female before spawning, Shrivastava (1999). In the present communication the term fecundity is referred to the number of eggs borne in the brood pouch during a single spawning act.

M. lamarrei matures in freshwater condition and the female deposits the eggs in brood pouch, beneath the abdomen formed by the long pleura of the abdominal segments and the first four pairs of pleopods. The number of eggs shed by different species of the prawn may vary considerably (Bhattacharjee and Dasgupta, 1989; Kurian and Sebastian 1986; Manna and Raut, 1991; Sakuntala, 1976). Individual of the same species produces varying number of eggs depending on their age, length, weight and environmental condition (Bal and Rao, 1990).

The trends of relationship between fecundity and body length and fecundity and body weight were estimated by the formula: F= a + bX

For body measurements *i.e.* X (L and W) Where, F= fecundity, L= body length, W= body weight, a and b= constants. The coefficient of correlation (r) of each of the relationships was also assessed. In addition to the above studies, the relative fecundity was also calculated by using the following formula: Total number of eggs in the brood pouch Total weight of prawn (Jay Chandran and Joseph, 1992)

Development in Prawn

The *M. lamarrei* has four phases in its lifecycle: egg larva, post-larva and adult. The fertilization is external and takes place as soon as the eggs are extruded. The newly fertilized eggs are homogenously granulated. All the larval stages are completed inside the freshwater. Histology of the mature eggs were studied which helped in the study of the development of the embryonic as well as the larval development of the prawn which can help to know the nature, habit and habitat for the proper management of the prawn.

Stages of Development

The fertilized females with dark green coloured eggs in their brood pouch were reared. The dark green colour of eggs turned into light green after seven days and

Stage I: Sessile eyes, uropod present

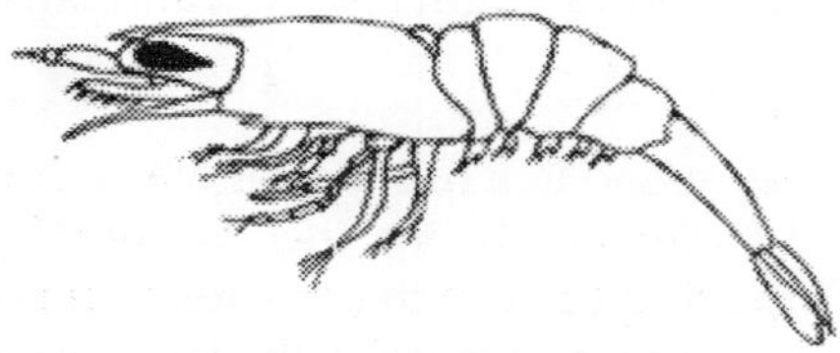

Stage II: Pleopod developed, telson present

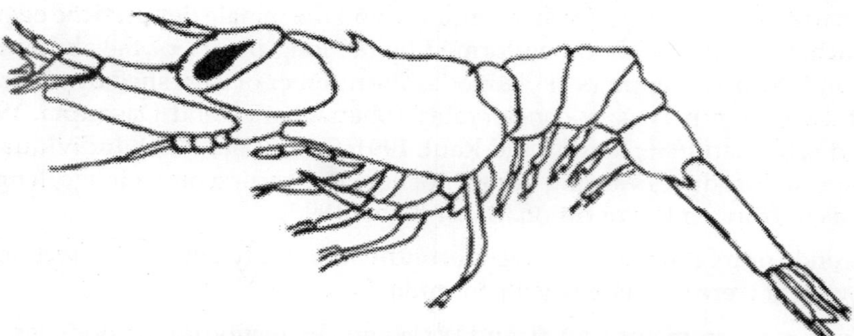

Stage III: Stalked eye, one dorsal teeth present

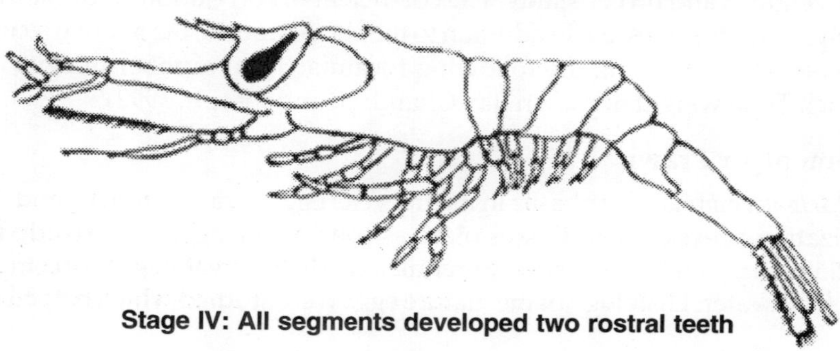

Stage IV: All segments developed two rostral teeth

Figure 5.164: Larval Stages of *Macrobrachium lamarrei*

then again turned into transparent white in the next 6-7 days. The transparent eggs possessed two black spots of eye which hatched after 2-3 days. The eye spots were clearly visible externally before hatching. The mature eggs were slightly elliptical and measured 0.54-0.64 mm on their long axis. The eggs hatched completely after 5-6 hrs of laying. At the time of laying the eggs, the female vigorously move their pleopods so that the eggs dispersed evenly into the water. The temperature measured at the time of egg laying was 30 ± 20C, Dissolved oxygen measured was 10 mg/l and pH was 7.75-8. The hatching is accomplished with increased internal pressure due to increased size of the larva and by the movement of the appendages. The vigorous movement of the mouth parts and thoracic appendages cause rupture of the egg membrane resulting in emergence of the telson first, followed by the head of the larva. The mother prawn moves the pleopods rapidly during hatching to disperse the new hatchling. The newly hatched larva swims actively. The larvae swim close to the surface of water. During the developmental study of the prawn, altogether five different larval stages were recorded The first stage larvae were collected after 6 hrs of hatching. The first stage larvae measured 8 mm in length. They possessed sessile eyes, very small pleopods, unsegmented antennae and uropods (Figure 5.165).

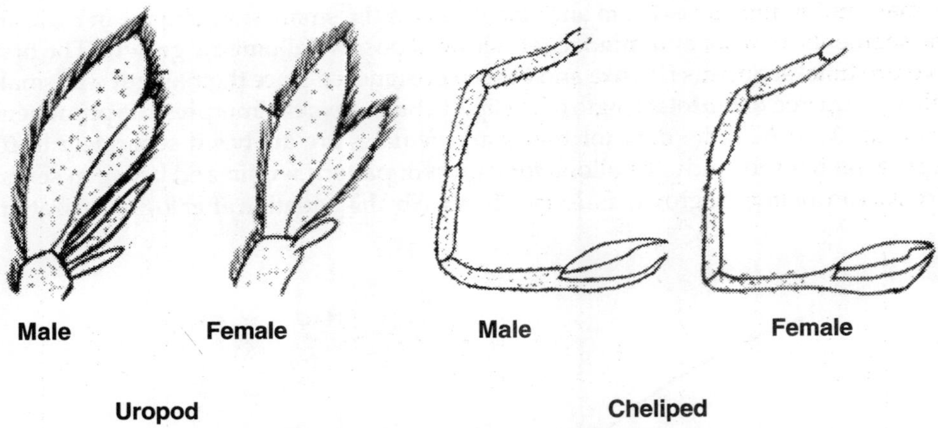

Figure 5.165: *Macrobrachium lamarrei*

Macrobrachium veliense (New species)

Macrobrachium veliense sp. nov. is described from Veli lake and Kuttiyadi river, on the south-west coast of India. This species is closely related to *M. nipponense* and *M. equidens* but is separated from them by the lanceolate shape of the rostrum, number of teeth of both dorsal and ventral margins, the almost smooth nature of the carapace, the slender telson and the ratios of the fingers to the merus, carpus, propodus and palm of pereipods 1 and 2. The carpus of the second pereiopod is equal to or longer than the chela and is a diagnostic character of the new species. (Jayachandran, K.V.; Joseph, N.I.1985)

Macrobrachium kulsiense (New Species)

Macrobrachium kulsiense sp. nov. is a very small sized prawn (maximum size–34.5 mm in total length), exhibiting species-specific characters, such as highly elevated and moderately long rostrum with 9–12 dorsal teeth, a single ventral tooth, and percentage ratios of ischium, merus, carpus, palm, dactylus of first and second chelipeds (19.05:28.57:33.33:09.52:09.52 and 21.43:25.00:21.43:14.28:17.86, respectively). The species shows close similarity with *Macrobrachium mirabile* (Kemp). The females are larger than males, eggs are large in size (1.2×0.9 mm) and fecundity is low (15–20). In the Kulsi River, a major tributary of Brahmaputra River, N. India prawns were collected. On identification, one of the prawns was found to be undescribed, and hence is described herein. The ecology of the river consisted of: temperature fluctuating widely from 15 to 28 °C, depth from 0.8 to 10 m, turbidity of 11–19 cm, sand mining @ 12,500 MT annually, and fish catch of 300–800 kg (from 1.5 km area). All these factors pose a great threat to the fish and prawn wealth of the river. (Jayachandran K. V, *et al.*, 2007)

Macrobrachium nobilii (Henderson and Matthai, 1910)

Propodus was the longest segment followed by carpus and merus in both chelae of male and in female; ischium and dactylus was the smallest segments. In male all the segments in major and minor chela showed positive allometric growth. The first two proximal segments *i.e.* coxa and basis were ignored since their length was small when compared to the total length of chelipeds. Since sexual dimorphism is prominent in adult, *M. nobilii*, the data for males and females are analyzed separately to fit regression lines to study the allometeric growth pattern within and between sexes. Crustacean allometric growth is defined by y=axb; this equation after log linearization

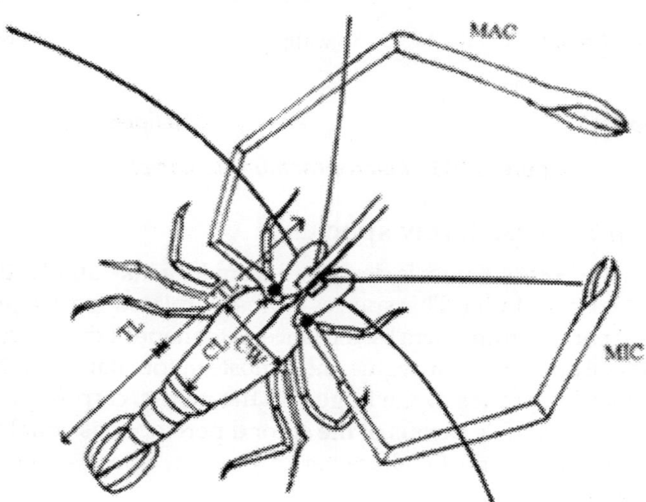

Figure 5.166: *Macrobrachium nobilii*

yields log y = log a + b. x which eases the application of the results (Hartnoll, 1982). Male *M. nobilii* had a minimum body weight of 1449 mg and a maximum of 12676 mg. The body weight of female ranged from 336 mg to 7178 mg. In general, growth was manifested through size increment, which could be measured by increase in volume or weight.

In *M. nobilii,* the carpus is shorter than the propodus while the level of allometry is higher when compared to propodus. In male *M. nobilii,* the growth of propodus and carpus had positive allometry. In females it is isometry for major and minor chela. The maximum allometry level in *M. nobilii* was recorded in carpus in both chelae and both sexes.

In general, growth was manifested through size increment, which could be measured by increase in volume or weight. The increase in weight of both sexes of *M. nobilii* as measured by growth constant 'b' was higher for males (4.28) than females (3.02), indicating the positive allometric growth in this species.

Chela Growth

In females, the growth pattern of major and minor chela length was isometric. In males, there was a remarkable difference in the pattern of length increment between major and minor chela. For both sexes the ratio between major and minor chela length and weight were calculated. The length ratio varied from 1.15 to 1.58 and that of weight varied from 0.87 to 2.42 for males. For females, the chela length ranged from 1 to 1.43 and the weight ratio ranged from 1 to 2.08. The change in the chela length (major/minor) remained constant ($P>0.05$) in male and female. For males there was a significant inverse relationship between major chela weight/minor chela weight ratio and carapace length. Elevations (intercept of minor and major chela) differed significantly ($P<0.001$). Regarding the growth of segments of major and minor chelae in male and female, the length increment of each segment differed one from another ($P<0.001$). Among sexes, carpus had a maximum allometry level, which declined to the minimum in dactylus and schium. (Pitchaimuthu Mariappan and Chellam Balasundaram, 2004).

Macrobranchium rude (Heller)

Branchiostegal spine is absent; hepatic soine present; dactylus of last 3 legs simple. Carpus of 2nd pereopod is longer than merus. Rostrum without a distinct elevated basal crest. Larger chela of 2nd leg of adult male is with tubercles at both sides of the cutting edges; carpus of 2nd leg in adult male shorter than chela; all joints of 2nd legs in adult male pubescent.

Macrobranchium idella (Heller)

Branchiostegal spine is absent; hepatic soine present; dactylus of last 3 legs simple. Carpus of 2nd pereopod is longer than merus. Rostrum without a distinct elevated basal crest. Large chella of 2nd leg of adult male is without tubercles at each side of the cutting edges. Rostrum is with 9–11 teeth corsally, 3 of which generally placed behind the orbit; carpus of 2nd leg in adult male larger than chela.

Macrobranchium equidens (Dena)

Rostrum curved upward and lower margin is with 5–7 (Seidom 4) teeth; fingers covered with stiff or velvety hairs on the entire surface or in the proximal part.

Branchiostegal spine is absent; hepatic soine present; dactylus of last 3 legs simple. Carpus of 2nd pereopod is longer than merus. Rostrum without a distinct elevated basal crest. Large chella of 2nd leg of adult male is without tubercles at each side of the cutting edges. Rostrum curved upward.

Macrobranchium scabriculum (Heller)

Fingers of 2nd legs of adults male with more than 4 teeth placed at regular intervals, sometimes restricted to the proximal part: teeth are generally of equal size, but one of the proximal may sometimes be larger; fingers with a velvety pubescence in their basal portion; dorsal teeth of the rostrum being in the distal third of the carapace. Fifth legs are of about the same length as the fourth. Carpus of 2nd pereopod is about as long as or shorter than merus. Branchiostegal spine is absent; hepatic soine present; dactylus of last 3 legs simple.

Class	Crustaceans
Family:	Palaemonidae
Scientific Name	*Palaemon serenus*

Figure 5.167: *Palaemon scabriculum*

Figure 5.168: *Palaemon serenus*

Branchiostergal spine present; mandible with a palp; eyes distinctly pigmented; first pleopod of male with or without a rudimental appendix interna on the endopod; branchiostergal spines present as a sharp line; propodus of 5th pereopod with transverse rows of setae on the distal part of the posterior margin; the two median hairs of the posterior margin of telson are slender.

The Red handed shrimp (*Palaemon serenus*) belongs to the Palaemoid shrimp family Palaemonidae. Palaemoid shrimps are part of the shrimp referred as carids, which are characterized by piners on the first two pairs of legs, a second abdominal segment that overlaps the first and eggs that are carried under the abdomen. *Palaemon serenusis* can often be found in large numbers hiding in cracks or under debris. They are small shrimps that can be recognized by the bright red banding on the front claws and the small yellow and brown spots that cover the translucent body. This species grows to a maximum body length of 6 cm.

Palaemon tenuipes

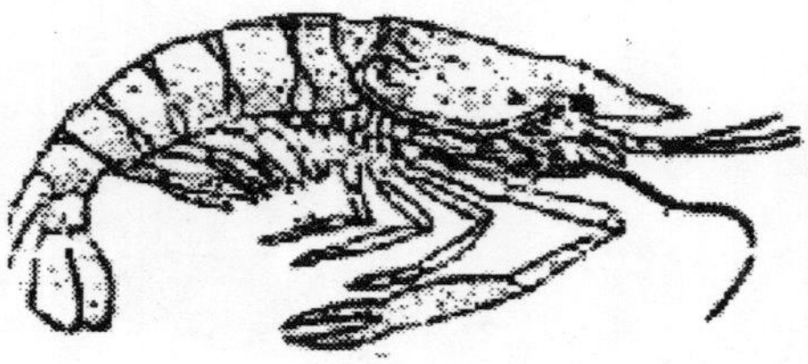

Figure 5.169: *Palaemon tenuipes*

Figure 5.170: Rostrum of PL-20

This is one of the most important fisheries in Bombay and Gangetic Delta. The maximum length of this species is 80 mm. The species is available in Malaysia to New Zealand coastal water up to 20 m depth and in estuaries. In India it is available in northern regions of both coasts.

Palaemon styleferus

This is one of the most important fisheries in Gangetic Deltas. This species is also available in West Pakistan and India to Malay Archipelago, in shallow coastal waters, brackish and some times freshwater areas. In India it is available in Northern regions of both coasts. Maximum size of this species is reported as 90 mm.

Palaemon serratus

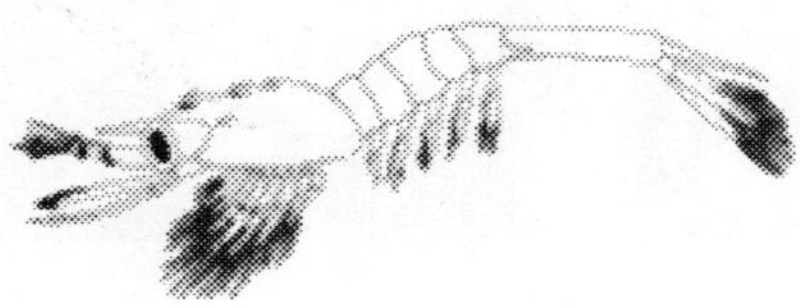

Figure 5.171

Figure 5.172

Palaemon or *L. natator*

They are circum tropic pelagic species associated with floating algae, especially *Sargassum*, has been reported from our waters and should be watched for in the north. Its rostral teeth 8-14/5-7 with ventral teeth concealed by double row of setae and no rows of setae posterodistally on propodus of fifth leg. Their Length is up to 2 inch. Posterior margin of telson are with 3 pairs of spines. (Sub-family Pontoninae)

Figure 5.173: Rostrum of *Palaemon natator*

Large group of small, secretive, often commensally forms, mainly Indopacific in distribution. Many genera, those in New Zealand with rostrum compressed and toothed; all maxillipeds with exopods; pleura of first to fifth abdominal segments rounded or bluntly pointed; dactyli of fifth legs without basal protuberance (mandibular palp absent).

Palaemon affinis M.-Edwards, 1837

Palaemon affinis M.-Edwards, 1837 with rostral teeth 5-10/2-5 and propodus of fifth leg with transverse rows of setae posterodistally, is our endemic, common, intertidal shrimp. Found throughout New Zealand, being extremely abundant in the brackish waters and mangrove swamps of the north. They are transparent with longtudinal wavy red and green bands and a prominent diagnostic orange and black spot laterally on sixth abdominal segment. Their length is up to 3 inch.

Palaemon longirostris

Figure 5.174: ***Palaemon longirostris*** **(Top View)**

Figure 5.175: ***Palaemon longirostris*** **(Side View)**

Size

Total length up to about 70 mm

Ecology

In esturies of large and small rivers.

IDENTIFICATION ON THE BASE OF CHEMICAL CHARACTERISTIC

Isozyme Patterns of *Penaeus* sp.

Because of the ambiguity in morphology seen in some samples, the species of individual specimens were further determined by isozymes analysis. Proteins were separated by electrophoresis and gel was incubated in staining solutions containing appropriate substrate and cofactor for the activity staining. The zymograms of alcohol dehydrogenase (ADH), glucose-6-phosphate ehydrogenase (G-6-PDH), succinate dehydrogenase (SCDH), 6- hosphogluconate dehydrogenase (6- PDH), were obtained (data not shown). The isozyme pattern of ADH showed three isoforms, ADH1, ADH2 and ADH3 (in order of increasing mobility in the gel). Individuals which were identified by morphology as *P. indicus* had isoform ADH2 and ADH3 whereas individuals of *P. merguiensis* had ADH1, ADH2 and ADH3. *P. merguiensis* also had extra slow-moving isoforms of SCDH, G-6-PDH and 6-PDH. Therefore, prawn samples were classified into *P. merguiensis* and *P. indicus* according to the presence and absence of slow moving isoform of ADH, SCDH, G-6-PDH and 6-PDH. Haplotypes 1 and 2 were only found in *P. indicus*, whereas, haplotypes 3-7 were found in *P. merguiensis* and samples with mixed morphology. (Satyal Nadlal and Timothy Pickering.)

Macrobrachium nipponense (De Haan, 1849)

Molecular Analysis

Three specimens were collected were preserved in 70 per cent EtOH. Genomic DNA was extracted using a modified version of a CTAB-phenol/chloroform extraction (Doyle and Doyle 1987). Three fragments of mitochondrial DNA (mtDNA) were amplified using the polymerase chain reaction (PCR). These were two fragments (5′ and 3′) of cytochrome oxidase subunit I (COI), and one fragment of 16S ribosomal DNA (rDNA). The 5′ COI fragment (popularly known as the "DNA Barcode" sensu (Hebert *et al.*, 2003) was amplified using universal COI primers LCO-1490 (5′-TGA TTT TTT GGT CAC CCT GAA GTT CA-3′) and HCO-2198 (5′- GGT CAA CAA ATC ATA AAG ATA TTG G- 3′) (Folmer *et al.*, 1994) and the 3′ fragment of COI was amplified using CDC0.La (5′-CCN GGG TTY GGR ATA ATT TCT C-3′; Page *et al.*, 2005) and COIa.H (5′-AAG CAT CTG GGT ART C-3′). Primers for the 16S PCR were 16S-F-Car (5′-TGC CTG TTT ATC AAA AAC ATG TC-3′) and 16S-R-Car (5′-AGA TAG AAA CCA ACC TGG CTC-3′). PCR amplifications were 12.5µl reactions on a Geneamp PCR System 9700 (Applied Biosystems, Foster City, CA, USA) of 0.5µl template DNA, 0.4µM primers, 0.1µM dNTPs, 2µM MgCl2, 2.5µl 10X PCR Buffer, 0.5 units of Taq polymerase (Bioline Pty Ltd, Alexandria, NSW, Australia) and the rest ddH20. The following cycling conditions were used for the COI primers: 15 cycles of 30 s at 94°C, 30 s at 40°C, 60 s at 72°C; 25 cycles of 30 s at 94°C, 30 s at 55°C, 60 s at 72°C. The 16S primers had the following cycling conditions: 40 cycles of 30 s at 94°C, 30 s at 50°C, 30 s at 72°C. (Salman D. Salman *et al.*, 2006)

From the molecular analyses, the final sequences produced were 602 base pairs (5′ COI), 557 base pairs (3′ COI) and 534 base pairs long (16S rDNA) lodged in Genbank under the accession numbers DQ656414–DQ656416. The haplotypes for each specimen were identical for each respective gene fragment. A BLASTN search of Genbank (www.ncbi.nih.gov) failed to find *Macrobrachium* sequences that were closely related (*i.e.* within 10 per cent) to either of the COI fragments. There are currently no *M. nipponense* COI sequences on Genbank. However, our 16S sequence closely matched (99 per cent similarity) seven 16S sequences of *M. nipponense* from Japan (Genbank AY282771, Murphy and Austin 2004) and China (Genbank DQ462406-411, Sun, Li and Feng, unpublished). *Macrobrachium nippponense* inhabits fresh and brackish waters throughout much of East Asia (Holthuis 1980).

DNA "barcoding" (Hebert *et al.*, 2003) also provides a potentially effective method to supplement morphological analysis in species identification, as in the present study. A further factor that requires consideration when identifying *Macrobrachium* is that environmental rather than genetic factors can determine expressed morphological characters (Dimmock *et al.*, 2004). Nonnative species are unlikely to appear in locally appropriate morphological identification keys and because invading species may not be morphologically distinct, biosecurity can be significantly improved by a combined morphological and molecular approach (Armstrong and Ball 2005).

Part II
CULTURE

Chapter 6

Culture System and Guidelines for Culture

History of Prawn (Shrimp) Culture

Shrimp seed production technique and the shrimp culture system were developed by *M. Fudinaga* and his colleagues in the *early 1960's* in Japan after many ground-breaking studies over the preceding 30 years. Their techniques have been spreading through out the world and have helped the world shrimp culture industry grow enormously.

Prawn Culture Status

Introduction

The disproportionate growth of shrimp culture in the vast coastal regions of South and Southeast Asia has given rise to several conflicts in the micro as well as macro sphere of shrimp industry chain.

Global Level

Focus on the role of Multi Lateral Institutions (MLIs) in the process policy formulation and implementation for the promotion of shrimp culture in the third world countries thereby creating a 'politicized environment' of change in the patterns of resource use at micro level. Additionally focus on the role played by corporate capital, consumer demands and resistance of international civil society.

Findings at Global Level

The study finds that global actors like M L Is, Corporate worlds, Consumer countries of the first world play significant roles in the promotion of shrimp culture

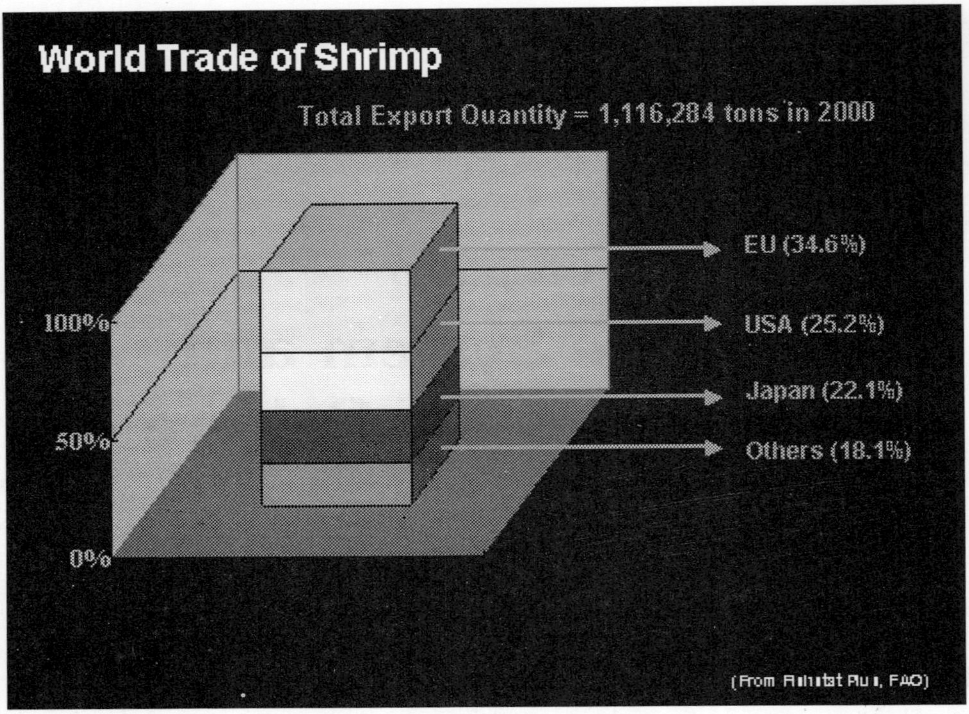

in the third world countries such as in India. The main observations and findings of the study on the global level are briefly as follows:

- ☆ The FAO and the WB have formulated policies to introduce the process of shrimp culture in the global fishery sector
- ☆ MLIs like the FAO, the World Bank and the IMF have directly contributed to shrimp culture development in the third world countries (e.g. India) by providing multilateral and bilateral assistances.
- ☆ MLIs and corporate sectors in shrimp culture have often cited global food security and poverty alleviation needs as justification of shrimp culture development in the producing countries, which in most cases have proved to be a myth.
- ☆ The main aim of the MLIs and the corporate world is maximization of the production of shrimp for the global market by initiating a process of control over the natural resources of third world countries and utilizing the same according to the global market demand.
- ☆ The geographic distribution of global shrimp production sites clearly indicates the dominance of Southeast Asian and South Asian countries in the business whereas the marketing trend in the global sphere indicates a total dominance of USA, Japan and EU countries as consumers of cultured

shrimps. · Within a decade shrimp has established its position strongly in the export basket of third world countries of Asia. Correspondingly, an increasing dependency of these states on shrimp culture as a source of earning considerable foreign exchange is observed.

- ☆ In the era of globalization and open market economy, the third world countries have a huge burden of debt, compelling them to accept the structural adjustment suggestions put forward by MLIs. Structural Adjustment Programme emphasis the promotion of export-oriented production like shrimp cultivation.
- ☆ A global concern regarding aquaculture activities started taking shape since the mid 90s, which has forced the producer as well as consumer countries of shrimp to come together to discuss the issues concerning externalities of shrimp culture in several international forums. Thus, a number of agreements, plans of action and guidelines including the Code of Conduct for Responsible Fisheries (CCRF) have been developed, although the implementation of these are left on voluntary willingness of the respective states.
- ☆ The third world countries mostly with a huge burden of debt do not have enough incentive to implement CCRF and have not made enough serious efforts to control the unabated growth of shrimp culture, which ensures earning of foreign currency for the state.
- ☆ Several serious socio-economic and ecological crises have started taking shape in connection with the shrimp culture in the shrimp producing countries.
- ☆ A growing awareness of the international civil society regarding the impact of the shrimp industry in the third world countries is gaining momentum, facilitated by international NGOs.
- ☆ Sustainability in shrimp culture has become the topic of international debate and all the global actors associated with the shrimp culture are concerned about sustainability, though often with different perceptions.

(Ujjaini Halim, 2004).

Findings at National Level

The post independence agrarian as well as fishery policies of India are embedded in resource management patterns in the primary sector, introduced during colonial era, which was marked by the abolition of land rights of the traditional users, introduction of private property and integration with the capitalist market of the west. This was reflected in greater degree of commercialization of agriculture and greater penetration of commodity money relations in the Indian agrarian system on the one hand and increasing poverty, hunger and loss of livelihood security of millions of traditional resource users on the other. Thus a structure of exploitation was imposed on Indian traditional primary resource users, which continues even today though in a changed form, in independent India.

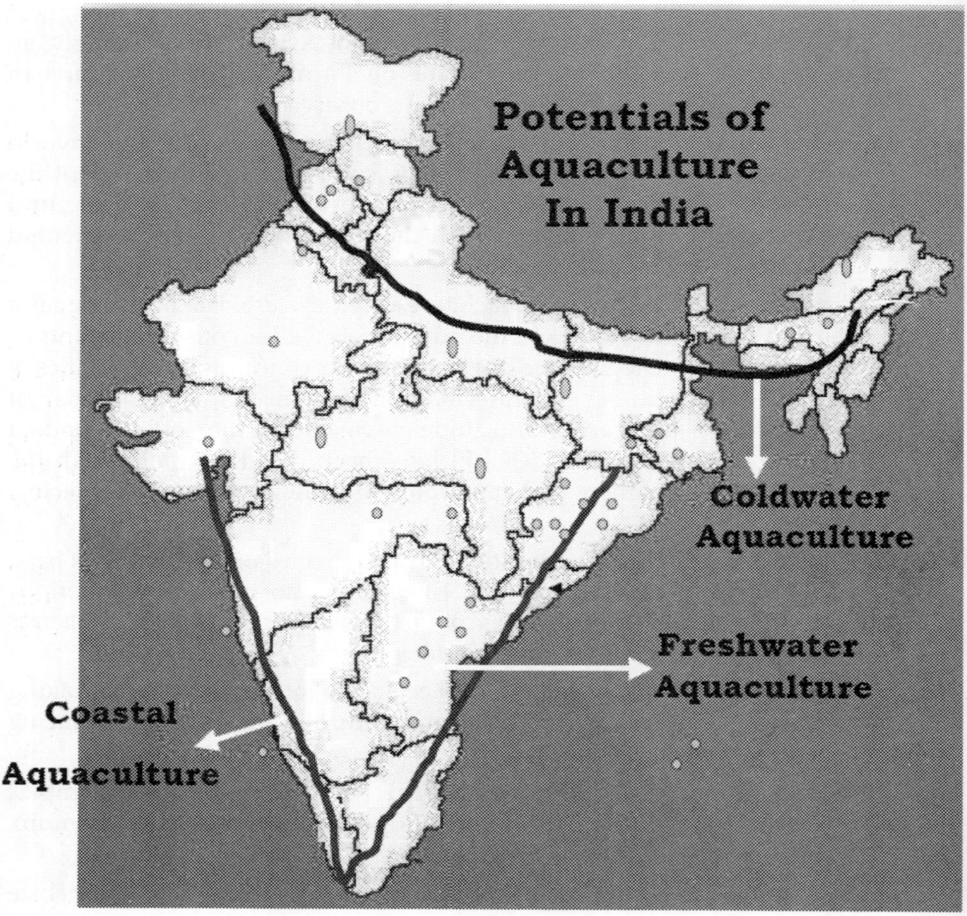

- ☆ After independence a few measures had been taken to reduce the extreme economic disparities in rural India albeit only with limited success. The basic exploitative structure has remained the same and the new policies of the government introduced strong capitalist trend in primary sector resource use and production.
- ☆ To encounter ever increasing poverty at the prescription of MLIs, Government of India engineered Green Revolution in agriculture in the 1960s and 70s, which though initially increased food production but in the long run led to centralised control of the trade in food grains and made farmers dependent on corporate sectors for various inputs. Green Revolution thus paved the way for introducing similar packages of technocratic solution of production and food security problems in other fields of primary resources *i.e.* forest, live stock and water management which are well known as Social Forestry, White Revolution and Blue Revolution, in Indian economy.

- At the same time the policy of the Indian government to promote massive industrial development resulted in the abolition of several restrictions on production to liberalise the entire economy of the country. The new liberal policy of India introduced in the early 80s is well known for opening up the Indian market for foreign investment. On the one hand subsidies and several other incentives were introduced to lure private entrepreneurs into investing in export oriented production and on the other hand huge reductions of subsidies were announced in traditional agriculture and fishery sectors.
- This policy of liberalisation in the 80s was followed up by the new economic policy of the 90s, the aim of which was to generate growth, relying on market forces or, in other words depending on resource mobilisation and investment.
- Extreme focus has been given to expansion of export oriented production during this phase as the government had to accept huge loans from the International Monetary Fund (IMF) with certain conditions of 'Structural Adjustment' of India's economy in order to deal with the Balance of Payment Crisis of the nation. Special schemes have been announced for EOU (Export Oriented Units) and Aquaculture/shrimp culture has been identified as a major thrust area.
- The Coastal Aquaculture Project of 1986, first launched shrimp mono culture officially in the country in the sixth five year plan (implemented during 1988-93) and offered special infrastructure facilities to Multinational Corporations (MNCs), Transnational Corporations (TNCs) and the national corporate sector to invest in shrimp culture.
- The Brackish Water Fishery Development Authority (BFDA) and Marine Product Export Development Authority (MPEDA) have been established to ensure smooth development of shrimp culture.
- Government of India also received multilateral assistance from the World Bank and development aid for aquaculture projects from consumer countries like Britain. Different federal governments also came forward to welcome this foreign exchange earning industry and introduced special facility packages for attracting investment in their respective states.
- Thus shrimp monoculture which is an input intensive, species specific commercial culture of brackish water shrimp gained momentum in India in mid 90s and total potential area for brackish water shrimp culture was estimated to be 1,190,800 hectares among which 13,816 hectares are under cultivation in 2000.
- With an annual production of 70,000 metric tons of shrimp India became fifth in shrimp production in the world in 1998-99. The contribution of shrimp to India's export reveals why the government has taken such a promotional role to develop the industry. By 1997-98 India had registered the export turnover of Rs. 4,120 crores 3 of which shrimp alone contributed 2,700 crores.

- ☆ The cultured shrimp hence contributed 43 per cent in quantity and 60 per cent in value of total aquaculture exports of India. The foreign exchange realised registered a growth of 14 per cent, which indicated a 23 per cent growth in three years. Major markets of Indian shrimp are Japan, USA and EU countries.
- ☆ Thus within a decade commercial shrimp monoculture has replaced the traditional brackish water shrimp culture system in India which were practised by coastal communities for generations. These traditional practices were different from one coastal state to another and were low input, natural shrimp production system that was often carried out in rotation with paddy or other types of aquaculture.
- ☆ Though the production in traditional farming was not enough for earning huge amount of foreign currency but was sufficient to meet local demands and needs of the producers and consumers of the coastal belt.

(Ujjaini Halim, 2004)

In 2005, according to MPEDA, India had 43,433 hectares of prawn ponds that produced 42,820 tons of prawns, and 140,000 hectares of shrimp ponds that produced 143,000 metric tons of shrimp. The bulk of this production came from the states of Andhra Pradesh, Tamil Nadu and West Bengal.

Establishment of Shrimp Seed Hatcheries

For the projected development of shrimp farming on 10,000 ha area the requirement of the seed will be around 600 million annually. It is also recommended that establishment of Backyard hatcheries of production capacity 1-5 million shrimp seed should be encouraged in fisheries cooperative sector. Establishment such hatcheries are relatively lesser capital intensive and can be established with an investment of about Rs. 3 to 5 lakh each. The coastal marine fisheries of the State, alike mechanized fishery in nearly every coastal State in the country and several other places in the world, is presently facing economic crisis and there has been an insistent demand to declare fish famine. The two month's fishing ban and rising costs of fuel have added to the woes of coastal fishermen. While a fishing ban is meant to be a necessary measure of conservation of stocks, a simultaneous positive step will be 'Sea ranching', *i.e.,* artificial seed stocking in estuaries by the State government. Financial assistance for establishment of such Backyard hatcheries to fisheries cooperative societies and purchase guarantee of the produced seed by the State government will provide a shore-based livelihood opportunity to coastal fishermen will assist in stock replenishment and can increase marine shrimp production and the earnings of the coastal fishermen.

It is therefore recommended that the State government may initiate a 'Sea ranching' programme for the next decade and undertake artificial stocking of shrimp seed in the 30 major estuaries in the State by involving the concerned fisheries cooperative societies to produce shrimp seed through Backyard hatcheries. Financing of such Backyard hatcheries by the financial institutes will be feasible if the State government becomes a partner in the project by providing seed capital and purchase guarantee. The State Agricultural University of the coastal region can provide the technical consultancy and assistance in this programme.

Guidelines for Sustainable Aquaculture

Rapid expansion of aquaculture in India is likely to lead to a number of social and environmental side effects which have been witnessed elsewhere in the world. Therefore, the basic objective of laying down guidelines for sustainable aquaculture is to reduce and eliminate any adverse impact of aquaculture on the environment keeping in view the experience of South–East Asian countries in this regards and developing it as an eco-friendly activity. These guidelines are expected to be useful in formulating appropriate shrimp farming management practices and adopting measures for mitigating the environmental impact for management of shrimp pond wastes and utilization of the land/water resources in a judicious manner.

The activities commonly associated with shrimp farming and their likely impacts on the environment are listed below. The norms to be adopted by the shrimp aquaculturists in particular and the regulatory agencies in general in the country are indicated.

1. Mangroves play an important role in soil binding, as a source of nutrient cycling, as a breeding ground and nursery areas for many important fin and shellfishes. There is evidence that removal of mangroves leads to a decline in finfish and shellfish recruitment to the open waters through reduced availability of post-larvae.

   Large concentration of shrimp farms in mangrove areas has not proved sustainable else where in the world. Poor soils and deteriorating water quality have lead to abandonment in several areas. Government permission should not be given for any construction activity within the natural mangroves areas, or ecologically sensitive wet lands, swamps, etc.
2. Construction of shrimp ponds on marginal land not fit for cultivation alone should be permitted by the States.
3. A vast majority of problems affecting the shrimp culturist as well as the environment could be avoided by better site selection and improved culture management. Site selection process should include proper environmental impact assessment. The existing criteria for site selection also need to be reviewed and consideration should be given to long term capacity of the area to sustain aquaculture development.
4. Detailed master plans for development of aquaculture through macro and micro-level surveys of the potential areas and zonation of coastal area delineating the land suitable and unsuitable for aquaculture using the Remote Sensing data, ground truth verification, Geographical Information System (GIS) and socio-economic aspects should be considered.
5. The shrimp culture units with a net water area of 40 ha or more shall incorporate an Environment Monitoring Plan and Environment Management plan covering the areas mentioned below.

   ☆ Impact on the water sources in the vicinity;

- ☆ Impact on ground water quality;
- ☆ Impact on drinking water sources;
- ☆ Impact on agricultural activity; Impact on soil and soil Stalinization;
- ☆ Waste water treatment;
- ☆ Green belt development (as per specifications of the State Pollution Control Board)
- ☆ All farms of 10 ha and more but less than 40 ha shall furnish detailed information on the aforesaid aspects.

6. An Environmental Impact Assessment (EIA) should be made even at the planning stage by all the aquaculture units above 40 ha. size For 10 ha and above statement will be required to be given in the detailed plans. The State Pollution Control Boards should ensure that such a EIA has been carried out by the aquaculture units seeking No Objection Certificate from them, before giving clearances for such projects.
7. The quality of feed plays an important role in waste output in shrimp culture, and there is scope to improve pond environment by good feed management. Nutrients and organic loads are higher in ponds where shrimps are fed with trash fish and fresh diets than where palletized moist or dry feeds are used. Fresh diets infrequent feeding and high stocking densities, increase nitrogen loads in the waste water form the shrimp farms. A considerable amount of detritus and wastes often accumulate on the pond bottom, in areas where water circulation is slow, leading to increased BOD and release of harmful gases, which could cause stress on bottom leaving shrimps. On the contrary, regular feeding with palletized diets is known to maximize the growth of shrimps and minimize the nutrient enrichment of the waste water.
8. Feed waste plays an important role in the total waste loadings in the environment. This is because; feed settles directly on the pond bottom and the feed wastage can have a significant effect on sediment quality and ultimately the health of the shrimp which normally live at the bottom.

   Hence the use of wet diets such as fresh fish and invertebrates has to be reduced and preferably avoided in shrimp aquaculture systems/Feed Conversion Ratio (FCR) should also be optimized. Monitoring of feed input is required to keep feed wastage to the minimum. Similarly, careful monitoring of standing stock in the ponds will also help to ensure that correct feeding levels are observed.
9. Chemicals should be avoided in shrimp culture ponds for prevention or treatment of disease, as feed additives, disinfectants, for removal of other fish or for treatment of soil or water. However, chemicals may be required in hatcheries. Entry of such chemicals into the natural waters from the hatcheries should be carefully monitored and steps should be taken to remove such materials from the waste waters.
10. Both organic and inorganic fertilizers are used widely in semi-intensive culture systems for promoting the growth of fish food organisms,

particularly for the early post-larval stages. This may contribute to the nutrient load in receiving water. Therefore, as far as possible only organic manure/fertilizers and other plant products should be used for such purposes.

11. Piscicides and molluscicides are widely used for removing predators and competitors from shrimp ponds. It would be advisable for aquaculturiest to use only the biodegradable organic plant extracts for this purpose as they are less harmful than the chemical agents. Use of chemical piscicides in culture systems should be avoided.
12. Some of the chemotherapeutants such as formalin and malachite green which are commonly use as disinfectants are known to be toxic and may affect adversely the pond ecosystem, the external water, etc. and hence their usage in culture system should be avoided.
13. A number of antibiotics used in shrimp culture for preventing outbreak of disease are harmful and may result in development of shrimp pathogens resistant to such drugs and the transfer of these pathogen's into human beings might result in development of resistance in human pathogens. The use of antibiotics/drugs in culture system, therefore, should be avoided.
14. Outbreak of disease in shrimp culture system is related to the environmental factors such as deterioration of water quality, sedimentation and self–pollution. The production losses are also linked to rthe acid sulphate soil particularly in the areas which have been developed by destruction of mangrove areas for shrimp farming.
15. Since the introduction of imported shrimp seed may bring with it a number of problems including diseases, disease producing pathogens etc., the use of exotic seed in the culture systems should be prohibited.
16. Access to the sea front and other common resources to the coastal communities by the aquaculture units should be ensured. The interest of the communities and organizations in the area should be safeguarded.
17. Care should be taken to see that the natural drainage canals which are used as water source for aquaculture units are not blocked so as to avoid flooding of low lying areas and villages.
18. Channelisation of saline water supplies for shrimp culture can lead to changes in the water quality and soil characteristics. Salinisation of ground water might render drinking water sources non-potable. Salt water intrusion into some of the freshwater aquifers may also be accompanied by soil salinisation in the coastal areas resulting in further devaluation of the marginal agricultural land and conflicts with local farmers and residents.
19. To avoid problems of ground water salinisation, drawl of ground water should be prohibited for shrimp aquaculture.
20. An appropriate legal framework should be evolved by the State for developing shrimp culture taking into account technical, environmental and social issues arising from development of this activity.

21. Spacing between adjacent farms may be location specific. In smaller farms, at least 20m. distance between two adjacent farms may be considered, particularly for allowing easy access to the common public, to the fish landing centers and other common facilities. Depending upon the size of the farms, a maximum of 100–150 m. between two farms could be fixed. In case better soil texture, the buffer zone for the estuarine based farms could be 20–25m. A gap having a width of 20m for every 500m.distance in the case of sea 300m.distance in the case of estuarine based farms could be provided for easy access.
22. A minimum distance of 50–100 meters shall be maintained between the nearest agricultural land, canal or any other water discharge/drainage source and the shrimp culture unit.
23. Agriculture units shall be located at least 100m away from any human settlement in a village/hamlet of less than 500 populations and beyond 300m from any village/hamlet of over 500 population. For major and heritage towns it should be 2 km. All aquaculture units shall also maintain 100m distance from the nearest drinking water sources.
24. Water spread area of a farm shall not exceed 60 per cent of the total area of the land. The rest 40 per cent could be used appropriately for other purposes. Plantation could be done wherever possible. For the existing farms, these provisions should be made within one year of issuance of appropriate orders by the State concerned.
25. The unit has to abide by the rules and regulations notified by Ministry of Environment and Forests, Government of India on Water Pollution, Coastal Regulation Zone, etc. The units with 10 ha. or more water spread area shall obtain No Objection Certificate of the State Pollution Control Boards (SPCB) before commencement of Aqua-farming. In addition all units established under Export Oriented Unit/Joint Venture guidelines will be required to obtain such certificate. The units of 40 ha, and more water spread area shall obtain regular consent of the SPCB under Sec. 25/26 of Water (P and CP) Act 1974. The shrimp culture units should obtain all the requisite clearance from various Departments concerned, before approaching the Government for issuance of permit/license.
26. Improved pond water quality and soil quality by aeration of the pond and fertilizing it with organic manures/fertilizer should be resorted to for preventing serious disease outbreaks and more effective environmental management. For maintaining good water quality in shrimp farms, periodic water exchange besides aeration is very desirable. Daily water exchange of 10–30 per cent may be affected, keeping in view the stocking density, the quality of feed administered and the phase of culture operation.
27. The seed collection from the natural sources should be discouraged. There is an apparent relationship between stocking density and disease problems. Farms stocking less than 25 PL/m^2 of shrimp have some advantage in avoiding disease outbreaks and in reducing the build up harmful organic

matter on pond bottoms. As far as possible only hatchery produced seed should be used in culture operation.

28. The solid waste of the farms, including sludge and scrapped soil from the ponds should not be disposed off into the water ways. The waste shall be disposed off within the premises of the farm after adequate treatment without allowing it to get into waterways. Not less than 10 per cent of the total pond area should be provided/equipped with waste water treatment or sedimentation pond.
29. The waste water should be discharged into receiving water bodies only after checking the quality for specified parameters and standards.

Chapter 7
Site Selection for Shrimp Farming

Introduction

Selection of suitable site is the most critical step in the development of successful shrimp farming venture, as this can have a significant impact on the profitability of the business. Hence, prior to establishment a careful analysis of the hydrology and hydrography of the site is to be carried out to ensure satisfactory supply of the required water for farming. Since no site will have the entire desired characteristic, a number of judgments have to be made for every site like:

1. Can shrimp be farmed profitably
2. The type of appropriate management
3. System which should be taken up at the location.

The main criteria for an ideal shrimp farming site depends up on:

1. The variety of species that are desired to be cultured
2. The anticipated production expected from a given area
3. Physiological, hydrological and other parameters desired to achieve the targeted production.

Factors to be Considered for Selection of Site

The major factors to be considered for identifying a site for shrimp farm are classified into three different heads namely:

1. Ecological Aspects
2. Biological Aspects
3. Social and Economic Aspects

Details are given as under.

Ecological Aspects

The following are considered under ecological aspect during selection of suitable site.

1. Water sources and quality
2. Salinity
3. pH
4. Tidal characteristics
5. Environmental conditions
6. Temperature
7. Soil

Water Sources and Quality

The site should have brackish water of desired quality for most part of the year. This is possible only in estuaries with an open month. Alternatively, direct seawater pumping could be resorted to. If the site is situated away from the sea, the salinity variation data for one full year or at least for the different seasons should be studied. If the salinity range is found in the acceptable range, such sites can be selected.

Salinity

Salinity is the total concentration of dissolved ions in water which is expressed either in mg/lit or parts per thousands (ppt or per cent o). Salinity of normal seawater will be around 35 ppt where as that of freshwater is zero. For selecting a site for shrimp farm the salinity of water during high tide in different seasons should be tabulated. This is especially important in rivers and canal where there is mixing of salt water and freshwater. The frequency of floods and the duration of freshwater flooding is to be ascertained.

pH

It is the hydrogen ion concentration in the water. By this parameter, it can be ascertained whether water is acidic or alkaline in nature. This is an important aspect since the water pH at the site should fall in the range of 7.8 to 8.3. Extreme values of pH (below 5 and above 11) are harmful to many species in decreasing their growth and productivity significantly.

Tidal Characteristics

Tidal characteristics in relation to land elevation at the site should be determined. It is critical to determine and decide whether the proposed ponds are to be filled by tide or by use of pumps. This is also important in fixing the pond bottom, dike height etc. If the fluctuation is too large, heavy embankments will have to be made which becomes too expensive, thus making the project financially non feasible. So for an ideal site the amplitude should be about 2 meters.

Actual measurements of the tidal aptitude should be made at the pond site throughout the year or at least in the different seasons to determine the high and low water marks. Since the tidal amplitude fluctuates ion different seasons, high tide levels during the flood seasons are also to be ascertained through local enquiries and with the date available with meteorological departments. Wave actions during normal tides, storms and monsoon also help in understanding the tidal characteristics of a site.

Environmental Conditions

Currents Prevailing in the Area

Prevailing currents and wind direction in the area is an important aspects to be studied for an effective planning of the farm bunds so as to minimize erosion and also to protect from excess sediment deposition in the water channels. Data on these aspects over a period of years is essential.

Rainfall

This is an important factor in the area of the construction. Data on average annual distribution of rainfall is necessary for knowing the runoff in relation to the site under selection and to design discharge sluices.

Evaporation Rates

Evaporation rates at the proposed site is to be checked for a period of time as this may pose a great problem if freshwater supplies are lacking in the area through which the required salinity of the pond can be maintained.

Pollution

Determination of harmful substances which are used nearby the area and which are released in the upstream is necessary. Harmful substances like pesticides generally used for agricultural purposes, industrial and urban wastes are likely to be released in the nearby areas. Also anticipation of future pollution problems form industrial estates or a city is to be ascertained from the local authorities. Hence such a site may be rejected.

Temperature

Water temperature range of 26–33°C is considered to be optimum for good growth and survival of shrimp. Whether there is a possibility of occurrence of higher water temperature, providing an increased water depths by way of digging trenches will be helpful. This is important for good growth and survival of the shrimp. The range of temperature during different period of times is to be taken into account.

Soils

Soil characteristics of the site should be carefully studied. Samples should be taken at least at ten different places per hectare from ponds that are likely to be constructed. Soil core samples should be taken at least to a depth of 0.5 mtrs below the proposed pond bottom. This is because a good soil might overlay an unsuitable soil and a surface soil analysis may not be sufficient. The soil samples, thus taken are then analyzed for particle size, texture, permeability, shear and compaction tests. An ideal soil texture for construction and water retention are clay loam or loam. The pH of the soil should also be measured both initially and by disturbing, to detect any acid sulphate conditions.

(*a*) Type of Soils

Soil classification is normally based on particle size or sometimes on other properties like plasticity and compressibility. The system of classification based particle size is widely used. General classification of soil is given below:

Table 7.1: Classification of Soil Based on Particle Size

Texture Class	*Sand*	*Particle Size in Per cent*	
		Silt	*Clay*
Sand	80-100	0–20	0–20
Sandy loam	50–80	0–50	0–20
Loam	30–50	30- 50	0–20
Silty loam	0–50	50–100	0- 20
Sandy clay loam	50–80	0- 30	20–30
Clay loam	20–50	20–50	20–30
Silty clay loam	0–30	50–80	20–30
Sandy clay	55–70	0–15	30–45
Silty clay	0–15	55–70	30–45
Clay	0–55	0–55	30–100

Source: MPEDA, 1996.

Many coastal soils are usually high in sand content and will not hold water. The potential soils must have reasonable clay content (Particle size less than 0.002 mm) so that the pond can hold water a good field test used to determine water retention is to shape a handful of moist soil into a ball. If the ball remains intact and does not crumble after considerable handling, there is enough clay in the soil. Sandy clay or sandy loam types of the solids are best for dike constructions as these are hard and do not crack on drying. Another type of sample test usually done is to take 5 cms of soil into a measuring cylinder with water. Vigorously stir and allow the sample to settle. After an hour the larger particles like sand will settle to the bottom followed by silt and clay. By measuring the depth of each of these, the proximate analysis of the type of soil is defined.

(*b*) Soil Acidity

Soils having a pH value equal or less than 4 are termed as acid sulphate soils which are also easily identified by the presence of mottles of the pale yellow mineral jarosite. In the drained areas, these conditions are characterized by the presence of red colouration on the soil surface.

These are much more difficult to determine, since they do not become acidic until after oxidation. Soils of this nature are acidified by exposure to air but the extent and rate of the acidification process are regulated by tropic bacteria. This activity is low in dry soil, so it is best to keep the soil moist.

(*c*) Determination of Acid Sulphate Soil

The following procedure is adopted to identify a potential acid sulphate soil.

A soil sample is made into a 1 cm thick cake and sealed in a thin plastic bag. The preserves the soil moisture and if this is permeable will allow oxidation of the pyrite to proceed rapidly. The pH of the soil should be reduced to below 4, within a month period if it to potential acid sulphate.

Considering the problems associated with acid sulphate soils, a detailed survey of the soils is necessary before the construction work starts. Ponds on acid soils required limiting for improving the soil and water quality. Common liming compounds used are calcium carbonate ($Ca\ CO_3$), Calcium magnesium carbonate ($CaMg(CO_3)_2$, Calcium hydroxide ($Ca\ (OH)_2$) or Calcium oxide (CaO).

(*d*) Construction in Areas of Acid Sulphate Soils

Special procedures are necessary to ensure pond fertility and to prevent mortalities due to low pH. Some of the aspects which can be considered are:

1. Excavation of only the minimum required soil from internal canals for construction of dikes without disturbing the top soil. If this type of construction is taken up, ponds can be filled only with the help of pumps.
2. The top soil from the pond bed is scrapped off. Once the pond has been excavated, this top soil is replaced over the bond bottom.
3. Otherwise alternate strips can be excavated twice as deep as necessary and then the good soil from the unexcavated portion may be placed in the deep portions.
4. Sometimes alternate 10 to 20 m wide strips are excavated between the undisturbed soils. This pond is used for culture and after few years, the other strips (undisturbed soil) should be excavated.
5. If the sub soil is not very acidic, but of high potential acidity, the soil should be kept moist so that it will not be oxidized.
6. In case of construction and for use in dikes, the dikes should be surfaced with good top soil, otherwise runoff during rains will have greater effect in the pond.
7. During the construction of dikes, designing should be made carefully so as to ensure minimum amount of seepage during runoff into the ponds. This can be maintained by keeping the pond water at a higher level than that of outside water.
8. A beam can be constructed near the water edge to catch acid runoff during rains and preventing it from washing into the ponds.

9. In acid sulphate areas, turfing with acid tolerant grass of *Cynadon* sp on the bunds is worth trying to control erosion and acid run off during rains.

(*e*) Percolation Rates

Knowledge of the percolation rate of the soils under investigation will help in determining the extent of water loss through the pond bottom or dikes which can affect the design and management. To check the percolation loss a small pit is made and filled with water. After 24 hrs the water loss is measured.

(*f*) Load Bearing Capacity

This is especially important if heavy equipment is to be used. It also will help to determine the number of pilings required under the gates and the need for special foundation under dikes.

Biological Aspects

Under the biological aspects survey of seed resources, predators, competitors, boring animals and the type of vegetation available in the area are to be examined.

Seed Resources

Availability of fry either from the wild stock and supply through seed suppliers or from the hatcheries is to be ascertained. The local sources of the variety of seed available and the seasonality of abundance should also be determined.

Predators, Competitors etc.

As the predominant varieties of predators vary from area to area, the kind present in a given area, may have an effect on the management, construction and estimates.

Boring Animals

Presence of these animals in the area and also the extent of damage that is likely to occur are to be known. To find out this, old wood present in the area or the wooden boats of the local areas are to be examined. This gives an idea as to the type of wood that can be used for sluice gate constructions.

Vegetation

The type of vegetation in the area gives an idea about the soil type. Some of the common mangroves in the tidal zones are *Avicennia, Rhizophora, Excoecaria.* Mangrove areas are generally acidic. Areas with grown up trees are not recommended for farming.

Social and Economic Aspects

The various social and economic aspects to be considered are: Land cost, Accessibility, Availability of man power, Marketing outlets and prices, Social and political factors, availability of technical assistance, Use of land and nearby water.

1. Land cost should be determined first so that economic viability of the project can be evaluated.

2. Accessibility to the site is importing of construction material, equipments, feed, seed and daily operations to the farm are essential. It should be accessible even during monsoon season. A good approach and communication will greatly help in reducing the cost of transport of inputs. If the accessibility is by water, it should be made sure that even during monsoon it is possible.
3. Local labour will work out to be cheaper than hiring from some other place. By employing local people cost on housing, transportation, food etc can be saved. Local customs, traditions and agriculture activity are to be known as this will help in planning labour requirement for construction work. The above information will help in planning for the maintenance, culture work such as stocking, harvesting etc in season when sufficient labour is available.
4. Marketing and outlets and prices will have an impact on management. Market with cold storage facilities near by will be of use for getting better price for the product. Harvesting as per the local market demand or as per the processing capacity of the processing plants in the area will help in getting better price for the harvest. If it has to be marketed to a greater distance, it is better to harvest completely at a time. However, it is better to start harvesting after negotiating the price.
5. Social and political factors includes the licensing requirements, land ownership laws, legal restrictions, governmental regulations etc are to be studied.
6. Availability of technical assistance can be from governmental extension services, university research stations or from financing agencies and input suppliers.
7. The use of nearby land and water should be assessed to determine as to what impact they will have on the project. Activities like fishing, industry, public utilities, nursery areas, agricultural lands may create problems if their activities are disrupted. It is to be made sure that the project does not block a traditional right of way or interfere with other's work.

Type of Area for Coastal Aquaculture

Potential area for coastal aquaculture are broadly grouped into the following classes like areas which are located in coastal zones and the low lying areas along the periphery of shallow backwaters.

Intertidal Zone

For rearing to marketable size and also to drain the pond completely for harvesting it is better to have ponds in the intertidal zone. Though a sluice gate is required to drain water it is advisable to pump water into the ponds without depending on tidal amplitude.

Intertidal zone areas are submerged during high tide and exposed during low tide. These are usually covered with mangroves and grassy swamps which generally contain clay loam mixture. High organic matter content may sometimes make it unsuitable for dike construction. Land clearing and excavation require higher expenditure. For areas having low tidal amplitude of less than I meter, pumping may be necessary for draining and drying ponds.

Super Tidal Zone

This is usually located above high water area. Such areas are not normally acid sulphate and commonly have sandy clay, sandy loam type of soil. Developmental cost is usually less when compared to the intertidal zones. Operation can start immediately after construction of the pond. Water management is usually by pumps regardless of the type of culture management. Since the bottom elevation is above mean high water level, it helps in faster drainage by gravity. Hence, these areas are usually preferred for semi-intensive and intensive system of practice.

Low Lying Areas

The shallow water bodies swell up during monsoon periods due to rain water in the catchments areas. During summer due to evaporation there is considerable reduction of depth in the lake thereby exposing a vast low lying area. These areas are also suitable for culture practices of traditional and extensive types. Soils of these areas are highly fertile and water management in the ponds constructed at these sites is easy due to less seepage from the lakes adjacent of the pond.

Site Requirements and Construction

Site selection has been covered earlier in this manual. Having selected the site you will need to thoroughly survey it to determine the best layout for water intake, ponds, access roads, and effluent discharge. These topics are not specific to freshwater prawn farms, so there is no attempt here to duplicate the FAO manuals already available, which have been mentioned above. The development of sites for freshwater prawn farming is discussed in detail in Muir and Lombardi (2000).

Measuring Soil pH

Take 10-12 samples of the upper 5 cm layer of the soil, before any soil treatment has been applied, dry them in an oven at 60°C, and pulverize them to pass a 0.085 mm screen. Bulk the samples together and mix 15 g of the pulverized soil with 15 ml of distilled water. Stir occasionally for 20 minutes and measure the pH, preferably with a glass electrode. The hand-held pH-soil moisture testers used by some farmers are not accurate enough (Boyd and Zimmermann, 2000).

Chapter 8

Pond Preparation

Objective

The five main objectives of pond preparation are as under:

1. To eradicate predators and competitors.
2. To degrade organic matter accumulated in the pond soil.

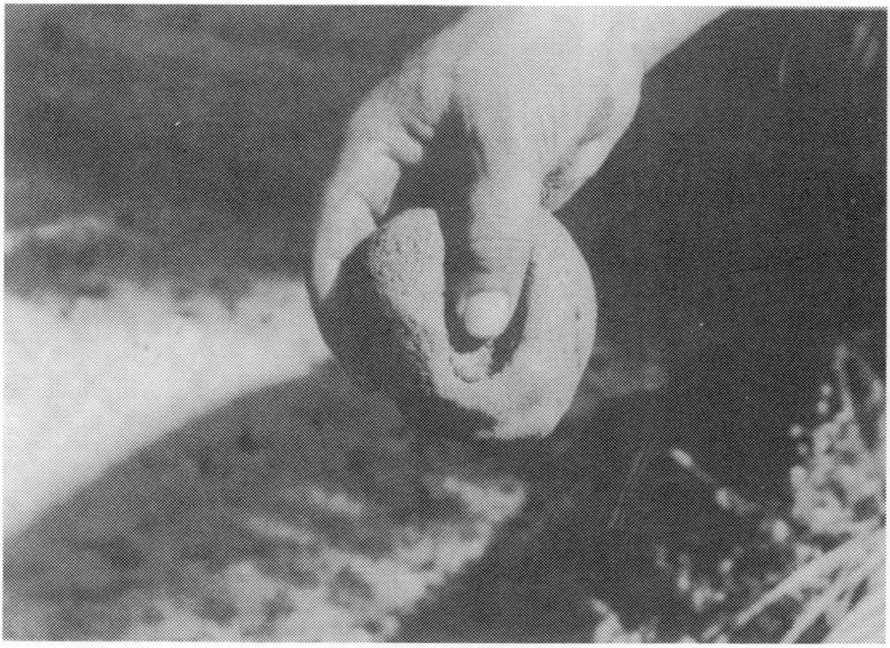

Figure 8.1: Soil Test for Water Retention

3. To oxidize the accumulated hydrogen sulfide, ammonia, nitrite, ferrous ion, methane in the soil.
4. To promote plankton by the addition of organic and inorganic fertilizers.
5. To adjust the pH of the soil and water of the pond.

Table 8.1: Chemical Parameters for Shrimp Farm

Sl.No.	*Parameter*	*Limits*
1.	Soil pH	7.5–8.7
2.	Soil salinity (ppt)	0.18–1.5
3.	Cat ion exchange capacity (per 100gm)	18–26
4.	Convertible Potassium	0.20–0.65
5.	Convertible Sodium	3.5–8.8
6.	Convertible Calsium	10.0–14.0
7.	Convertible Magnesium	3.5–4.5
8.	Available Phosphorus	0.0046–0.0092
9.	Soil Carbon	0.50–1.5
10.	Total Nitrogen	0.040–0.095

Source: N.D. Chhaya, 1993.

Pond Preparation

The success of shrimp production through aquaculture system relies heavily on the design and construction of the system. A well designed and good constructed pond will respond to good operation and management. If the design and construction is poor, even the best management techniques can do very little to optimize production.

In extensive system the ponds are large in size and water is taken by gravity through sluice gate. In semi-intensive farms, the ponds vary between 0.5 and 1 ha. Though sluice is required for draining, water is taken by pumping. The water depth in the ponds is about 1.2 meters. In intensive system the ponds vary between 0.25–0.5 ha in size with a depth of not less than 1.5 meters with adequate number of pumps for exchange of water.

A careful layout of the described facilities and support will ensure a smooth and trouble free operation allowing proper management of production activities.

Pond Layout

Layout of the shrimp farm depends mainly up on the topography and shape of the area. Pond orientation should be depending on the direction of prevailing wind. The longer sides of the pond should be oriented parallel to the prevailing wind to increase aeration and circulation of pond water by way of diffusion and turbulence. Ponds with internal canals are better in places where silt or mud content is high, as these get deposited in the canals which can be removed easily. If the area is facing a mud flat a sedimentation tank is necessary so that the sediment gets deposited in the

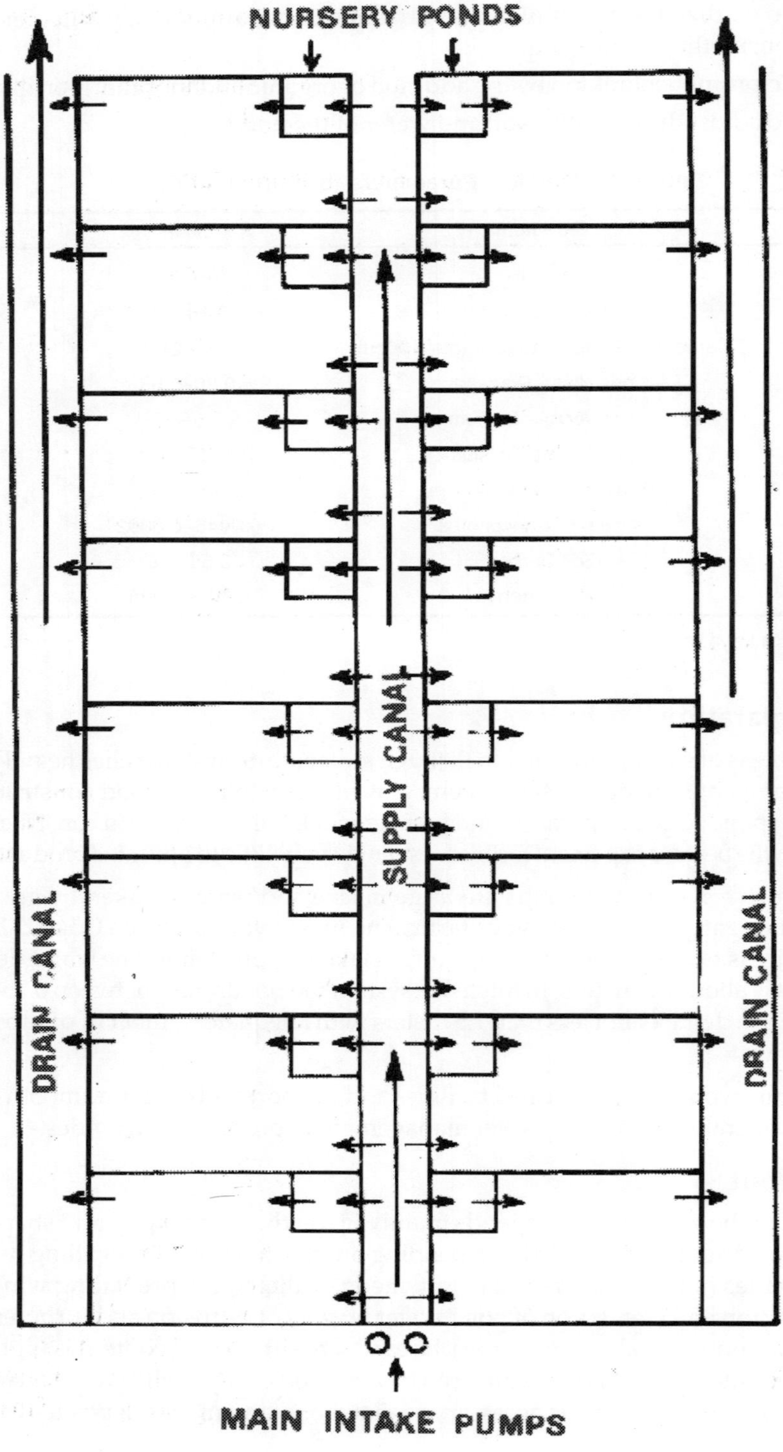

Figure 8.2: A Plan of Shrimp Farm
***Source:* MPEDA, 1999**

tank before water enters the pond. The main water supply gate should not be located at the bends of a canal or facing the open sea, as these areas are subjected to strong currents and wave action which may cause damage to the gates resulting in costly repairs. The layout of channels and dikes should be made based on the existing land slop and undulations. The discharge from the farm should be at the down stream of the intake system so that it is not taken back in to the ponds. Deep wells for freshwater should be located at a reasonable distance to avoid intrusion of sea water due to pumping of freshwater in to ponds to control salinity. Size of the pond area should be optimum for easy operation and management of the farm.

While designing for intensive and semi-intensive systems, a flat bottom with slight slope is preferred whereas in the case of extensive type the bottom may be designed either flat or with trenches covering about 20 per cent of the total area. Trenches are made in such a way so that they can hold a water of 1 meter and the other areas can hold a depth of 0.60 to 0.75 meters. The bed level of the trenches should be in such a way so that all the water can be drained off to facilitate drying. By providing trenches, protection from birds and shelter during high temperature are achieved. The pond bottom should preferably be above the low water mark of spring tide.

Construction of Dikes

Pond dikes are considered to be the most important structures in a pond. The main dikes (Figure 8.3) are those which enclose the entire farm area and render over all protection to the stock and other related structures. They are wide and of considerable height. Secondary or partition dikes are those which separate one pond from the other and are of smaller in width and height depending on the water height preferred in the pond.

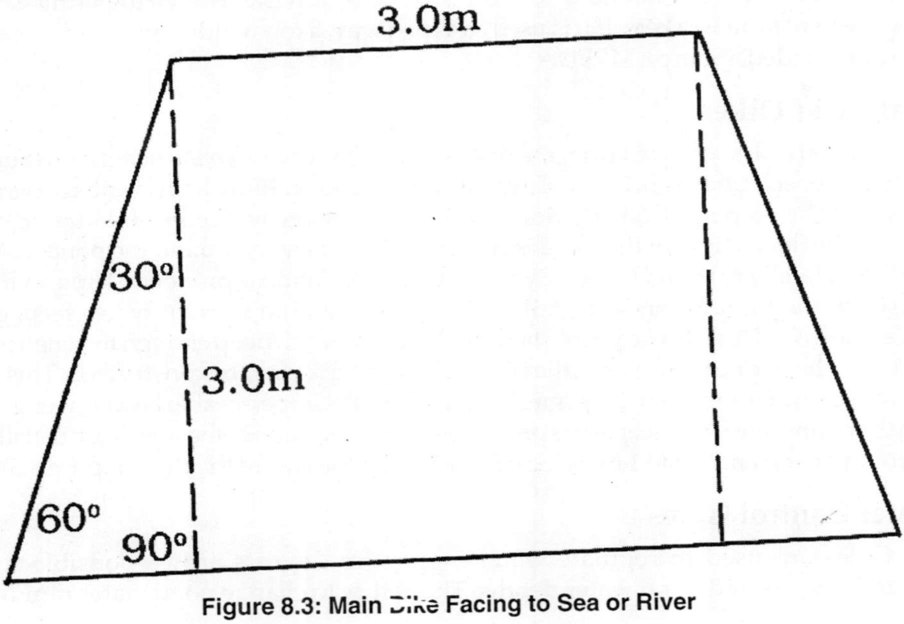

Figure 8.3: Main Dike Facing to Sea or River

The primary dike (Figure 8.4) is to be constructed considering the type of soil. Construction should be in stages with consolidation. First 1/3 rd of the final height of the bund is constructed all the way around the pond then 2/3 rd and final to its full height. By this way consolidation is proper to support the weight of the top portion.

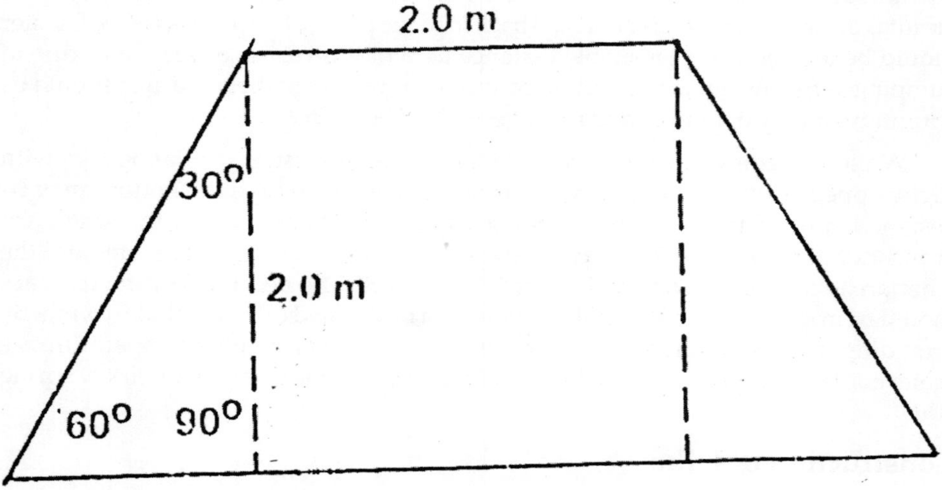

Figure 8.4: Dike Adjacent to Pond Dike

Dike top width or commonly known as crown depends up on the type of structures which are likely to present on the dikes. Structures such as elevated canals, electrical pots and service roads take considerable amount of space and should be given attention. A minimum standard of 2 to 2.5 meters is advisable so as to permit easy movement of farm workers. If it is used as a road way, a top width of at least 4 meters is to be provided. (Source: MPEDA, 1998)

Feature of Dikes

The top soil is removed from the dike area and the dike is constructed on the underlying consolidated soil. The soil must have low permeability to prevent excessive water loss by seepage. Construction should be in a stage with consolidation. First 1/3rd of the final height of the bund is constructed all the way around the pond, then 2/3rd, and finally to its full height. By this way consolidation is proper, to support the weight of the top portion. A puddle trench is essential to prevent water seepage under the dike. This is an accomplished by digging a trench deep enough to penetrate well in to the permeable layer in the place where the dike is to be constructed. This is then filled with compacted impermeable soil. Since the impermeable layer is wedged into the permeable layer, seepage is prevented. The dimension of the trench are usually 0.5 to 1 mt deep and 0.5 to 1 mt wide depending on the size of the dike (Figure 8.5).

Water Control Gates

Gates are used to regulate water ingress and egress are responsible for maintaining desired pond water depths. The ability to change pond water rapidly

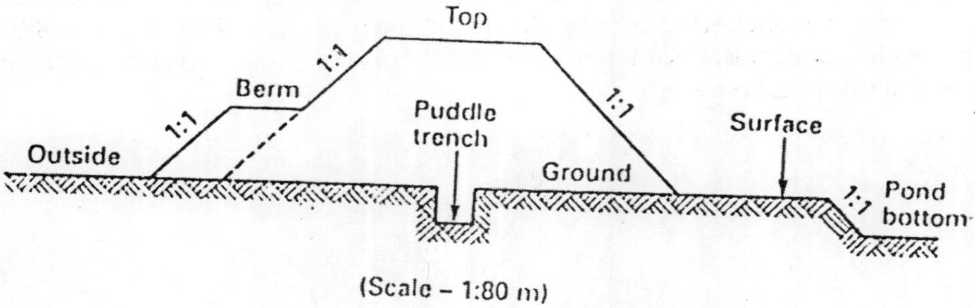

Figure 8.5: Main Dike

and efficiently depends on better design and construction of sluice gates. In semi-intensive and intensive systems of farming separate inlet and outlet gates must be provided.

Type of Gates

There are different types and size according to their function to control water, and their construction can be of various materials. The basic requirement of the gates should be as under.

They should be of adequate dimension for the amount of water required to be taken or drained. Construction should be in such a manner that water can be taken in and discharge from the bottom. Provision for draining surface water also should be provided. The bottom of the gate must be at an elevation which can drain the pond completely. Gates should be provided with slots for placement of screens and should be water tight. Provision for installing the harvesting net is to be made. It should be durable and easy to operate and preferably made from the locally available materials. Its foundation should be adequate. Adequate reinforcement against side pressure from the dike and water should be arranged. Measures are taken to prevent under cutting by seepage of water along the sides and bottom of the gates.

Main Gate

The main gate connects the main channel of the pond system to the source of water. These are generally located on the perimeter dike. They are usually big enough for filling the ponds and draining and generally or reinforced cement concrete.

Secondary Gates

These are located in the partition of the dikes and are used to regulate water level in the ponds. Usually of smaller size and are made of Ferro cement, RCC, brick masonry or wood.

Inlet Gates

In case where secondary gates are used to regulate water from primary feeder channel to secondary feeder channel, pond inlet gates are separately used for controlling water level in the ponds. This is mostly either open type or closed type. In case of open type, either RCC or wood is used. Grooves are provided for screens and shutters to control water flow into the pond. In case of closed type either PVC or cement hume pipes are used.

Figure 8.6: Inlet with Filter Bags

Outlet Gates

This regulates water flow from the pond into the drainage canal. These gates are open type. Usually it is constructed with RCC or wood (Figures 8.7).

Pond Bottom Conditioning

Conditioning of a pond bottom must be done from the very next day after harvest:

1. The entire pond bed and all inlet and out let canals are flushed thoroughly, so that all accumulations of undesirable materials are removed.
2. 40–60 mesh per inch (1.0 mm) nylon screens are installed in the gates and water is taken into the pond to a depth of 10–15 cms.
3. This water is kept for 1 or 2 days and then drained out. Preferably two gates are installed diagonally opposite (one inlet and one outlet) are found to be more effective for each pond.

Figure 8.7: Drainage/Outlet

4. This is done so that sediments are not solidified.
5. Rakes and chains are used to loosen the solidified organics.
6. In case of small pond area, earth moving equipments like spades may be used.

When a pond cannot be drained out fully and dried or preparation is needed in rainy season, following method is generally applicable.

1. One mm nylon screens are to be installed in the gates and water is allowed into the pond to a depth of 10–15 cms.
2. After allowing the water to remain for 1 or 2 days, dredging and physical removal of reduced sediments is to be carried out along with flushing of water as far as possible.

Procedure for Drying the Pond

1. Once the flushing is completed, all inlet and outlet water gates will have to be closed.
2. If there is any water remaining in the deep areas it has to be pumped out as these deep areas contain the most anaerobic organic matter.

3. Remove the sediments from the pond.
4. The pond bottom is allowed to dry for 7 to 10 days, until cracks appear at least 1–2 cms deep or the pond bottom surface can support the weight of a person without sinking.
5. Do not allow the soil to become too dry (rock hard) as it may destroy the microorganisms necessary for decomposition of organic matter.
6. If the soil is too dry it crumbles or may become dusty and is more vulnerable for wind erosion.
7. After the pond bed is dried as per requirement, plough or till the soil to a depth of 5–15 cms so that iron content in ponds with acid sulfates is reduced.

Pond Bottom Sterilization

1. If pond is to be sterilized to get free from pathogenic organisms (Bacteria, fungus, virus etc) 1000 to 2000 kgs/ha of hydrated or burnt lime is applied. This will raise pH of soil and kill the pathogenic bacteria which reside in soil between crops.
2. Application of potassium permanganate ($KMnO_4$) is another method. If $KMnO_4$ is used, it is applied at a rate of 1–2 gm per M^3 of water by mixing in water and all aerators are operated for proper mixing.
3. If Benzal Konium Chloride (50 per cent) is used it is applied at a rate of 0.5–1.0 ppm level into the pond.
4. If organic iodide (10 per cent available iodine) is used, it is at a rate of 0.5–1.0 ppm concentrations. Iodine required is diluted and applied evenly in the pond.
5. Measure the pH of the supernatant water by the regular procedures using lovibond comparator, pH paper, pH pen, pH meter etc.

Liming the Pond

Lime is used for oxidation and the activity adjustment of pond soil.

Application of Lime Dosage

1. If the pond soil pH is 6.5 or above, lime is required in small dosages.
2. If even after drying and flushing, the pH is below 6.5, agricultural lime stone ($CaCO_3$) is applied in following quantities.

Soil pH	Lime ($CaCO_3$) (MT/ha)
3–4	2.0–4.0
4–5	1.0–1.5
5–6	0.5–1.0
6–7	0.3–0.5

Procedure to Overcome the Acid Sulphate Problem

For reclamation of acid sulphate soils, the procedure should be started at least 3–4 months before stocking and during the dry season as follows:

1. The ponds are drained if any water is present.
2. Dried till the pond bottom cracks up to 10–15 cms deep.
3. pH is monitored regularly. Bottom is flushed with water and submerged.
4. Flushing and draining the ponds are continued until the pH becomes more or less constant at or above 5.
5. Per hectare 500 kgs of agricultural lime stone is broadcasted.
6. Construct small trenches with levees on top of the dikes.
7. Water is pumped in to the trenches to flood them to more than 10 cms deep. This should be done simultaneously with the submergence of pond bottom, so that the acid leaches can directly seep into the pond which can be flushed out during draining.
8. Pumping is topped and the trenches are allowed to dry when the pond is to dry. pH is monitored regularly.
9. Submergence and drying procedure are repeated for a few times as that of the pond.
10. The levees are removed and lime is broadcasted on the slope of the dike @ 1 kg/10m and above the dike @ 1 kg/20m.
11. pH is monitored regularly and apply agricultural lime immediately into the pond water and along the dikes.

Source: MPEDA, 1996.

Eradication of Predators

The first thing has to do about predator control is screening of the water through gates, so that entry of predators, competitors are effectively controlled and chemicals etc is applied in to the pond water as the required dosage to kill predators and pests, present inside the pond water.

Table 8.2: Chemicals Commonly Used as Piscides with Dosages

1.	Rotenone (5 per cent)	-	40 mg/lit.
2.	Tea seed cake (5 to 7 per cent saponin)	-	20 mg/lit
3.	Mahua oil cake (4–6 per cent saponin)	-	200 mg/lit
4.	Lime + Ammonium sulphate (5–10:1)	-	1100–1200 kg/ha
5.	Tobaco dust	-	20 mg/lit
6.	Anhydrous Ammonia	-	10 mg/lit
7.	Sodium hypochlorite (5.25 per cent conc)	-	10–300 mg/lit
8.	Calcium oxide	-	1000–1500 kg/ha
9.	Calcium hydroxide	-	1000–2000 kg/ha

Source: MPEDA, 1996.

Deodar in Pond Water and its Removal

Unconsumed feed and decomposition of soil creates deodar. This deodar adversely affects the growth life system of shrimp. It is therefore necessary to maintain clean and clear water. To maintain proper algal broom Urea, super phosphate etc chemicals can be used depending upon availability of Nitrogen and Phosphorus, Proper growth of algae and microbes can be manually calculated by observing the colour of water. Colour of pond water changes to greenish brown or yellowish, when sufficient quantity of algae and microbes are present in water. Some time people use Formalin before water is added to pond to kill un-needed insects and removal of deodar. Recommended quantities of this are as under.

Table 8.3: Recommended Chemicals for Deodaring Pond

Available Nitrogen in Soil (mg/100gm)	*Urea*	*Available phosphorus in Soil (mg/100 gm)*	*Single Super Phosphate*
12.5	100	1.5	100
25.0	50	3.0	50
50	25	6.0	25

Source: Vyas, A.A. *et. al.*, 2004.

Chapter 9

Selection of Species for Culture

In commercial scientific shrimp farming practice it is better to select a native shrimp species which has a ready market and can grow fast in the prevailing ecological conditions. In India *P. monodon* and *P. indicus* have proved to be the most suitable species for culture activities. Other species which can be tried are *P. merguiensis* in the North West and North East coasts, *P. semisulcatus* and *P. latisulcatus* in suitable areas along Tamil Nadu coasts. *P. japonicus* along the Maharashtra, Goa, Tamil Nadu, Andhra and Kerala coasts.

We can broadly classify the major criteria under two categories, which are considered for the selection of a suitable species for culture:

- ☆ Biological
- ☆ Economical.

Under biological features, the following are the important aspects:

- ☆ Growth rate
- ☆ Ecological adaptability
- ☆ Distribution of species
- ☆ Reproductive characteristics
- ☆ Larval development
- ☆ Hardness of the animal
- ☆ Availability of seed.

The factors considered under economical aspects are:

- ☆ Market price
- ☆ Market demand.

Detailed information of each of above aspects is as under.

Biological Factors

Growth Rate

Selection of a suitable species largely depends upon its growth rate. The fast growing varieties, yielding short-term harvest are the most suitable species for scientific shrimp farming. The growth rates of important species discussed are as under:

Species	*Maximum Size*
P. monodon	320 mm
P. inducus	230 mm
P. merguiensis	240 mm
P. semisulcatus	250 mm

Source: Dholakia, A. D. 1994.

Ecological Adaptability

Success of a culture practice largely depends on the ability of the species to adjust to certain fluctuations in the culture medium. Adaptations to various ecological conditions differ considerably among different species. Some species prefer high salinity for their growth while others less. In case of temperature also, some type of disparity is found among different species. So according to the water and other environmental parameters, the species selected for culture under different regions of the country vary widely. *P. monodon* is preferred to culture in brackish water conditions due to its high degree of tolerance to varying environmental condition where as *P. indicus* is preferred for culture in normal saline conditions since sudden change in salinity brings heavy mortality. Another important aspect taken into consideration is the ability of the species to accept supplementary feed and its good food conversion efficiency.

Distribution of Species

Shrimps that are naturally occurring in the locality (Native species) will be the most suited species for culture. In India most of the culture operations are carried out at present in the coastal inshore regions and hence the species that occur around these zones of operations have greater advantages than those species that are distributed far way from the area. Most of the shrimps cultured at present are littoral in distribution having their bathymetrical distribution confined to the inner half of the continental shelf. Of the littoral species of shrimps, those having differential distribution are more suitable for culture than those restricted to a particular environment. The developmental stages of the penaeid shrimps of out coast have some different distribution, their eggs and early larvae occurring in the sea and the post larval and juvenile stages in the estuaries. It may be noted that the traditional culture practices prevalent in India is mostly based on the juvenile phase of these species that is being spent in brackish water environment.

Reproduction Characteristics

Reproductive characters are as under:

1. Availability of spawners close to the area of culture operations,
2. Their occurrence during greater part of the year.
3. High fecundity.
4. Capacity to breed more than once in a year
5. Spawning under captivity are the other important salient reproductive features of shrimps which contribute to their suitability for culture.

Larval Development

Duration of larval development in penaeid shrimps is relatively short when compared to that of Palaemonid pawns, as all the 13–15 stages in the larval development in the early life history of the former are being completed in 15–20 days. But in the case of Palaemonid shrimps the life cycle is fairly long and it takes about 40–50 days to complete the metamorphosis.

Hardiness of the Animal

The hardiness of shrimps makes them suitable for culture. Since the various stages in the life history of shrimps are distributed in widely fluctuating habitats, they show different tolerance and preferences to environmental factors. Among these, salinity and temperature are found to be the most important factors. The ability of larvae and juveniles to adjust to changes in salinity and temperature and the naturally evolved life history which is pre-adapted to different eco-systems makes most of the species ideally suitable for cultivation under controlled conditions.

Source: MPEDA, 1996.

Economical Factors

Market Demand

Recently the greater demand for shrimp products in internal and external markets has made shrimp farming attractive economic. In the context of greater demand for larger shrimps in the export trade, species that grow very rapidly and attain bigger size in the shortest period have to be selected for profitable commercial culture.

Market Price

Species having good market value both in the domestic and external markets should be selected for culture for the export point of view, the penaeid shrimps are more important than the non-penaeid shrimp on account of their larger sizes reaching within short duration and high unit value.

For selecting cultivable shrimp species you will have to know the advantages and disadvantages of each cultivable species, which is given in Table 9.1.

Table 9.1: Advantages and Disadvantages of Important Cultivable Shrimp Species

Sl.No.	Name of Species	Advantages	Disadvantages
1.	*Penaeus monodon* (Tiger shrimp)	1. It attains a very large size	1. Natural seed supply is relatively inadequate.
		2. It is the fastest growing of all shrimp tested for culture within short culture duration.	2. Females are more difficult to mature in captivity than many other species.
		3. High price and high market demand.	3. Gravid females are difficult to obtain from the wild along west coast in sufficient numbers to support a large hatcheries
		4. It can tolerate a wide range of salinity. It can survive up to 40 ppt salinity. Growth is reported to be slower at very high salinities and in freshwater.	4. The head to tail ratio is not as good as that of some other species. This could have an adverse effect on sales to the export market where only tails are desired.
		5. It can tolerate temperature up to 37°C. Mortalities occur at low temperature.	
		6. Very good food conversion ratio (FCR)	
		7. It is hardy and not greatly disturbed by handling.	
		8. It performs well at high stocking densities.	
2.	*Penaeus indicus* (White)	1. This shrimp grows to fairly large size and fetches good price.	1. Relatively high salinity (35ppt) is required for best growth.
		2. Fast growing, tasty meat, high market price and short growing period.	2. Mortalities occur at temperature above 34°C.
		3. Natural seeds and spawners easily obtainable.	3. There is a significant size difference between sexes.
		4. High survival rate for the first three months of growing period.	4. It cannot withstand rough handling as either a juvenile or as adult. Fry are weaker than *P. monodon* during transport.

Contd...

Table 9.1–Contd...

Sl.No.	Name of Species	Advantages	Disadvantages
		5. Females can be matured in captivity with relative case.	5. Higher mortality rate after three months growing period.
		6. Good growth rate in intensive culture.	
		7. The exoskeleton is relatively thin giving greater portion of edible meat to total weight.	
3..	*Penaeus merguiensis*	1. Fast growing tasty meat.	1. High mortality if cultured in salinity less than 5 ppt and higher than 40 ppt.
		2. High market price	2. Problem of cannibalism.
		3. Short growing period.	3. Dies in temperature higher than 34°C.
		4. Wild fry are usually abundant in estuaries near areas where the adults are present.	
		5. Gravid females are relatively easy to obtained from the wild	
		6. The exoskeleton is relatively thin giving greater portion of edible meat to total weight.	
4.	*Penaeus semisulcatus*	1. Artificial propagation of larvae is relatively easy.	1. It requires high salinity water
		2. High price and high market demand.	2. Slow growth rate
		3. Testy meat.	3. This species has not been successfully cultured to marketable size despite numerous attempts.

Chapter 10

Freshwater Prawn Farming

Freshwater prawn farming the world over has registered increase in the past decade. In India, a spurt in freshwater prawn farming activities can be seen in the recent years. The objective of the symposium was to evaluate globally the progress made and to critically analyze the constraints and shortcoming in freshwater prawn farming and research. K. V. Thomas, Minister for Fisheries and Tourism, Government of Kerala while inaugurating the symposium, stressed on the importance of freshwater prawn farming in India, especially in Kerala.

The current global status of freshwater prawn farming with comments on the statistical information was available. Going by statistical information of FAO for 2001, India produced 24,230 mt of the popular freshwater prawn variety *Macrobrachium rosenbergii* which is also known as Scampi, standing at the 3rd position after China and Vietnam, which produced 128,338 and 28,000 mt respectively. He predicted that national production of scampi in India will be more than 50,000 mt/year by 2010. (Michael New, European Aquaculture Society, UK, 2003).

Scampi is cultured in 34,630 ha area in the country. The average production per hectare ranges from 880 to 1250 kg. He has noted that 62 per cent of the scampi culture operation is in Andhra Pradesh. Presently 71 hatcheries are operating in various states supplying 183 billion scampi seeds to the farmers in India. The high cost of seed and feed is a problem facing scampi farming in India (J. Bojan, 2003)

The need for nation-wide promotion of giant freshwater prawn farming through survey of sites is assured aquaculture inputs and technical support for farming, processing and marketing. J. V. H. Dixtulu (Fishing Chimes, Visakhapatnam) The annual expansion rate of freshwater prawn farming in the world during the decade ending 2001 was estimated as 29 per cent and that between 1999 and 2001 as high as 48 per cent. He emphasized the requirement of establishing sustainable freshwater prawn farming systems as per the guidelines formulated by FAO and other agencies in order to prevent an unexpected collapse as in the case of shrimp farming.

Productivity varied from 1000 to 4500 kg/ha/year. Status of freshwater prawn culture in China with special reference to the high density culture of *M. nipponenus* prepared by Miao Weimin (Freshwater Fisheries Research Centre, China)

In India there was a spurt in the freshwater prawn in the freshwater prawn farming, activity in recent years resulting in production of 30,450 mt from 34,630 ha in 2002–03. He attributed this mainly to the availability of water bodies, establishment of hatcheries, production of low cost prawn feed and enthusiasm of entrepreneurs.(Bojan, 2003)

Regarding hatchery technology, Michael New, UK suggested the need for having small seasonal hatcheries on all farms or for co-operative seasonal hatcheries that serve a number of small local farms. Breeding programmes to improve performance, and possibly the production of hybrids that exhibit the favourable characteristics of more than one species, are desirable. There is also a need to protect the natural resources in a country like India where the major cultured species are indigenous.

In the case of *M. lamarrei* eggs were slightly elliptical and mature eggs measured about 0.54-0.64 mm in long axis. The colour of the eggs showed contrast from *M. rosenbergii* since they were dark green when immature and turned into light green and then transparent at the time of hatching. It was observed that the average incubation period in case of *M. lamarrei* was 16 days at 30°C which is 4-5 days lesser than that of *M.rosenbergii*. The larvae hatch as zoeae in the *Macrobrachium* spp. i.e., they typically utilize their thoracic appendages to swim, differing from panaeid shrimp which hatch as nauplii that are morphologically less developed and use their cephalic appendages to swim, Ismael and New (2000). The first stage zoeae are less than 2 mm long (from the tip of the rostrum to tip of the telson) in *M. rosenbergii* but the first stage larvae measure about 8 mm in *M. lamarrei*. The larvae of *M. lamarrei* possess sessile eyes and very small pleopod with unsegmented antenna and undeveloped uropod but in *M. rosenbergii* larvae possess only sessile eyes and all other characteristics are absent.

M. lamarrei does possess highly developed larval form at the time of hatching than in case of *M. rosenbergii. M. lamarrei* differs greatly with *M. rosenburgii* in the size of the larva and other characteristics. In case of *M.lamarrei* the second stage larvae collected after 12 hrs from the time of hatching, were having sessile eyes, more developed pleopods and also showed presence of telson and uropod but in *M. rosenbergii* eyes are sessile while the pleopods appear only after 7-10 days and uropod at 3-4 days and the telson at 5-8 days.

Dates of Beginning, Peak and End of Seasons

July to February is the main season for the species in the Gulf of Kutch area; its contribution to the fishery in different centers varies from 13.7 to 27.4 per cent. Slightly south of this area, along the Bombay coast, the species is available throughout the year and the peak season is from January to March. Here in 1952 to 1954 the species averaged 12.8 per cent of the prawn catch. In the Hooghly estuarine system the species is fished almost throughout the year. In the lower zone of the Hooghly and the lower Sunderbans, the bulk of the landings are during the winter months,

November to February. In the Matlah estuary, the fishery commences in August and continues until March (Rajyalakshmi, 1961). According to Subramanyam (1965), the season for the species in the Godavary estuary, on the east coast of India, is from March to June.

Polyculture and Integrated Culture

A considerable, but unquantified proportion of global freshwater prawn production comes from polyculture and integrated culture. No detailed recommendations for the polyculture of *Macrobrachium rosenbergii* with other species, or its integration into other farming activities, have been provided in this manual. This is because there is no single recommendable process. Many different management techniques are possible. It is hoped, however, that you will be stimulated by the examples given below to try polyculture with locally available species, as well as integration with other farming activities in your specific location. Further reading on this topic is available in Zimmermann and New (2000) and New (2000b).

Polyculture

Records exist of the polyculture of various Macrobrachium species in combination with single or multiple species of fish, including tilapias, common carp, Chinese carps, Indian carps, golden shiners, mullets, pacu, ornamental fish, and red swamp crayfish. Other combinations may be feasible.

The inclusion of freshwater prawns in a polyculture system almost always has synergistic beneficial effects, which include:

- More stable dissolved oxygen levels;
- The reduction of predators;
- Coprophagy (the consumption of fish faces by prawns), which increases the efficiency of feed;
- Greater total pond productivity (all species); and
- The potential to increase the total value of the crop by the inclusion of a high-value species.

However, the management of a polyculture system is more complex. This particularly applies to the harvesting of prawns. Some large fish can be cull-harvested from a polyculture pond but this interferes with the culture of the prawns. Prawn-fish polyculture systems are therefore normally batch-harvested. It is difficult to synchronize fish production with prawn production to achieve the maximum production of marketable animals. For this reason, most polyculture systems involving freshwater prawns concentrate management on the production of the fish and regard the harvested prawns as a high-value bonus.

The addition of prawns to a fish polyculture system does not normally reduce the quantity of fish produced. On the other hand, the addition of fish to a prawn monoculture system markedly increases total pond yield but may reduce the amount of prawns below that achievable through monoculture. Some problems have been reported. For example, tilapia which was inadvertently introduced into prawn ponds in Hawaii was described as a pest, causing serious competition for food. Escaped tilapia, which had been grown in cages in freshwater prawn ponds in Puerto Rico took years to eradicate. However, this problem could be avoided by the use of artificially incubated sex-reversed caged tilapia. The monoculture versus polyculture decision is site-specific and depends on economic factors, namely balancing the relative market values of the various species with the costs of a more complex management system.

Fish are faster than prawns in accessing any supplemental feed which is presented, so the feeding for polyculture systems is normally directed at the fish, not at the prawns. The prawns consume feed which falls to the bottom of the pond, as well as fish faeces and nutrients derived from detritus. Though commercial fish feeds are sometimes applied, tropical polyculture systems often use simple mixtures of rice bran with plant oilcakes, such as mustard and groundnut. Since there are so many potential combinations of fish and freshwater prawns, it is impossible to give firm guidelines on management in this manual. The culture cycles ranged from 3 to 6 months and the water temperatures were 26 ± 4°C. This table also gives an indication of the productivity obtainable. The results of other published studies on prawn and fish polyculture have been reviewed by Zimmermann and New (2000). Much of the output of *M. rosenbergii* produced in China comes from polyculture systems.

Outdoor (Secondary) Nurseries

Outdoor nurseries are similar to grow-out ponds and can be stocked with newly metamorphosed postlarvae (PL) from hatcheries, or with juveniles from primary nurseries. In the secondary nurseries you can rear them until they reach 0.8-2.0 g. This may take 4-10 weeks, depending on the source used for stocking.

General Management and Water Quality

Supplemental aeration is ideal but may be too expensive. If you use substrates you can improve performance. This is discussed in the section of this manual that deals with grow-out in temperate areas. Between rearing cycles, you should disinfect your nursery ponds by applying 1 mt/ha of burnt lime or 1.5 mt/ha of hydrated lime to kill unwanted pathogens. Water quality and its management in secondary nursery systems is similar to that employed for grow-out.

Systems of Management in Grow-Out Ponds for Freshwater Prawns

1) The Continuous System

This involves regular stocking of PL and the culling (selective harvesting) of market sized prawns. There is no definable 'cycle' of operation and the ponds are therefore only drained occasionally. One of the problems of this form of culture, which can only be practiced where there is year-round water availability and its temperature remains at the optimum level, is that predators and competitors tend to become established. Also, unless the culling process is extremely efficient, large dominant prawns remain and have a negative impact on the postlarvae which are introduced at subsequent stocking occasions. This results in a lower average growth rate. The decline in total pond productivity (yield) that has been observed when this system has been used for a long time is, however, not confined to this management system and may also be a function of genetic degradation, as discussed elsewhere in this manual. This results in less and less satisfactory animals being stocked. There are other major problems which occur when ponds are continuously operated.

The various real or perceived problems of the continuous management system were not obvious when the original FAO manual on freshwater prawn farming was revised. In its first English edition (New and Singholka 1982) the authors mentioned the continuous system but specifically omitted any details about it because they thought that it might be wrongly interpreted as a recommendation for application in all circumstances. However, following requests for details, the authors included detailed information on this topic in its revision (New and Singholka 1985); this information was also included in its French and Spanish editions. In view of the experience gained in the 17 years since this information was published, the long-term continuous management system is not now recommended and the annex providing details about it has therefore been omitted in the current manual.

2) The Batch System

At the other extreme to the continuous system is the batch system, which consists of stocking each pond once, allowing the animals to grow until prawns achieve the average market size, and then totally draining and harvesting it. This reduces predator and competitor problems. However, although dominant prawns cannot impact on newly-stocked PL (because there is only a single stocking), the problem known as heterogeneous individual growth (HIG) remains. This term (HIG) refers to the fact that freshwater prawns do not all grow at the same rate. Some grow much faster, tend to become dominant, and cause stunted growth in other prawns.

3) The Modified Batch System

This more complex management regime was developed in Puerto Rico (Alston and Sampaio 2000) and involved three phases. After 60–90 days in a 1,000 m^2 nursery pond stocked at 200 to 400 PL/m^2, 0.3–0.5 g juveniles were harvested and stocked at 20-30/m^2 into empty (without any existing prawns present) 'juvenile' ponds. After another 2-3 months, seine harvesting of these juvenile ponds began and was repeated every month after this. These harvests removed animals of 9 to 15g, which were then stocked into ponds with existing populations of small prawns. The juvenile ponds were themselves then either converted to adult ponds, to allow remaining animals to grow to marketable size, or were drained and refilled for further use. According to the owner of the farm (J. Glude, pers. comm. 1998), drain-harvesting into a catch basin, instead of seining, would have reduced labour costs and increased survival. Further advantages could have been obtained if postlarvae had been held longer in the nursery ponds and then graded into at least two size groups before stocking into juvenile ponds.

Monoculture in Temperate Zones

Special conditions apply to the culture of freshwater prawns in 'temperate zones', because of the short period during which the grow-out phase can be operated (usually about 4-5 months). A captive brood stock has to be maintained, an indoor heated hatchery operated, and postlarvae reared to juvenile size in indoor nurseries. This is necessary to provide larger animals for stocking grow-out facilities as soon as possible in the season, thus enabling the longest possible growing period. The highest possible average weight at harvest can be achieved in this way. These topics are fully discussed by Tidwell and D'Abramo (2000).

In the temperate zone culture of freshwater prawns, natural food, enhanced by feeding or fertilization, is used until the prawn biomass reaches about 200-250 kg/ ha. After that, supplemental feeding is essential. The use of a range of diets, both for initial fertilization and as a feed for prawns, is discussed later in this manual. Aeration may be necessary to maintain satisfactory levels of dissolved oxygen. Although average water temperatures during grow-out in temperate zones may be much lower than in the tropics, the maximum may become quite high (over 30°C).

Basic Requirements and Facilities for Freshwater Culture

The selection of sites for indoor nurseries should follow the same pattern as for hatcheries. Site selection for outdoor nursery facilities should be similar to that for grow-out ponds (Figures 10.1).

Figure 10.1: Holding Tank: Site for Outdoor Nursery

There are no special requirements, except that they must have supplies for freshwater and air. You can use branches and nets suspended from floats in the tanks (both referred to as 'substrates' in this manual) to increase the surface area available to the PL but this may make the normal operations of feeding, cleaning etc. (similar to hatchery operation) more difficult. Figure 10.2 shows PL utilising a nylon screen inside a holding tank.

After rearing your freshwater prawns in your hatchery, you need to be able to hold them until you stock them in your ponds or sell them to other people. 50 m^3 concrete tanks are convenient for holding postlarvae (PL) prior to transport for stocking in ponds. However, you can also use other sizes and types of tanks, similar to hatchery tanks. There are no special requirements, except that they must have supplies for freshwater and air. You can use branches and nets suspended from

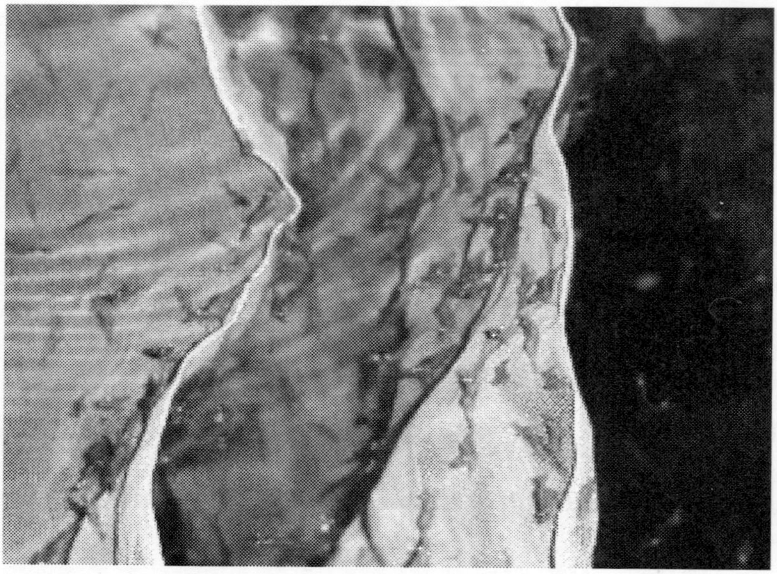

Figure 10.2: Postlarval Freshwater Prawns can Use a Nylon Screen as an Additional Surface Srea in Holding Tanks

floats in the tanks (both referred to as 'substrates' in this manual) to increase the surface area available to the PL but this may make the normal operations of feeding, cleaning etc. (similar to hatchery operation) more difficult.(Figure 10.2) shows PL utilising a nylon screen inside a holding tank.

Indoor (Primary) Nursery Facilities

Tanks for indoor freshwater prawn nurseries can be constructed from concrete or fiber-glass. The use of asbestos cement tanks is *not* recommended. The shape of nursery tanks is not important and their size, usually from 10 to 50 m^2 with a water depth of 1 m, depends on the area of the outdoor ponds which you (or those you sell juveniles to) are eventually going to stock with your product. You can use artificial substrates of various designs and materials to increase surface area; these provide shelter and increase survival rates.

The water supplies for indoor nurseries can be flow-through or recirculating. For flow-through, allow the water to continuously enter from above the tank water level and exit from the lowest part of the tank through a vertical standpipe with an outside sleeve (pipe with a larger diameter) extending higher than the water surface. Cover the standpipes with a 1.0 mm mesh screen to prevent PL and juveniles from escaping. This drainage system draws water from the tank bottom where food waste and detritus settle (Figure 10.3).

If you wish to operate your primary nursery tanks on a recirculation system it can be similar to those used in recirculation hatcheries.

Figure10.3: Overhead Air and Water Distribution Systems are Used to Supply these Indoor Nursery Tanks (USA)

Prawns tend to use the edges of substrates, whether they be natural (*e.g.* leaves, branches) or artificial. Layers of mesh can therefore be used to increase the amount of surface edges available to the prawns in both vertical and horizontal planes. Plastic netting can be placed in several layers over wood, aluminium, or PVC pipe frames. Suspend these 10 cm above the bottom of the tank, so that it can be cleaned. Hanging the mesh vertically allows the prawns easy access to the tank bottom to search for feed and allows detritus to fall to the tank bottom, where it can be siphoned out. Other substrate designs are feasible but you must be careful to think about the effect of the substrates that you use on your ability to manage the tanks (feeding, observation, cleaning, etc.).

Nursery Cages

Postlarvae can also be nursed in cages but research on the best ways to manage cages for this purpose is not yet complete enough to recommend this in this manual. The scientific literature describes (for example) the use of 1 m^2 cylindrical cages constructed from 0.64 cm mesh galvanized hardware placed in the mud bottom of a pond (although these were described as cages, they are really pens). In this case, animals weighing almost 2 g were stocked. PL or juveniles would require a much smaller mesh size. 2 x 1 x 1 m cages made of iron with a 1.0 mm nylon mesh, suspended above the pond bottom (real cages) have also been described for stocking PL. Such fine mesh would need careful cleaning to ensure proper water exchange and the mesh size would need to be increased as the animals grow.

Outdoor (Secondary) Nurseries

Outdoor nurseries are similar to grow-out ponds and can be stocked with newly metamorphosed postlarvae (PL) from hatcheries, or with juveniles from primary

Figure 10.4: Nursery Tank

nurseries. In the secondary nurseries you can rear them until they reach 0.8-2.0 g. This may take 4-10 weeks, depending on the source used for stocking.

Filling and Stocking Ponds

Postlarvae are especially sensitive to the effects of algal blooms (excessive quantities of algae; the methods for controlling these are described in the grow-out sections of this manual) and high pH. Some operators allow natural food to build up, and pH to stabilize, over a period of 10-14 days after pond filling, before stocking. However, this causes predators and competitors to become established, with consequential effects on prawn survival. It is not recommended in this manual. Stock your ponds immediately (within 2 days) after filling them with filtered water, which has no predators and causes no photosynthetically-induced pH changes. You may not get quite such high initial prawn growth rates from ponds with little natural food but increased survival will outweigh this factor.

It is difficult to recommend exact initial stocking rate in outdoor nurseries because this is site-specific (*e.g.* temperature profile over time; prawn size at stocking; the length of time the prawns will remain in the nursery ponds; the presence or absence of substrates and aeration; the amount of predation and whether the ponds are covered to reduce this, etc.). If your nursery ponds have no substrates or aeration, do not exceed stocking rates of either 1 000/m^2 of PL, 200/m^2 of small juveniles (0.02 g), or 75/m^2 of 0.3-0.4 g juveniles. You can increase these stocking densities if you provide substrates, aeration and predator protection.

Feeding Strategy

Normally, outdoor nurseries use grow-out feeds, which may be either bought from commercial sources or made on the farm. Feeding once or twice per day is sufficient. You can also add some supplementary fresh feeds but you must be careful about water quality problems, as mentioned earlier in this manual. The quantity of feed should be adjusted after observing the actual consumption but should be about 10-20 per cent of the total weight of the prawns in the pond.

Survival and Growth Rate

Some mortality (10-20 per cent) will occur soon after PLs are stocked, even when the conditions are ideal. To determine the rate of survival, a sub-sample of the animals should be evaluated within a mesh bag (cage) suspended above the pond bottom. If the survival is poor after 24 to 48 h, stock more PL, unless poor water quality was the cause of the mortalities. Over-stocking is much easier to remedy later than under-stocking. Total survival from stocking (or re-stocking) until removal from the nursery ponds should be at least 75 per cent. The weight of the prawns at the end of the outdoor nursery period should be about 0.8-2.0 g, but the time taken to reach those sizes will depend on local conditions.

Harvesting, Grading and Transport

You can harvest juvenile prawns by seining your ponds two or three times with a 5 to 6 mm mesh seine, or by emptying them completely. If you use drainage, the juveniles should be trapped in a large catch basin or box at the end of the outlet. The catch structure should not stress the prawns. Polypropylene boxes or tanks filled with water from the nursery pond and kept aerated, can be used to transport the juveniles to the grow-out ponds if they are close by. More care needs to be taken if you are taking them to sites further away (see the section on transport of PL). You should estimate the numbers of juveniles harvested and transported to the grow-out ponds. There are some advantages in grading the juveniles into two or three groups, depending on their average weight, before stocking them into separate grow-out facilities. This decreases competition in grow-out ponds by reducing HIG and increases productivity. Grading is discussed in the grow-out section of this manual.

Other Systems

Multi-phase Nursery Systems

A number of multi-phase nursery systems have been developed for research and commercial systems. The simplest system, developed in Israel, involved stocking ponds with newly metamorphosed PL at 1 000 to 10 000/m^3 in the first phase. 15 to 30 days later they were transferred into second phase ponds at 100 to 200/m^2 for a further 60 day period. Survival rates of 92 per cent (phase 1) and 85 per cent (phase 2) were achieved. Other multiphase systems have been modelled or commercially applied but are not described here because they are complex and/or their true value has not been adequately demonstrated. Further details can be found in Alston and Sampaio (2000).

Nursing in Cages

Some research work on nursing postlarvae in cages has been carried out. These involved the rearing of newly metamorphosed PL (stocked at initial densities of 2-10 PL/L) in 2 x 1 x 1 m and 1 x 0.5 x 0.7 m cages for 20 days. The prawns grew to 50 mg at the lowest and 30 mg at the highest stocking densities. Survival rates were not significantly different up to a stocking density of 8 PL/L. Another experiment involved the stocking of 0.19 g juveniles at 50 and 100 prawns/m^2; the prawns grew to 3.2 and 2.4 g in 2 months, with survival rates of 86 per cent and 75 per cent, respectively. A further experiment stocked with 0.05g juveniles at 100-800/m^2 and reared them for 60 days in similar cages; the prawns reached 0.35-0.79g.

Research on this topic is sparse and the results need to be confirmed on commercial farms. At the time this manual is being prepared, the operation of cage nurseries is therefore not yet recommended for commercial practice. This does not mean that the practice has no potential value, simply that no clear recommendations can be made at this time.

Defining the Pond

Choosing its Area and Shape

If you are going to use seining for harvesting, which is often practised in freshwater prawn farming because of the necessity to cull out larger animals (and sometimes to separate females from males, when they have different values) before the final harvest, rectangular ponds are the most suitable shape. The maximum width for this type of management should not be wider than the space through which a seine can be conveniently drawn from one end of the pond to the other by manual labour. A convenient width is 30 m. In practice, of course, wider ponds can also be seined but not so efficiently as narrow ones. The length of the pond depends partly on the topography of the site and partly on the pond size and farm layout chosen. It is best to standardize the width of ponds; otherwise a range of different seine nets will be required for harvesting.

The most easily managed pond sizes range between 0.2 ha and 1.6 ha, with most farms having ponds around 0.2-0.6 ha. If kept to a 30 m width, a 0.6 ha pond will be 200 m long. Narrow ponds should be oriented so that the prevailing wind (which enhances the dissolved oxygen content of the water) blows down the long axis towards the drain end, to lessen the area of the pond bank subject to wave erosion.

Large ponds are normally wider than 30 m and often drained for harvesting. If the total harvest is going to be taken at one time, the size of the pond should be influenced by the maximum weight of prawns that the market will accept at one time without price deflation. For example, if a quantity greater than 300 kg of freshwater prawns would swamp your market and reduce prices it would be pointless to have a drainable pond greater than 0.15 ha in area (assuming a productivity of 2 mt/ha/crop).

The bottoms of grow-out and nursery ponds need to be sloped towards the drainage point and to be smooth; this increases the efficiency of both drain-harvesting and seine-harvesting

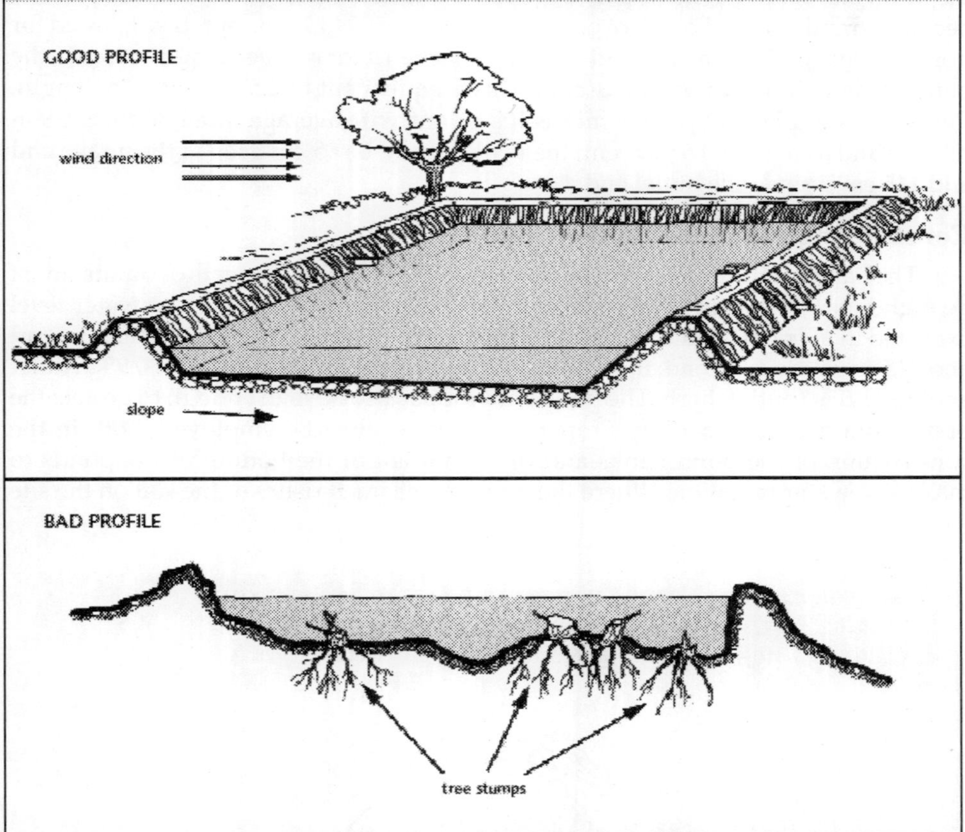

Figure 10.5: Bottoms of Grow-Out and Nursery Pond (*Source*: Emanula D'Antoni)

Choosing its Depth

The average depth of water in freshwater prawn ponds in tropical areas should be about 0.9 m, with a minimum of 0.75 m and a maximum of 1.2 m. Deeper ponds (an average of 1.2-1.4 m) are used in colder areas to maintain more stable water temperatures. However, deeper ponds are difficult to manage. Even if you have ponds of the recommended average depth you may have to drain or pump out part of the water to facilitate seining operations at the deep end. In the cool season, the temperature of the water at the bottom of deep ponds may become low enough the reduce feed consumption by the prawns. On the other hand, the water in shallower ponds may become too hot for the prawns in the hot season and the water may be quite clear, exposing the prawns to greater predation. Shallow ponds also tend to support the growth of rooted aquatic plants and are not recommended.

The bottom of the pond must be smooth (Figure 10.6). There must be no projecting rocks or tree stumps in it; these would prevent efficient seining and damage nets. The pond bottom must slope gradually and smoothly from the water intake end towards the drain end so that, when drained, pockets of undrainable water in which prawns become stranded and die do not occur. A slope of 1:500 (0.2 per cent) is suggested for ponds 0.4 ha or more in area and 1:200 (0.5 per cent) for smaller ponds towards the outlet, where drain harvesting occurs. This is equivalent to 2-5 cm per 10 m length. Thus (for example), in a pond which is 100 m long with average water depth of 0.9 m (90 cm) and a slope of 0.5 per cent, the water would be 65 cm deep at the intake end and 115 cm deep at the outlet end.

Constructing the Pond Banks

The banks of the ponds (sometimes referred to as embankments or bunds) must be high enough to provide a freeboard of 30-60 cm above the highest water level expected in the pond. Thus, in a pond with a water depth of 65 cm at the shallow end and 115 cm at the deep end, the total bank height must be a minimum of 0.95 m (inlet end) to 1.45 m (outlet) high. The pond banks must also be high enough to protect the pond from exterior flooding. Proper compaction must be employed, both in the construction of the pond banks and the treatment of the bottom of the ponds to maximize water retention. Where the retention characteristics of the soil on the site

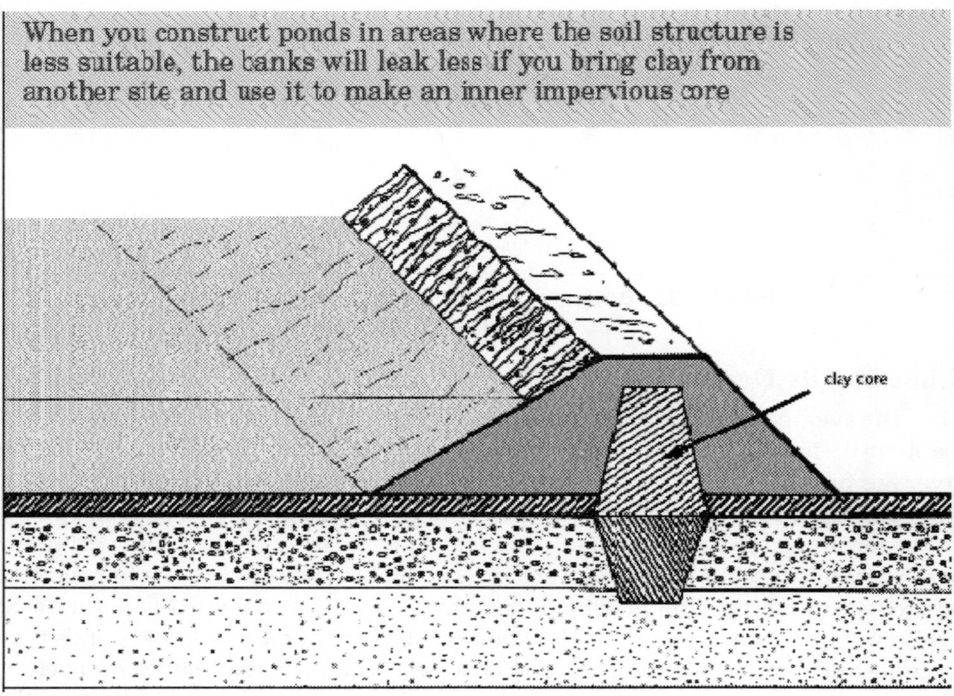

Figure 10.6: Construction of Pond
(*Source*: Emanula D'Antoni)

are not good, a core of impervious material brought from outside the site must be provided during pond bank construction. This core should extend below the level of the bottom of the pond

For ease of management the internal slope of the pond bank should be 3:1 but it may need to be as high as 4:1 in sandy areas to minimize erosion (and the consequential need for repairs). In highly stable soils the inner slope should not be less than 2.5:1 (Figure 10.6). Very small ponds with almost vertical sides may be constructed for artisanal purposes in floodplains having very sticky and impermeable clay soils. Fruit trees or other crops may be planted on the pond banks. Sometimes attempts are made to protect the banks from excessive erosion by stakes (for example). Having vertical or near-vertical pond banks almost certainly leads to rapid erosion problems. This means that a lot of maintenance will be required and they are certainly not recommended for larger commercial-scale farms.

The external pond bank slope should preferably be at least 2.5:1 and never less than 1.5:1, even for highly stable soils. Properly constructed pond banks are more expensive and use more land but failure to build them correctly may result in severe erosion (Figure 10.7). After construction, you should plant the banks with fast growing grass (*e.g.* Phyla nodifera), kudzu (a woody vine) or taro (dasheen), to help prevent erosion.

Figure 10.7: Pond with Grass Turfs

The banks of these ponds have had grass turfs laid on them.

Figures 10.8 and 10.9 illustrate pond banks overlaid with turf.

Pond bank planted with coconut, grass, and banana; besides stabilising the bank this is a form of integrated farming.

Figure 10.8: Pond Banks Over Laid with Turf

Figure 10.9: Pond Banks Over Laid with Turf

It is not normal to treat the water entering freshwater prawn ponds except to screen it to prevent entry of predators. Screening is not necessary where the water supply is piped from a well or a spring but is essential where surface water or open channel distribution is used. Well water requires aeration by cascading (Figure 10.10) or by supplying it above pond water level to re-establish gas equilibrium, as it is often initially very low in dissolved oxygen content. There are many alternative methods of screening. Crude screening excludes adults and fingerlings of unwanted species but not their eggs or larvae.

The dissolved oxygen levels of incoming pond water can be increased by rippling and cascading

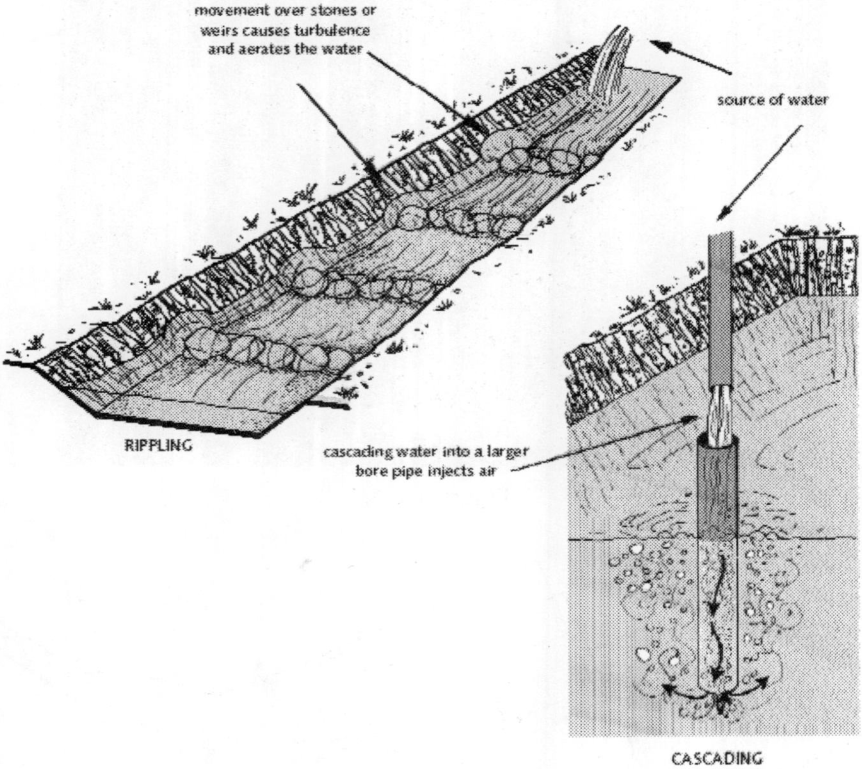

Figure 10.10: Gravel Filter for Exclude Fish Eggs and Larvae (*Source*: Emanula D'Antoni)

Figure 10.11 shows a simple gravel filter which will exclude fish eggs and larvae as well.

The flow of water into each pond must be controlled by valves, weirs, stop-logs or plugs. Many farms (due to site, technical, or financial limitations) have water inlets below the pond water level and receive water from an inlet channel (or a low-lying area such as paddy fields) with the same water level as the pond. In some cases ponds are directly interconnected. These farms produce freshwater prawns, often profitably. However, the use of such water supplies increases risk substantially; a proper water distribution system is essential for reliable high production.

There are many ways to control entering water in your farm. Some of them are shown in Figures 10.13 and 10.14.

Figure 10.11

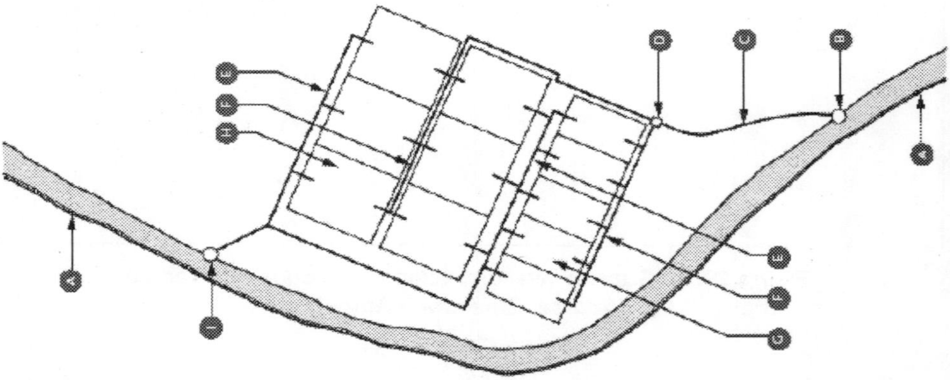

Figure 10.12: Water Management in Pond.
A: River; B: Main water intake; C: Main water feeder canals; D: T-box division; E: Drainage canal; F: Inlet feeder canal; G: Series of small ponds; H: Series of larger ponds; I: Drainage canal to discharge to river

Discharging Water from the Ponds

It is preferable to be able to drain ponds by gravity than to have to pump the water out and, where this is possible, you should construct a 'monk' or sluice gate outlet structure. These structures (Figure 10.13 and 10.14) will allow you easily to control water depth and drainage speed and can be screened to prevent the loss of stock. In flow-through water management, water is continually flowing through this

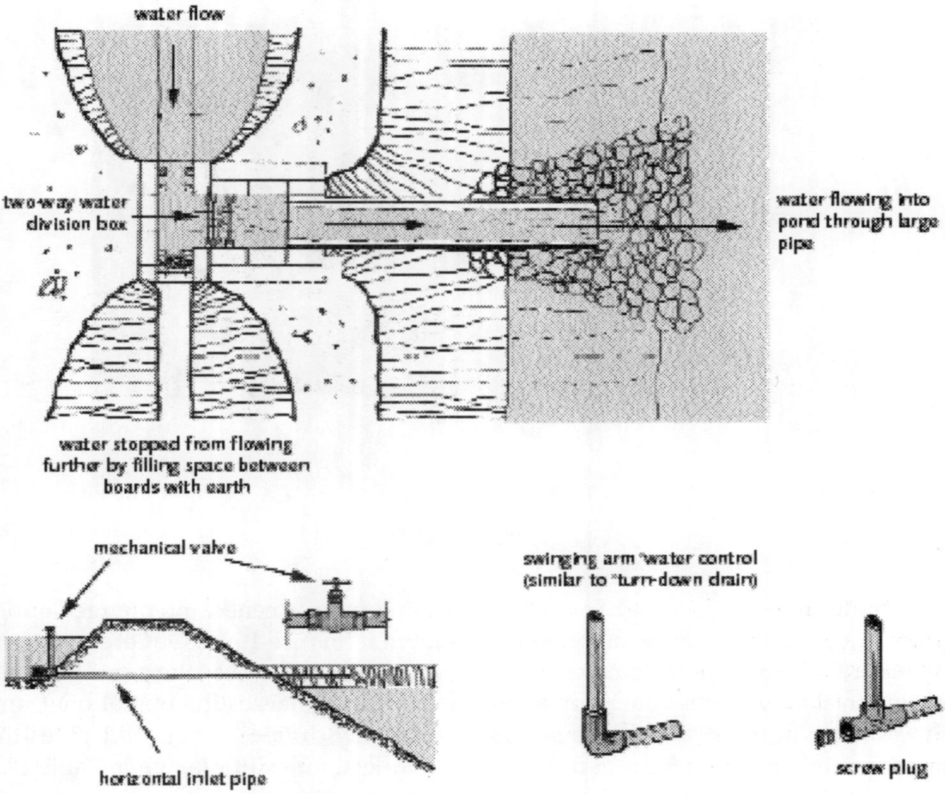

Figure 10.13: Control on Entering Water

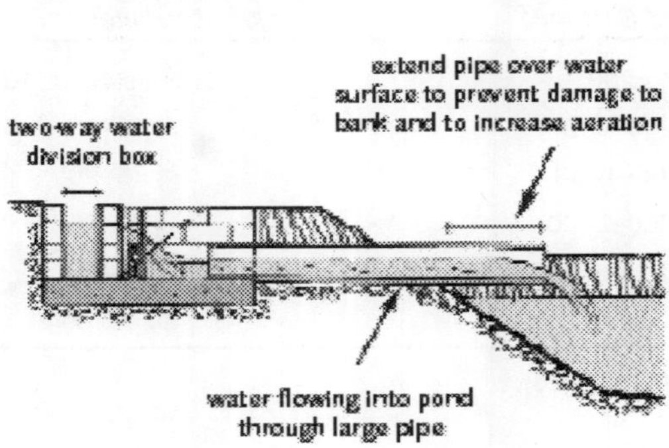

Figure 10.14: Control on Water Flowing

structure. The monk allows you to totally drain your pond and, more importantly, enables you to control the water level during seine harvesting operations, flushing and water circulation.

Figure 10.15

Static (non-flow-through) ponds can have a simple screened and plugged outlet pipe or a turn-down drain, as previously shown in Figure 10.15. Outlet structures, whether they are pipes or monks, must be carefully sized so that the pond does not drain too slowly (to prevent poor water quality during harvesting operations), and they should be sited so that the pond can be totally drained. Table 10.1 gives the appropriate pipe sizes for ponds with monk outlets, for example, while Table 10.2 gives the time to drain a pond under various circumstances.

Table 10.1: Sizes of Outlet Pipes for Ponds with Monks

Pond Size (m^2)	*Inside Diameter of Pipe (cm)*
<200	Not less than 10
200–400	10–15
400–1000	15–20
1000–2000	20–25
2000–5000	25–30
>5 000	40 or more

Source: Derived from FAO (1992b).

In addition to its ponds and its water distribution systems a freshwater prawn farm has the following equipment and facility requirements:

- ☆ Power;
- ☆ Roads and access paths;
- ☆ Accommodation: every farm should have accommodation for some of its workers to live on site;

- ☆ Fencing: a perimeter fence and, on larger farms, lighting, to deter poachers (human predation);
- ☆ Storage facilities: dry storage is needed for feeds (or ingredients), chemicals, nets, etc.;
- ☆ Feed distribution and monitoring equipment;
- ☆ Nets;
- ☆ Water quality monitoring equipment;
- ☆ Predator protection; and
- ☆ Transport: larger farms will need trucks for prawn distribution and feed collection.

Table 10.2: Time taken to Drain Ponds (in hours) with Different Drain Pipe Sizes

Inside Diameter of Pipe (cm)	*Pond Area (ha)*						
	0.1	*0.2*	*0.5*	*1.0*	*2*	*5*	*10*
10	96	192	480				
20	15	30	75	150	300		
50	1.5	3.5	8	16	32	80	
100			2	3.5	7	17.5	35

Note: These figures assume an initial water depth of 1 M with pipe velocity limited to 1 M/second; If you have two pipes, the time for draining in each case would be halved.

Source: Derived from FAO (1995).

Preparing your Pond

Before you stock your pond you need to prepare it. After the final harvest of the last batch of prawns that you reared, the pond should be drained to remove all predators. Make any necessary repairs to the pond banks and the major structures at this time. Check all inlet and outlet screens. Completely dry the pond for 2-3 weeks (this may not be possible between every cycle, for example in the rainy season, but should be done at least once per year). It is not normally necessary to remove pond sediments from freshwater prawn ponds after every cycle. The sediment consists of particles contained in the incoming water, the effects of erosion, the remains of dead pond organisms, prawn faeces, remnants of feed, and exoskeletons cast during prawn moulting. One of the effects of a heavy sediment build-up is a decrease in the volume of water available for the stocked prawns to occupy. Scraping the bottom of the pond can be used to remove sediment but care must be taken not to place the excavated sediment where it will wash back into the pond or supply/discharge canals when it rains, or cause a local environmental problem. Site-specific means of sediment removal need to be developed. However, if there is no opportunity to place the sediment elsewhere, it can be spread in a thin layer over the pond bank surfaces and allowed to dry until it cracks.

You should till (harrow) the bottom of your ponds during the drying period to increase the oxygen content of the soil, especially if it has a heavy texture (clays and clay loams). A disc harrow is the best equipment to use and tilling should take place while the soil is still wet but is dry enough to support the weight of the tractor. Where there has been a severe disease problem in the previous crop, you should spread 1 000 kg/ha of agricultural limestone ($CaCO_3$) or 1 500 kg/ha of hydrated lime [sometimes called slaked lime–$Ca(OH)_2$]. It is better if you use agricultural limestone. The use of slaked lime, or quick lime (CaO) may increase the subsequent pH of the water above tolerance limits if prawns are stocked (as is recommended for other reasons later) soon after the ponds are filled.

Aeration

Most prawn farms use water exchange to keep dissolved oxygen levels high, as well as curing other water quality problems. When the original FAO manual on freshwater prawn farming was written in 1982 it was pointed out that the dissolved oxygen level of incoming water can be enhanced if ripples are built into gravity inflow channels and water is injected into the ponds above water level. It was also noted that permanent aeration equipment was not normally provided for many freshwater prawn grow-out ponds but that equipment for emergency aeration was useful in times of oxygen depletion (perhaps caused by an algal crash). However, since that time, aeration has become more common place in freshwater prawn farming because the higher stocking densities that are used in some grow-out systems and nurseries increase oxygen demand. Paddlewheels (Figure 10.16) are the most efficient method of increasing dissolved oxygen levels in pond water.

Figure 10.16: Using Paddlewheel Aerators keeps the Dissolved Oxygen Level High Enough to Increase Stocking Levels

Table 10.3: Oxygen Transfer Efficiencies of Basic Types of Aerator

Type of Aerator	*Average Oxygen Transfer Efficiency (kg O_2/kWhr)*
Paddlewheels	2.13
Propeller-aspirator pumps	1.58
Vertical pumps	1.28
Pump sprayers	1.28
Diffused air systems	0.97

Source: Boyd (1990).

If your pond has previously been stocked with fish and you want to convert it to freshwater prawn culture, or if a lot of fish were present during your last prawn grow-out season, treat it with a piscicide after harvesting and while it still has water in it. Rotenone or tea seed cake are commonly used for eradicating unwanted fish between cycles. They are effective if spread evenly throughout the pond. However, the use of rotenone is banned in some countries because of environmental concerns: check before you use it.

Acid soils may cause your pond rearing water to be too acid for good prawn productivity. These soils need treatment to improve the alkalinity of pond water. Liming will be necessary if the water in your pond is pH 6.5 or below at sunrise. If it is necessary to treat the soil, you must apply the lime before the ponds completely dry out, so that it dissolves and penetrates the soil. Routine liming should be sufficient to increase total alkalinity to about 40 mg/L. The quantity of lime required depends on the type of soil and the pH. Agricultural limestone is the best compound to use for increasing alkalinity.

Table 10.4 shows the quantity of lime to use in treating pond bottoms between cycles. Spread the limestone uniformly before fertilizers are applied. Liming may be necessary every time the pond is drained if it is managed with a rapid water exchange. Judge the need by testing the water before draining. If the pond water contains less than 30-40 mg/L of alkalinity it will be necessary to lime. If it is more than 60 mg/L it should not be limed.

Some soils may benefit from the application of nitrates to oxidize the soil and aid the decomposition of organic matter where pond bottoms cannot be completely dried out. For most ponds 150-200 kg/ha of sodium nitrate would be sufficient. Calcium peroxide is also sometimes used for this purpose but is less efficient and is not recommended.

Some farms use organic fertilisation. Manure is used for fertilising ponds, before and during the rearing cycle, where freshwater prawns are grown with silver and bighead carps in China. In Brazil, freshwater prawn ponds are often fertilized between cycles using 1000–3000 kg/ha of cattle man-urea or other organic material. This increases the benthic fauna, which become an important feed for PL and juveniles. However, this practice is not encouraged. If you are really convinced that organic fertilisation between cycles is helpful, use plant meals, such as soybean meal or rice bran, not animal manures. Generally, the productivity of ponds improves as they get older and as a rich bottom area and grassy banks are established.

Table 10.4: Lime Requirements for Treating the Bottom of Ponds Between Cycles

Soil pH	*Agricultural Limestone Requirement (mt/ha AS $CaCO_3$)*		
	Clays or Heavy Loams	*Sandy Loams*	*Sand*
<4.0	14.32	7.16	4.48
4.0–4.5	10.74	5.37	4.48
4.6–5.0	8.95	4.48	3.58
5.1–5.5	5.37	3.58	1.79
5.6–6.0	3.58	1.79	0.90
6.1–6.5	1.79	1.79	nil
>6.5	nil	nil	nil

Source: Derived from Boyed and Tucker, 1998.

Stocking

It is better to stock ponds immediately after filling them with filtered water. This has no predators and causes no photosynthetically-induced pH changes. There may be a slight reduction in growth from the initial lack of natural food, but increased survival will outweigh this factor. Stocking the ponds quickly reduces the amount of competitors and predators, which have less time to become established. Often postlarvae (only about a week or two old after metamorphosis) are used to stock grow-out ponds, where they will remain until harvesting. Some farmers prefer to use PL reared in a simple (in contrast to a sophisticated) hatchery, believing them to be hardier because the strongest have been naturally selected. Juveniles are more tolerant of high pH and ammonia than PL and there are some advantages in stocking juveniles instead of PL, even in tropical areas. Juveniles are more expensive to produce in nurseries, or to purchase from others, but the improved grow-out survival and shorter time to marketable size achieved should more than balance this out.

On arrival of PL at the pond bank you should take great care to acclimatize the PL to the temperature of the pond water by floating the transport bags in the pond for 15 minutes (Figure 10.17) before emptying them into the water. Severe mortalities can be caused not only by thermal shock but also by sudden changes in pH. You should measure the pH of the pond water before stocking. If it is more than 0.5 pH units different from the pH in the PL holding tank or the nursery ponds, acclimatize the PL to this pH level slowly (over a one-day period) in the hatchery-nursery before transporting and stocking them at the grow-out site.

Due to the aggressive nature of *M. rosenbergii* and the hierarchy between males, stocking densities are much lower than in penaeid shrimp farms. Intensive farming is not possible due to the increased level of cannibalism, so all farms are either stocked semi-intensively (4 to 20 postlarvae per square meter) or, in extensive farms, at even

Figure 10.17

lower densities (1 to 4/m²). The management of the grow out ponds must take into account the growth characteristics of *M. rosenbergii*: the presence of blue-claw males inhibits the growth of small males, and delays the metamorphosis of Orange Claws males into blue-claws. Some farms fish off the largest prawns from the pond using seines to ensure a healthy composition of the pond's population, designed to optimize the yield, even if they employ batch harvesting. The heterogeneous individual growth of *M. rosenbergii* makes growth control necessary even if a pond is stocked newly, starting from scratch: some animals will grow faster than others and become dominant Blue claws, shunting the growth of other individuals.

The stocking rate you need to use depends on the size of the animals you will eventually be selling (and thus on the demand of the local, national, or international market that you are targeting), on the length of the growing season (determined by water availability and temperature), and on the management system you are using. Older ponds tend to be more productive than new ones. Your decisions about stocking rate should consider all these factors. Specific stocking densities are not recommended in this manual because no guarantee can be given that a certain quantity of prawns will be produced. Semi-intensive stocking rates vary between 4 and 20 PL/m² (40 000-200 000/ha). The lower stocking rates will tend to result in prawns of a larger average size. Higher stocking rates tend to result in greater total productivity (mt/ha/crop) but smaller average prawn size. The stocking rate you choose should therefore be adjusted according to your previous experience in your farm or locality, and the size of marketable animals desired. If you are stocking juveniles, there are some advantages in grading them before stocking, as discussed later.

The postlarvae (PL) you have purchased and brought to your pond-site will have been counted into the transport bags at the hatchery. You may wish to be present at that time to ensure fairness. Normally, hatcheries will put more PL into the bags, rather than underestimate them. However, if you are receiving PL without having seen them packed, it is advisable for you to count the contents of one or two bags at random to check the accuracy of the delivery. If a standard number of PL are packed into each transport bag the stocking procedure will be easier because it is only necessary to count the number of bags to achieve the desired density.

Holding Postlarvae before Sale

You should not retain PL in your holding tanks for more than a week or two prior to stocking in nursery facilities, grow-out ponds, or sale. The length of time you hold them depends on the demand for PL. If you have to retain the PL longer, you must reduce the density of animals. You can then sell the prawns as juveniles, which have a higher value than PL, reflecting the increased costs of holding them longer. Whilst the PL are in the holding tanks you must continue to exchange the rearing water (40-50 per cent every 2-3 days) and provide aeration. You can maintain PL at densities of up to 5 000 PL/m^2 for one week (Note: Once animals become PL, it is normal to refer to density on an area rather than a volume basis, that is per m^2, not per m^3), or up to 1 500-2 000 PL/m^2 for one month under these conditions. If you need to hold them for one month, you could improve survival if you reduce the density to 1000/m^2. Using substrates can help you maximize the stocking density, thus reducing other equipment and labour costs.

Transporting Postlarvae

Cooled and aerated fish transport tanks are ideal for transporting freshwater prawn postlarvae (PL) from the hatchery holding tanks to the pond site but they are rarely available or affordable. For journey times of up to one or two hours to the pond site, you can use aerated garbage cans. A 100 L trash can, holding 50 L of water, will hold 50 000 PL. You should insert baffles in the container to prevent excessive water movement during transport. Larger, open plastic tanks (1 m^3), containing about 500 L of aerated water, can hold about 500 000 PL during a short journey.

For longer distances you can use the method employed for transporting aquarium fish. Place them in double plastic bags containing 1/3 water and 2/3 air or oxygen. You can put about 250-400 PL in each liter of water. A 45 x 80 cm bag holding 8 L of water, for example, will take 2 000-3 000 PL. Higher or lower transport densities are used by some hatcheries. It is suggested that, if you have not done this before, you carry out some simple experiments to determine the optimum density for your conditions (length and duration of journey, climatic conditions, etc.). Round off the corners of the bags with rubber bands to prevent animals getting trapped there. Twist the top of the bag and bend it over, sealing it tightly with a rubber band after you have inflated it with air or oxygen.

These inflated bags can be used to transport PL very long distances (up to at least 16 hours travelling time by road). If you put them into insulated 'styrofoam' boxes you can ship PL by air most effectively. If they are in non-insulated boxes you can send them on night (cool) journeys by rail, for example. For long day-time (hot) journeys, you can stack the plastic bags on shelves in a home-made transport box mounted on a truck.

Lowering temperature during transport reduces metabolic activity and improves survival. You should also use water from the holding tank to fill the plastic transport bags. If you place the PL into 'new' water for transport, many will moult during the journey and many with be lost through cannibalism. Some hatcheries add a very small amount of seawater to the transport bags, claiming that survival rates are better in brackish water than in freshwater. You can try this for yourself but remember that increasing salinity will lower the dissolved oxygen content of the water and, if extreme, may make it necessary to adjust the salinity again before stocking the animals in the rearing water. A transport temperature of about 20-25°C is recommended for journeys of less than 6 hours. Keep it down to 20-22°C for longer trips. A one ton truck can transport up to 500 000 PL in plastic bags. You can keep the temperature steady throughout the truck by using electric fans driven from the truck engine or batteries. You can get very good transport survival in this way.

Pond Size

The most easily managed pond sizes range between 0.2 ha and 1.6 ha, with most farms having ponds around 0.2-0.6 ha. If kept to a 30 m width, a 0.6 ha pond will be 200 m long. Narrow ponds should be oriented so that the prevailing wind (which enhances the dissolved oxygen content of the water) blows down the long axis towards the drain end, to lessen the area of the pond bank subject to wave erosion.

Large ponds are normally wider than 30 m and often drained for harvesting. If the total harvest is going to be taken at one time, (batch management), the size of the pond should be influenced by the maximum weight of prawns that the market will accept at one time without price deflation. For example, if a quantity greater than 300 kg of freshwater prawns would swamp your market and reduce prices it would be pointless to have a drainable pond greater than 0.15 ha in area (assuming a productivity of 2 mt/ha/crop).

General Management

Tanks for holding postlarvae (PL) are a form of indoor nursery. However, their purpose is not really to grow the PL to a larger size before stocking but simply to be able to maintain them before sale. Sometimes hatcheries use holding tanks to acclimatize their PL to the pH and temperature of the rearing facilities where they are to be stocked. True indoor nurseries contain tanks where PL are intentionally reared to a larger size before transfer to outdoor nurseries or grow-out ponds.

Nursery tanks require aeration and may be operated as flow-through or recirculating systems, like hatcheries. Siphon the tank bottoms regularly to remove food wastes, faeces, and decomposing organic matter. Some nurseries allow organic matter to accumulate to enable PL to graze on 'lab-lab' but this may be difficult to manage without getting into water quality problems. Between cycles, you should dry out the tanks, disinfect them (the same way as hatchery tanks), and leave them to dry out for at least 48 h to minimize problems with pathogens. Do not forget to flush them out well to remove all traces of chlorine.

Construction

The external pond bank slope should preferably be at least 2.5:1 and never less than 1.5:1, even for highly stable soils. Properly constructed pond banks are more expensive and use more land but failure to build them correctly may result in severe erosion. After construction, you should plant the banks with fast growing grass (e.g. *Phyla nodifera*), kudzu (a woody vine) or taro (dasheen), to help prevent erosion.

The tops of the pond banks between ponds should be a minimum of 1 m wide to allow workers to walk round the ponds carrying feed and harvesting gear. Narrow pond banks with almost sheer sides are sometimes staked to prevent collapse but they need constant maintenance particularly if the site is sloping and the water level in adjacent ponds is different. Make sure that you have a pond bank width of at least 2-3 m at one side of the pond (usually the drain end or where harvest nets are to be beached) so that trucks can be brought to the pond side for delivering PL and feed and picking up harvested prawns, especially live prawns. On larger farms, particularly where mechanical broadcasting of feed is employed, you must provide a wide pond bank top (usually 3.5-4.0 m) on one of the long sides of the pond as well as at one end.

Semi-Intensive Monoculture in Tropical Zones

Although this section concentrates on the management of prawn monoculture in tropical zones, it also contains information which is equally applicable to other types of freshwater prawn rearing. Freshwater prawn monoculture can be extensive, semi-intensive or intensive but the definition of these terms is rather vague (Valenti and New 2000).

Stock Estimation during the Grow-out Period

Once the prawns have been put into a pond, it is extremely difficult to estimate growth rate or survival. Multiple seine and cast net samples seem the only reasonable method of following the growth rate of a crop of prawns. At least this enables a

Figure 10.18: Substrates have been Placed Vertically in this Temperate Zone Rearing Pond for *Macrobrachium rosenbergii* Culture (USA) (*Source*: Charles Weibel)

comparative estimate to be made. It is important that the method of sampling on each occasion is exactly the same (the same net; the same time of day; the same areas of the pond sampled; the same method of casting or pulling the net through the pond; and preferably the same person doing the sampling).

Another means of improving results in temperate freshwater prawn culture is to place artificial substrates in the ponds, which makes it feasible to increase stocking rates above the level recommended earlier for ponds without substrates. PVC fencing (such as is used to close off areas when roads are being resurfaced) forms an ideal substrate. This material can be expensive in some countries but the investment should be worthwhile, as the following information indicates. Substrate provision on a commercial scale has resulted in production and mean harvest size exceeding 1 800 kg/ha/crop and 35 g respectively, from a stocking rate of 4 PL/m^2, while yields exceeding 2,500 kg/ha/crop with average weights of >40 g have been consistently achieved at a stocking rate of 64,500/ha (Tidwell and D'Abramo, 2000). It is therefore suggested that you increase the stocking rate of juveniles from the 4/m^2 (40 000/ha) recommended earlier for use without substrates to 6.5/m^2 (65 000/ha) when you use either horizontal or vertical substrates. No extra labour (apart from its initial installation) is necessary if this form of substrate is used because it can be permanently installed in ponds equipped with catch-basins at the drain end. As the water is drained, prawns abandon the substrate and follow the water flow to the catch basin. You can spread the cost of the labour for installation, as well as the substrate material itself, over several production cycles. This new technology is still being developed but it clear that the use of substrates can markedly increase the productivity of freshwater prawn farming.

Without using substrates to increase productivity, a stocking rate of about 4 juveniles/m^2 (40 000/ha) is recommended for the monoculture of *Macrobrachium rosenbergii* in temperate zone ponds. There are some advantages in using larger juveniles for stocking. For example, it has been demonstrated that increasing the average stocking weight at 4 animals/m^2 from 0.17 g to 0.75 g increases production at harvest by nearly 30 per cent. However, this stocking size advantage does not apply indefinitely; research has shown that stocking 3 g animals did not improve production because the animals matured too rapidly.

Grading nursed juvenile prawns before stocking also has significant advantages. In temperate zones it has been found to increase average harvest size and total pond production. Size grading is a way of separating out the faster growing prawns and lowering the suppression of growth that they cause to other prawns; it can also result in improved feed conversion ratios (FCR). You may need to experiment to refine the technique.

Some notes on size grading are given hereunder, but you should note that this procedure is still in the developmental stage.

Size Grading

Place a floating grader box (these are commercially available for finfish grading) into a holding tank.

Trial and error is necessary to select the size of the grader bars to use. Your choice will depend on the size of the animals you want to grade. The efficiency of the procedure is a function of the average size of the population to be graded, how variable the size range of that population is, and the average weight and proportion of the total that you wish to achieve in the two graded portions. For example, prawns with an average weight of about 0.6 g can be separated into two portions with size #13 bar graders (13/64 inch; 5.16 mm) and #14 bar graders (14/64 inch; 5.55 mm).

Net the juveniles from the nursery tanks and pour them through the grader. Smaller animals will pass through the parallel bars of the grader and the larger ones will be retained above the bars.

The grading process can be speeded up by causing water movement (water flow, moving the box, air stones) but it is important not to overload the box because this will cause the juveniles to stack up and they will not actively try to swim out of the grader. Over-crowding may also cause mortalities to occur.

It is recommended that the juveniles be graded into equal (50:50) numbers of upper and lower sized individuals. These should be reared in separate ponds to achieve the best average yield of marketable prawns from the total area of the two ponds.

Culture Technology of Freshwater Prawn, *Macrobrachium rosenbergii*

The giant freshwater prawn, *Macrobrachium rosenbergii,* known as scampi in commercial parlance, is a highly valued delicious food and commands very good demand in both domestic and export market. Twelve species of freshwater prawns were identified; the largest among these was giant freshwater prawn *M. rosenbergii*. It is fast becoming one of the most important cultured species in the inland aquaculture

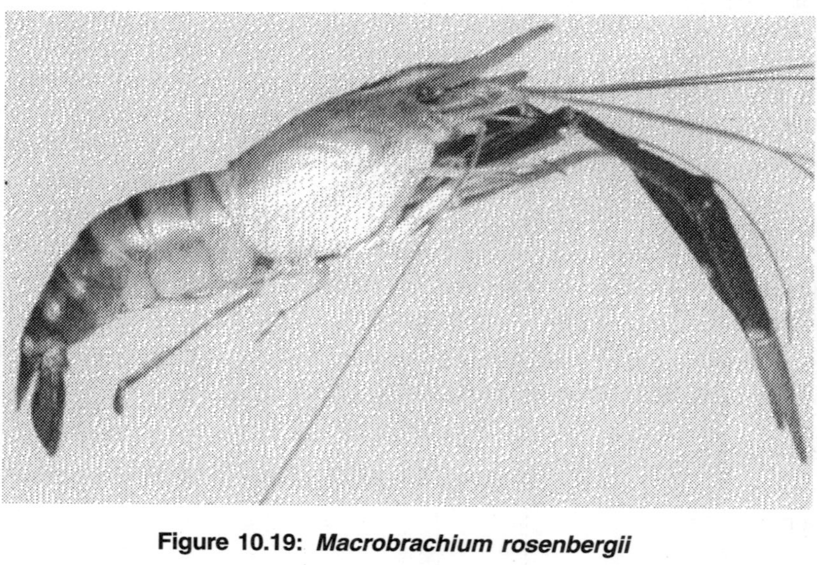

Figure 10.19: *Macrobrachium rosenbergii*

system in the country due to its fast growth rate, high market demand, attractive price and its compatibility to grow with carps. This species can also be cultured in low saline brackishwater areas (salinity <10%) and can be cultured either alone (monoculture) or in combination with carps, tilapia and chanos (poly-culture). It is also a suitable species for incorporation in the paddy-cum-fish culture. The culture of *M. rosenbergii* can be carried out in earthen ponds, cement cisterns, in pens or in cages. However, most of the operations are being carried out in earthen ponds.

Biology of *Macrobrachium rosenbergii*

Giant river prawns live in turbid freshwater, but their larval stages require brackish water to survive. Males can reach a body size of 32 cm; females grow to 25 cm. In mating, the male deposits spermatophores on the underside of the female's thorax, between the walking legs. The female then extrudes eggs, which pass through the *spermatophores*. The female carries the fertilized eggs with her until they hatch; the time may vary, but is generally less than three weeks. A large female may lay up to 100,000 eggs, from these eggs hatch *zoeae*, the first larval stage of *crustaceans*. They go through several larval stages before metamorphosing into postlarvae, at which stage they are about 8 mm long and have all the characteristics of adults. This metamorphosis usually takes place about 32 to 35 days after hatching. These postlarvae then migrate back into freshwater.

There are three different *morphotypes* of males. The first stage is called "small male" (SM); this smallest stage has short, nearly translucent claws. If conditions allow, small males grow and metamorphose into "orange claws" (OC), which have large orange claws on their second *chelipeds*, which may have a length of 0.8 to 1.4 their body size. OC males later may transform into the third and final stage, the "blue claw" (BC) males. These have blue claws, and their second chelipeds may become twice as long as their body.

Grow-out technology of giant freshwater prawn, *Macrobrachium rosenbergii* developed at CIFA aims at a sustainable production 1.5 t/ha/crop. The technology involves an initial nursery phase for a period of two months followed by a grow-out phase of six months. The culture technology involves the following steps: (*a*) preparation of pond, (*b*) eradication of competitors and predators, (*c*) fertilization of pond with organic and inorganic manures, (*d*) provision of hide outs, (*e*) stocking of ponds with juveniles, (*f*) feeding, (*g*) management of water quality, (*h*) pond aeration, (*i*) sampling of prawns for growth measurement, (*j*) disease control, and (*k*) harvesting.

Preparation of Nursery and Grow-out Culture Ponds

- ☆ Prior to initiation of culture the ponds should be well prepared. The pond bunds/dykes should be repaired and strengthened.
- ☆ Ponds should be drained and the pond bottom should be exposed to sun for a week to kill all predatory fishes.
- ☆ Lime may be applied as per the requirement after testing the soil pH. It can be applied @ 200 kg/ha, if the soil pH is between 6.5 and 7.0. Higher dose will be required in case of soil with low pH values.

Figure 10.20: View of a Freshwater Prawn Culture Pond

- ☆ Water should be let into the pond up to two feet using nylon mesh nets to prevent the entry of eggs and larvae of predatory fishes and competitors.
- ☆ Pond should be fertilized with raw cow dung/poultry manure and super phosphate as per the requirement. In general for a pond of medium nutrient contents the fertilizers may be applied at the rate of 5 tonnes raw cow dung, 200 kg urea and 300 kg/ha/crop super phosphate.
- ☆ After a week of fertilization the pond should be filled up to 4 feet water level.
- ☆ Transparency of pond water should be checked after 2-3 days using a Secchidisc
- ☆ Ponds can be stocked with post-larvae in case of nursery pond and with juveniles in case of grow-out ponds once the transparency is 30-35 cm during early morning or late evening hours.

Nursery Pond Management

- ☆ The preferred stocking density in the nursery pond is $20/m^2$
- ☆ Post-larvae (8-10 mg) may be fed with pellet diet (crude protein 35 per cent; lipid 8 per cent) in crumble form @ 100 per cent of the biomass during the first fortnight and further reduced to 50 per cent in subsequent period. In the absence of pellet diet a mixture of groundnut oil cake (powdered) and

rice bran may be given as feed. The feed should be broadcasted in the pond twice daily preferably in the morning and in the late evenings.

- ☆ In nursery ponds approximately 10 per cent of the pond surface may be covered with floating weeds with dense root system such as *Eichhornia* sp. to improve the survival rate of post-larvae. The weeds should be kept inside a PVC or bamboo frame to avoid their spreading in the pond.
- ☆ Aeration is provided for ~8 h/day.
- ☆ A fortnight after stocking sampling of post-larvae may be done to observe the growth using cast net or fry net.
- ☆ During nursery rearing water temperature may be checked twice daily. pH, dissolved oxygen, transparency and depth may be checked once every week and to be maintained in optimum ranges.
- ☆ Loss of water due to seepage and evaporation should be compensated by water addition at least once every fortnight.
- ☆ Nursery rearing may be done for 45-60 days. At the end of rearing period the juveniles (>1.0 g) are collected by dewatering the pond and transferred to grow-out ponds.

Grow-out Pond Management

- ☆ Once the pond is ready for stocking the juveniles collected from the nursery pond are stocked in grow-out pond for further growth.
- ☆ The preferred stocking density in grow-out pond is 40,000/ha.
- ☆ The prawns are fed daily with formulated pellet diet (2-3 mm size) @ 10 per cent of the biomass initially and then reduced to 3 per cent of the biomass towards the end of the culture period.
- ☆ The feed should be broadcasted in the pond as mentioned above. Check trays 3-4 nos may be kept in different corners of the pond to check the consumption of food.
- ☆ During the course of culture the water quality need to be maintained at optimum levels for good growth by routine monitoring of important water quality parameters such as temperature, pH, transparency and dissolved oxygen content.
- ☆ Pond depth should be preferably maintained at four feet.
- ☆ Regular partial harvesting/cull harvesting of bigger size specimens using a large mesh cast net from the fourth month of culture will increase the yield and improve the growth of smaller prawns.
- ☆ Major problems that may arise during culture are mortality of the stock due to low dissolved oxygen in the pond water. Heavy plankton bloom, very low water level and lack of water exchange leads to low dissolved oxygen levels. Continuous rainy/cloudy days precipitate this problem. Immediate

water exchange or aeration of ponds during night hours prevents this problem.

- ☆ Development of bottom algae due to high transparency of water is another problem during monoculture of prawns. To avoid this problem always maintain transparency in 30-40 cm range by frequent fertilization.
- ☆ After 8 months of culture the ponds may be harvested by complete draining.
- ☆ Prior to harvesting arrangements for marketing the prawns should be made.

Figure 10.21: A Haul of Giant Freshwater Prawn

During the south-west monsoon season (June-September), all penaeid prawn (shrimp) hatcheries along the west coast of India have to be shut down due to sudden drop in salinity which is not suitable for rearing penaeid prawn larvae. However, salinity and other climatic conditions during this season are conducive for rearing larvae of the giant freshwater prawn *Macrobrachium rosenbergii*, commonly known as "scampi".

M. rosenbergii is a commercially important farmed species of freshwater prawn. The life history of this species is quite interesting because, while the adults inhabit the freshwater bodies, the gravid females migrate to the low saline estuarine region for spawning and the post-larvae migrate back to freshwater areas for growth and maturity. The catadromous behaviour of this species facilitates its spawning and larval rearing in low saline waters in the hatchery. Therefore, the salinity reduction occurring during the monsoon season in the coastal waters and estuaries along the west coast of India can be beneficially used for rearing the PL of this species under captivity. In the inland areas, on the other hand, saline water has to be prepared by adding common salt or brine to freshwater, or sea water has to be transported over long distances prawns/(shrimps). This enabled the hatchery to carry on with production during the monsoon season and also earn additional revenue. The culture of *Macrobrachium* in temperate zones offered positive opportunities despite the inability to culture year round.

Table 10.5: Economics of Grow-out Production of Prawn

Sl.No.	*Item*	*Amount (in Rupees)*
I.	**Expenditure**	
A.	**Variable Cost**	
1.	Pond lease value	10,000
2.	Prawn seed @ 60,000/ha @ Rs. 600/1000 nos including transportation cost	36,000
3.	Fertilizers and lime	6,000
4.	Supplementary feed (pellet form @ 3 t/crop @Rs. 20/kg)	60,000
5.	Wages (One @ Rs. 2000/month for 9 months)	18,000
6.	Electricity and fuel	3,000
7.	Harvesting charges	5,000
8.	Miscellaneous expenditure	3,000
	Sub-total	**1,41,000**
B.	**Total Cost**	
1.	Variable cost	1,41,000
2.	Interest on variable cost (@ 15 per cent per annum for 6 months)	10,725
	Grand total	**1,51,725**
II.	**Gross Income**	
	Sale of big size prawn (@ Rs. 175/kg for 1000 kg)	1,75,000
	Sale of small size prawn (@ Rs. 70/- kg for 500 kg)	35,000
	Grand total	**2,10,000**
III.	**Net Income (Gross income–Total cost)**	**58,275**

Eyestalk Ablation

Eyestalk ablation is a frequently adopted procedure for induced maturation of gonads and spawning. This method has also been tried on a few occasions to enhance growth in some crabs and lobsters. The crustaceans eyestalk is known to have a neurohaemat function. This method is also most frequently performed for inducing maturation of gonads and spawning of prawns. Extirpation of an eyestalk removes the endocrine system of prawns located in the eyestalk which influences growth, reproduction and other metabolic activities. Besides, eyestalk ablation is reported to influence lipid metabolism, protein metabolism, carbohydrate metabolism, hydromineral regulation, gonad inhibition and limb growth. Moulting in Crustacea is part of the mechanism of growth. Change in form and increase in size can only occur when the hard calcareous exoskeleton is shed and before the new cuticle is hardened.

Males can reach a body size of 32 cm; females grow to 25 cm. In mating, the male deposits spermatophores on the underside of the female's thorax, between the walking legs. The female then extrudes eggs, which pass through the spermatophores. The female carries the fertilized eggs with her until they hatch; the time may vary, but is generally less than three weeks. A large female may lay up to 100,000 eggs.

Protocols for Rearing *M. rosenbergii* Larvae

Brood Stock Collection and Maturation

Healthy adult females and males are collected from the wild or from farms. They are disinfected in aerated freshwater containing 15-20 ppm formalin for half an hour before transferring into the maturation tanks. Berried females (carrying orange or grey coloured egg mass) are kept for incubation or hatching (depending on colour of eggs). Artificial shelters are provided at the bottom of the maturation tanks. Female:Male ratio of 4:l is maintained in a 1000 L rectangular FRP tank (maturation tank) and usually 8 females and 2 males are kept for maturation. Dechlorinated and aerated freshwater is used in the maturation tanks. Before introducing the animals, conditions are created for rich natural algal bloom (mostly *Chlorella* spp.), rotifers and zooplankton to grow as they provide a healthy environment and food for the brooders. In addition, the prawns are also fed on feed of about 5 per cent of body weight every morning consisting of an assorted variety of fresh food consisting of mussel meat, lam meat, worms, small shrimps, pieces of potatoes, formulated feed, etc. The tank bottom is cleaned by siphoning out very morning unconsumed food, faecal matter and moults. About 20-30 per cent of water is exchanged to compensate for the loss. Occasional sprinkling of freshwater or rain water speeded up the maturation process. Spawning occurs within 8-12 hours after mating. The extruded eggs get deposited on the ventral side of the abdomen and the egg mass is firmly held by the pleopods. Under captivity a female matures 2-3 times in a season. The interval between two successive puberty moults is about 20-30 days. In some females the next maturation process may start while the eggs are still incubating and in such cases the puberty moult occurs within 10 ways.

Incubation, Hatching and Fecundity

Fecundity

Males can reach a body size of 32 cm; females grow to 25 cm. In mating, the male deposits spermatophores on the underside of the female's thorax, between the walking legs. The female then extrudes eggs, which pass through the spermatophores. The female carries the fertilized eggs with her until they hatch; the time may vary, but is generally less than three weeks. A large female may lay up to 100,000 eggs. Initially, the egg mass is range in colour. The berried female is transferred into a 1000 lit FRP tank containing 8 psu water for incubation. Feeding, cleaning of tank and exchange of water are done as mentioned earlier. Artificial shelters are provided for protection and 3 to 4 females are kept in each tank for incubation. The incubation period is 18-20 days at 26-28°C. During this period, the colour of the egg mass gradually turns to grey/dark grey. Extreme are is taken while handling the incubating females so as not to cause damage to eggs or loss of eggs. The females remove dead eggs and extraneous materials with the first pair of pleopods and they periodically ventilate by vigorous movement of pleopods. Females carrying dark grey egg mass are carefully transferred into a dark coloured cylindro-conical FRP tank (110 lt) containing aerated 8 psu water for hatching of eggs. Only one female is kept in one tank. Feeding is not done at this stage as the female avoids food. This helps in maintaining good water quality, latching normally occurs in the night and may last for 2-3 days. This result in laving 2 or 3 different stages of larvae form a single spawn. The maximum latching occurs in the second day. The newly hatched larvae are dispersed by the mother prawn with the help of its pleopods. The spent females are transferred directly into the maturation tank containing freshwater having thick algal bloom. Abrupt exposure to freshwater enhances re-maturation process. The larvae are removed by opening the ball valve provided at the bottom of the tank after two days. A 25 g female produces >3,000 eggs in the first brood. It increases with the size of the female and a 60 g female carries about 25,000 eggs. The fecundity, however, decreases in subsequent broods. Eggs of *Macrobrachium rosenbergii* are slightly elliptical, long axis measures 0.6-0.7 mm, bright orange in colour, until 2-3 days before hatching when they become grey-black.

Larval Rearing

Larva undergoes 11 distinct zoea stages before metamorphosing into a postlarva. Each stage can be recognised by certain distinct morphological characters. Microscopic as well as visual observation of the larvae is made every day to assess the quality of larvae, health, survival, growth stage, size, deformity (if any), gut state, ectoparasitic infestation, swimming behaviour, etc. Moulting sometimes gets delayed in zoea stages III, V and VIII due to protozoan infestation. This can be overcome by application of ~20ppm formalin for 6 hours for 2-3 consecutive days. Type and size of the rearing tanks have to be changed according to the stage of the larvae. The freshly hatched zoeae are grown in the hatching tank itself for 2 days until they reach zoea I or II and no feeding is done. Later they are shifted to rectangular or circular FRP tanks of 500 or 1000 lit capacity depending on the density of larvae. They are stocked @ 60- 80 larvae/lit. Water depth in the tanks is maintained around 70 cm. Feeding is done from this stage onwards. It takes 2- 3 days for one stage to moult to the next stage. They are reared in these tanks until they grow to Mysis VI-VII stage. Since synchronised growth is lacking, 2 or 3 stages may exist in the tank at a time. From stage VII onwards, they are reared in large oval shaped LRTs of 2000 lit capacity made of FRP or in plastic pools (4' diameter, 2' depth) until they attain postlarval stage. Water depth of 70 cm is maintained in LRT and in pool around 60 cm. The stocking density is reduced to 30-40 lit. After attaining PL stage, they are transferred to 20,000 lit concrete nursery tanks containing dechlorinated freshwater to grow for a further period of 15 to 20 days before supplying to farmers. Rich algal growth and zooplankton will improve the health and enhance the growth of postlarvae. Weaning of postlarvae from brackishwater to freshwater is done gradually and with care as sudden change will cause mortality. Salinity reduction is done by 1 psu at a time and brought down to freshwater condition over a period of 12-15 hours. Vigorous and uniform aeration is given in all LRTs and nursery tanks to keep the feed in suspension for the larvae and to maintain dissolved oxygen level high and uniform. Survival rate improves by providing artificial floating shelters in LRTs. Artificial shelters are placed at the bottom of nursery tanks which increase the surface area for the postlarvae to settle and also act as hiding places for PL.

Salinity

Zoea I to early stages of PL survive and grow faster in 12 psu water. Sudden and wide variations affect the survival rate. Dechlorinated freshwater only is used for dilution. Sea water passed through a sand filter and cartridge filters improves the water quality.

pH

Wide variations in pH are not conducive for larval growth. They will affect moulting and growth. Extreme levels cause mortality. An optimum range of 7.3- 7.8 is conducive for larval growth.

Temperature

Growth is adversely affected if the temperature of water drops <26°C. Lower temperatures delay metamorphosis and cause mortality. A sudden change in temperature is avoided while exchanging water as they affect the survival rate.

Light

Larvae are phototactic and exposure to sunlight tends to aggregate them to the brighter part of the tank resulting in cannibalism and poor survival. Even in well aerated tanks, clumping of larvae occurs in the area with sunlight. Therefore, tanks are kept in shaded area away from direct sunlight.

Aeration

Survival is high if the aeration is uniform and vigorous. Vigorous aeration keeps the food particles in suspension and helps in dispersion of larvae in the tank besides providing the much needed dissolved oxygen. Aeration is stopped only for short durations while cleaning the tanks and observing larvae.

Nutrition

Nutrition is an important criterion for larval survival and growth. For the first 2 days no feeding is done as the zoeae survive on the yolk. From 3rd day onwards, freshly hatched brine shrimp nauplii (BSN) and formulated feed are given. Feeding is done seven times a day. The first and the last feeding are exclusively on BSN. Quantity of feed given for different larval stages is worked out based on visual observation of the gut content as well as feed utilisation by the larvae. Larvae feed continuously. However, excess feeding is avoided to reduce feed cost and to maintain good water quality.

Depending on the larval stage, the feed is sieved through appropriate mesh as shown below:

Larval Stage	*Mesh Size (mm)*
Zoea II-III	0.1
Zoea IV-V	0.15
Zoea VI-VIII	0.2
Zoea IX-X	0.3
Zoea XI-PLI	0.4
PL1-P15	0.5-1.0

A combination of three prepared feeds and BSN yields the best survival and growth. PL stage is attained within 21-28 days with a survival rate of >60 per cent.

Brood stock feed of high level of 18: $2n$–6 and n–3 HUFA (13 and 15 mg/g DW) has been found to improve fecundity, egg hatchability and overall quality of the larvae.

Water Quality Management

Water Quality for Indoor Nurseries

General water quality requirements for indoor nurseries are similar to those for freshwater in hatcheries. Maintain the optimum temperature (27-31°C) by heating the water in the system or the building in which they are housed, if necessary. If you are operating a recirculation nursery system, a turnover rate through the biological filter of 12 times a day is suggested.

Maintenance of high quality water in the hatchery is very important for the successful rearing of the freshwater prawns. Freshwater is dechlorinated before use. Sea water is filtered through a sand filter to eliminate all abiotic and biotic particles>50mm. Since the hatchery requires saline water of 8 and 12 psu besides freshwater, it is prepared in advance and stored in overhead tanks after treating with 1.5 mg of free chlorine or 15-20 ppm formalin to eliminate all microorganisms and parasites. The tank bottom is cleaned before the last feeding so that the quality of water remains good during the night when the moulting and metamorphosis actively take place. Daily water exchange of 30-50 per cent ensures good quality water.

Stocking Rates

The best stocking density for indoor nursery tanks depends on the length of time the animals will remain in the tanks before transfer to an outdoor nursery or grow-out facility. You are recommended not to exceed a stocking density of 1 000 PL/m^3 in tanks without substrates. You can stock 2 000 PL/m^3 in tanks where substrates are provided. These stocking densities assume that the indoor nursery rearing time is not more than 20 days. You would need to reduce the density if you are going to keep the prawns in your indoor nursery for longer periods (*e.g.* in sub-tropical and temperate regions, where prawns are usually maintained in indoor nurseries until the grow-out ponds reach temperatures of at least 20°C). Maintaining prawns in commercial indoor facilities for periods longer than one month may prove too expensive, although they are often maintained longer than that for research purposes.(FAO, 1992)

Strict adherence to the following criteria will help in successful larval rearing:

1. Maintenance of salinity, pH and temperature at the optimum range and avoidance of sudden variations.
2. Avoidance excess feeding.
3. Observation of health, growth, gut.
4. Checking for protozoan infestation and treatment in time.
5. Checking for contamination and quality of feed.
6. Cleaning and disinfection of all implements and equipments.
7. Maintenance of general hygiene in and around the hatchery.
8. Cleaning of tank bottom every day evening.
9. Cleaning sides of tanks once in 3 days and rotating (change) tanks every week.
10. Maintaining uniform and vigorous aeration.
11. Replacement of 30-50 per cent water every day.

Feeding Strategy

Feeding once or twice per day is sufficient. You should adjust the quantity of feed based on observing the actual consumption. It should normally be about 10-20 per cent of the total weight of the prawns in the tank. Grow-out feeds can be used but enhanced results may be obtained by supplementing them with other materials, such as beef liver, egg custard based diets (EC), or minced fresh fish. However, you must take great care if you use fresh feeds. Fresh feeds, which usually break down more easily than pelleted diets, may rapidly cause water quality problems. This could overload recirculation systems or mean that you would need to have a much greater water exchange in flow-through systems (this is not such a problem as in hatchery systems, because nursery water is not brackish; however, it would increase pumping and other costs). Adult Artemia (Artemia biomass) have also been used as a nursery feed for *Macrobrachium rosenbergii* in countries where it is readily available as a fresh (live) product from salt farms. Recently (2000), a freeze-dried version of this product has also become commercially available. (Alston, D.E. and Sampaio, C.M.S. 2000.)

Feed Formulation

Egg (97 per cent) +Yeast (3 per cent)

Egg(85 per cent) + Prawn(I2 per cent)+Yeast(3 per cent)

Egg (35 per cent) + Soyameal (25 per cent) + Mussel Meat

(25 per cent) + Skimmed Milk Powder (12 per cent) + Yeast (3 per cent)

Feeding Schedule (at hourly intervals)

Feeding rate of BSN (Brine Shrimp Nauplii) and feed varies with growth stage as shown in the Table below. The quantity of feed shown is for 10,000 larvae per feeding).

Larval Stages	*BSN per Larva*	*Feed (g) per 10.000 Larvae*
Mysis II-III	12	3
Mysis IV-V	28	5
Mysis VI-VIII	44	6
Mysis IX-X	54	8
Mysis XI-PL1	64	11

Using best management practices, production of 1500–1800 kg/ha has been achieved in commercial ponds on 110 days.

J. Bojan Marine Products Export Development Authority (MPEDA) presented the status paper on scampi farming in India. Scampi is cultured in 34,630 ha area in the country. The average production per hectare ranges from 880 to 1250 kg. He has noted that 62 per cent of the scampi culture operation is in Andhra Pradesh. Presently 71 hatcheries are operating in various states supplying 183 billion scampi seeds to the farmers in India. The high cost of seed and feed is a problem facing scampi farming in India.

The chemical contaminants and antibiotic residues were threats for scampi raised by aquaculture. However, microbial quality of farmed giant freshwater prawns in India did not exceed tolerance limit. Cadmium, lead, mercury in ppm were within limits.(Mukundan M. K., 2003)

Dealing with Problems of Predation

Predation is one of the greatest problems for any aquaculture enterprise, including freshwater prawn farming. Predation is caused mainly by other aquatic species, birds, snakes and humans. Two of the greatest sources of loss in freshwater prawn farming are human predation and operator error.

Freshwater prawn farms are more prone to human predation than many fish farms because of the high value of the product and because prawns are relatively easy to catch. The temptation to catch a few kilograms of prawns by cast-net at night (a kilogram of which may be as valuable as a tenth of a month's individual income to some) is sometimes too great to resist. You cannot eliminate any form of predation, including human poaching. However, you must minimize it by good management. Perimeter fences, dogs, lighting, and reliable watchmen help. If your farm is big enough to financially support it, you may be able to achieve some protection from human predation if you stock some PL into local public waters, thus generating a positive attitude towards your farm. If you own a small farm you may find it useful to form a cooperative with other farmers within the community. The activities of such groups are normally protected by the local community. You may also lose prawns through operator error and poor management. For example, water levels may be allowed to become too low and therefore temperatures too high, or CO_2 levels may be allowed to fall too low. Both errors will cause animals to die. Not maintaining outlet structures properly allows prawns to escape.

Normally, insects (mainly dragonfly nymphs), carnivorous fish and birds are the most serious predators in freshwater prawn farming. In the past, chemicals have been used to kill dragonflies and other insects but this is not recommended because it may negatively affect the pond ecosystem. Mosquito fish (*Gambusia affinis*) and related species were also once stocked in freshwater prawn ponds to control insects. *M. rosenbergii* postlarvae themselves, if they are stocked before the insects hatch, can control the dragonfly population. You can effectively control unwanted fish by using rotenone or tea seed cake between cycles, as discussed earlier in this manual. You can prevent the entry of fish and some insects by passing the intake water through suitable screens or gravel filters (Figure 10.22). Most commercial prawn farms rely on simple net filters. If fish eggs and larvae do get into your ponds (which they will!), it is not a complete disaster because, by the time they get to a dangerous size, many will be seined out during the cull-harvesting of prawns. The ideal would be to exclude all predators but this is not possible. The most important thing is to stock the prawns very soon after each pond is filled, so that predators and competitors have less chance to become established. The presence of many frogs and toads in a pond usually indicates that predatory fish have been fairly efficiently excluded.

Figure 10.22: Gravel Filter to Protect Predators

Pond Size

The most easily managed pond sizes range between 0.2 ha and 1.6 ha, with most farms having ponds around 0.2-0.6 ha. If kept to a 30 m width, a 0.6 ha pond will be 200 m long. Narrow ponds should be oriented so that the prevailing wind (which enhances the dissolved oxygen content of the water) blows down the long axis towards the drain end, to lessen the area of the pond bank subject to wave erosion.

Large ponds are normally wider than 30 m and often drained for harvesting. If the total harvest is going to be taken at one time (batch management), the size of the pond should be influenced by the maximum weight of prawns that the market will accept at one time without price deflation. For example, if a quantity greater than 300 kg of freshwater prawns would swamp your market and reduce prices it would be pointless to have a drainable pond greater than 0.15 ha in area (assuming a productivity of 2 mt/ha/crop).

Figure 10.23: Large Size Pond

Figure 10.24: Small Size Pond

Eradication of Predators

If your pond has previously been stocked with fish and you want to convert it to freshwater prawn culture, or if a lot of fish were present during your last prawn grow-out season, treat it with a piscicide after harvesting and while it still has water in it. Rotenone or teaseed cake is commonly used for eradicating unwanted fish between cycles. They are effective if spread evenly throughout the pond. However, the use of rotenone is banned in some countries because of environmental concerns: check before you use it. The quantities needed for treatment are shown as under:

Application of Rotenone and Teaseed Cake

Rotetone

20 g/m^3 (200 kg/ha when the water averages 1 m deep) of rotenone powder (which contains 5 per cent rotenone, usually from Derris roots, and thus equivalent to applying 1 g/m^3 of pure rotenone) is the normal dose. Rotenone needs to be mixed in water and the solution kept well-mixed while it is applied.

Teaseed Cake

The application of teaseed cake (containing 10-13 per cent of saponin) at a dose of 50-70 g/m^3 (500-700 kg/ha when the water depth averages 1 m) is adequate to remove unwanted fish. Teaseed cake needs to be prepared by drying and finely grinding the seeds, soaking the powder in lukewarm water for 24 hours, and diluting the suspension before mixing it evenly into the pond water.

More powerful chemicals, such as insecticides, are sometimes used for pest eradication (in severe cases, where there are very stubborn predators or competitors that resist other forms of treatments and/or because of their cheapness). However, the use of insecticides to remove unwanted fish is not recommended in freshwater prawn farms; they are potentially toxic to prawns and may accumulate in prawn tissues, with consequential dangers to human health.

Use of Artificial Substrate

Another means of improving results in temperate freshwater prawn culture is to place artificial substrates in the ponds, which makes it feasible to increase stocking rates above the level recommended earlier for ponds without substrates. PVC fencing (such as is used to close off areas when roads are being resurfaced) forms an ideal substrate. This material can be expensive in some countries but the investment should be worthwhile, as the following information indicates. Substrate provision on a commercial scale (Figures 10.18 and 10.25) has resulted in production and mean harvest size exceeding 1 800 kg/ha/crop and 35 g respectively, from a stocking rate of 4 PL/m^2, while yields exceeding 2 500 kg/ha/crop with average weights of >40 g have been consistently achieved at a stocking rate of 64 500/ha (Tidwell and D'Abramo, 2000). It is therefore suggested that you increase the stocking rate of juveniles from the 4/m^2 (40 000/ha) recommended earlier for use without substrates to 6.5/m^2 (65 000/ha) when you use either horizontal or vertical substrates. No extra labour (apart from its initial installation) is necessary if this form of substrate is used because it can be permanently installed in ponds equipped with catch-basins at the drain end. As the water is drained, prawns abandon the substrate and follow the water flow to the catch basin. You can spread the cost of the labour for installation, as well as the substrate material itself, over several production cycles. This new technology is still being developed but it clear that the use of substrates can markedly increase the productivity of freshwater prawn farming.

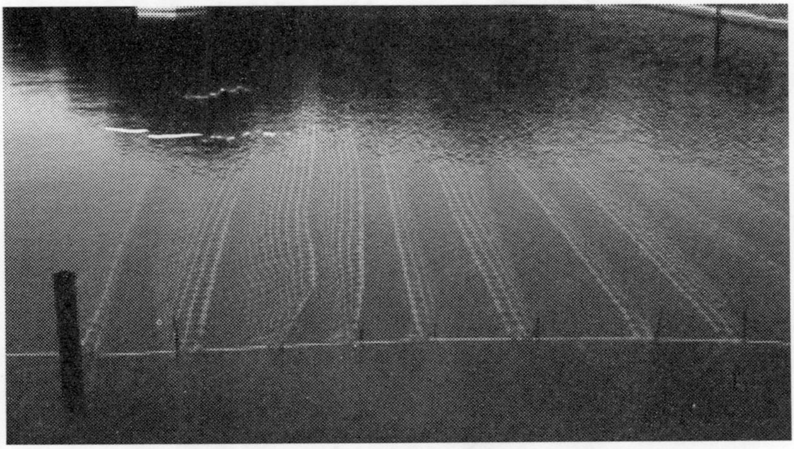

Figure 10.25: Substrates have been Placed Horizontally

These types of management make prawn production feasible in smaller, deeper ponds which were previously considered unsuitable. This is useful in hilly inland regions where suitable sites for large shallow ponds are very limited. Grading before stocking and the use of substrates has not been practised much in tropical monoculture yet but the advantages obtained in temperate culture should be transferable. One researcher believes that up to 9 mt/ha/yr of 20g prawns from three 4-month cycles might be achieved in tropical areas using the combination of grading and substrates.

Chapter 11
Marine Prawn Fishery and Culture

Before thinking culture of marine species, it is necessary to know some back ground of natural fishery of the respective prawn species. So that one can plan to get brooder from natural sources as well as plan culture.

FISHERY

Penaeus indicus

Penaeus indicus in India supports commercial fisheries in both marine and estuarine environments on the east and west coasts. The species is subjected to commercial exploitation at different stages of life from both estuarine and marine environments. The brood prawns, which come into backwater in November–December reach a size of 110 mm, in September-October of the following year, when they move out into the sea. The entire backwater fishery, therefore, are constituted by '0' year class prawns. Three-year classes (0,1 and 2) of this species are represented in the trawl fishery.

In the backwater of Kerala, the species is fished almost through out the year. The marine fishery is largely seasonal. On the west coast, the season generally coincides with the monsoon period, June-September, but the fishing season of the species in December-February have been observed both in east and west coast of India.

The estuarine and backwater fishery for the juveniles of the species is carried out in very shallow waters not exceeding 10 meters in depth. But the commercial fishery for adults is generally carried out in coastal waters up to a depth of 50 meters along Indian coast.

Penaeus monodon

P. monodon, the largest of marine prawns is known as "Jumbo tiger prawn" in most of the countries of Indo-Pacific region. The species is widely distributed in east and west coast of India and Sri Lanka. The species apparently prefers warm water habitats. It is recorded from seas, rivers, estuaries, brackish waters and even from freshwaters.

Like *P. indicus*, this species is also subjected to commercial exploitation at different stages of life from both estuarine and marine environments. The entire backwater fishery constituted by '0' year class. The species occurring in the trawl catches from both the coasts of India come under late 0-year to early 1-year class.

Specimens over 300 mm. in total length are common in the trawler catches landed from relatively deeper waters of the west coast.

In Kerala backwater fishery, the species is caught through out the season in small numbers. In Mumbai, they are found in commercial from August-October.

Penaeus merguiensis

Penaeus merguiensis is the major prawn species next to Tiger prawn in Gujarat. It forms about 4–5 per cent of 10.3 per cent of Jumbo varieties of total prawn landing of Gujarat. Local people know as white Jumbo. It was observed that maximum fertile eggs were found in September to October. It was between 1, 50,000 to 2,29,800 per female. Second breeding spell was observed during January to March. Average number of eggs were in between 1,01,260 to 1,05,000 per female.

Penaeus semisulcatus

In India *P.semisulcatus* is caught along with other prawns only occasionally. It is not known to contribute any significant proportion of the marine catch. However, the species is often well reported in the brackish water fishery of the west coast of India.

In India *P.semisulcatus* is caught along with other prawns only occasionally. It is not known to contribute any significant proportion of the marine catch. However, the species is often well reported in the brackish water fishery of the west coast of India.

The juveniles of the species have been observed to spend their life from late August to middle of October in areas of the sea, where Zostera marina are growing. After middle of October, the species seems to be fished only from the off shore areas, where the bottom is muddy. In the marine catches, the size composition varies from 150-180 mm. the largest recorded size being 222 mm. The species also form a significant portion of prawn catches of 'Bheris' of West Bengal, where they attain a length of 76-127 mm. at the end of the season.

Metapenaeus monoceros

Metapenaeus monoceros is distributed in South Africa, Mediterranean and Indian seas to Malaysia with the eastern limit a Malacca strait. Although it is a marine species, it is found in marine, brackish water and freshwater environments. In Indian waters, it occurs in the Juvenile stages in most of the estuaries and back waters with muddy bottom along the coast line and adults in the sea up to 50-60 meters depth both muddy as well as sand and silt bottom.

Only '0' year class contributes to the backwater fishery of Cochin. In the trawl catches, 3-year classes have been recorded. The bigger year class enters the fishery in November-December and the smaller size appears later. It is noticed that in some years, the bigger classes fail to appear in the fishery.

The backwater fishery constitutes 56-90 mm. the inshore fishery constitutes of the species of 40-120 mm, mostly juveniles. The adults are caught in the trawl fishery the size range being 90-175 mm. The maximum size attained is 180mm. in the deeper waters (50-60 meters).

The species are abundant in backwaters from March-June and in November. The season in the trawl fishery is November-December. In Bombay waters, the fishery commences during the rainy season, July-August. In Chilka Lake it is abundant in November-June.

Metapenaeus dobsoni

Metapenaeus dobsoni is distributed from Indian waters through Malaysia and Indonesia to Philippine islands. It is found in brackish water as well as marine environments. In Indian waters the species is present in the juvenile stages in most of the estuaries and brackish waters along the coast line and the adults in inshore areas up to 20 fathoms depth along the south west coast of India, where it contribute to a major fishery.

The fishery in backwater environments is constituted by the '0' year and 1 year classes by marine fishery.

During the monsoon months, when the mud banks occur, in various places along the coast, shoals of these prawns approach the shore in these areas, so close as to make it possible for fishermen to use for catching them.The population of the backwater and estuaries sizes ranging from 30-70 mm in the catches, the marine fishery size range from about 60-125 mm.

Juveniles are fished in backwaters, estuaries and paddy fields in shallow areas ranging from 1-15 meter depth. Young adults and adults are caught from sea in depths up to 25-30 metres. In marine inshore areas, the fishery is largely seasonal fron June to September. The off shore fishery extends from November to June. In brackish waters of Kerala, the fishery extends from middle of November to April.

Metapenaeus affinis

In Indian waters, the juveniles of the species are found in very small numbers in the backwaters and estuaries and adults occur in the inshore waters to a depth of about 45 meters.

In the backwater fishery only '0'year class (30-120 mm.) is represented. The inshore and off shore fishery is mostly represented by I and II year class (71-130 mm.)

In the trawl fishery, the II year class generally enters the fishery in earlier half of the season and one year in the later half (121-140mm.)

In the backwater fishery, the species is abundant from January-June. The peak season for the species in the trawl fishery is from December-February in Cochin, January-March in Mumbai and January-August in Calicut. The inshore fishery of the Kerala coast intensifies after the formation of mud banks (annual) on which the prawn concentrates.

Metapenaeus brevicornis

In the distribution of this species in Indian waters one difference noticed from other species like *M. monoceros* and *M.affinis* is that, it does not occur in the southern area, but contribute a good fishery in the northern region both on the west as well as east coasts. Well represented in estuaries and inshore waters especially in the east coast. In the Gulf of Kutch area the species is mostly distributed in areas with sandy bottom.

In Hooghly estuary I and II year groups of the species form the fishery. '0' and III year groups also contribute to the fishery. In the Hooghly estuary, the catches ranged in size between 15 and 115 mm. the inshore fishery for the species range from 40-110 mm. in length. They occur in shallow waters ranging 4-7 meters depth.

The species is found through out the year, the peak season is from January-March in Mumbai coast, July-February in Gulf Kutch area. In Hoogly estuaries it is fished through out the year with bulk landings in November-February.

Parapenaeopsis stylifera

Parapenaeopsosis hardwickii ranks third among the commercially exploited species of the genus Parapenaeopsis in the Indian region. The general distribution of the species is from the coasts of India through Malaysia to southern China. Although, the species occur on both the coasts of India, it supports a good fishery only in Mumbai and in lesser magnitude in Andhra coast.

In the inshore waters, the species is abundant up to 22 meters especially from the depth ranges of 12-20 meters. The population is composed of 0,1 and II year classes, having a size range from 10-145mm. At Veraval, the species support a good fishery during October-December period. In Mumbai coast they are caught through out the year. At Karwar, the species ranked second in the prawn catches landed by the trawlers. The peak season is from January-April in Mangalore, December-February at Cannarore and February-May in Malabar coast.

Although the species occurs all the year round in the west coast of India, it abounds the inshore waters from November-December to May-June and offshore waters in September-October.

Parapenaeopsis sculptilis

Parapenaeopsis sculptilis is widely distributed tropical specials found from west and east coast of India to Hong Kong through Malaysian waters and Indonesia to tropical Australia and New Guinea.

In India, commercial exploitation of the species is done at Kutch, Bombay in west coast and Hoogly estuarine system in the east coast. It is also reported from the Godavari river system through out the year.

Smaller individuals belonging to 0-1 year class contribute to the fishery of less saline areas and the larger sizes (I and II year's class) support the inshore fishery.

In the Gulf of Kutch area, the species contribute about 19 per cent of the total prawn catch during September-January period. In Bombay coast, the species occur throughout the year, but available in commercial quantities from October-May with peaks in December-February. In the Hoogly estuary, the species is dominant in sinter and monsoon months. The species is mainly confined to the fringes of the coast out to the four-fathom contour. But it may be found as deep as 7 fathoms. In India the inshore fishing seasons extend from October to May and the river systems during monsoon.

Parapenaeopsis hardwickii

The species forms about 0.6 per cent of the annual prawn landings of India. In Bombay coast, the species form 3.7 per cent of the total prawn catch, and the fishery starts in November and continues up to May, the peak season being November and January. The size ranges between 55-65 mm. in case of males and 80-100 mm. in case of females.

CULTURE

Introduction

Shrimps (prawns) are aquatic organisms inhabitating Sea, estuaries and backwaters. They are produced from these water bodies through fishing or farming. They breed and spawn in the sea and the young ones migrate towards coastal waters/ estuaries/backwaters for feeding and growth and return to sea for reproduction.

The best species for culture in India are the tiger prawn (*Penaeus monodon*), White prawn (*P. indicus*), banana prawn (*P. merguiensis*) and flower prawn (*P. semisulcatus*). Farming Techniques have been developed for three species and it is still in the experimental stage for flower prawn.

Types of Shrimp Farming

The process of growing the baby shrimp up to a marketable size in an enclosed water body can be termed as shrimp farming. It includes, at times, allied activities like seed production through hatchery, feed production, harvesting and marketing. According to the nature of scientific inputs and management, shrimp farming systems are termed as Traditional, Extensive, Semi-intensive and intensive in different parts of the world; although there is no clear cut demarcation between these systems.

Traditional System

In this system of shrimp farming, the incoming tides are trapped along with the young ones of shrimp, fishes and other organisms that co-existing the environment into the already existing impoundments adjacent to estuaries and backwaters. Escape of these organisms is prevented by fixing suitable screens in the sluice and the crop is harvested at frequent intervals. Owing to unavoidable and indiscriminate stocking of both desired and undesired varieties of shrimp and fishes, the production under this system is normally unpredictable and often low in quantity and quality.

Extensive System

This system of shrimp farming is an improved method of traditional farming, involving construction of new ponds ranging from 1 to 5 ha. In size, selective stocking with fast growing prawn seeds at a comparatively lower density ranging from few thousand to 100,000 seeds per hectare without much supplementary feeding and the water quality is maintained through the natural fall and rise of tide. The production under this system normally ranges from 2 to 3 tonnes per hectare per year in two crops.

The Semi-intensive System

The Semi-intensive System of shrimp farming involved construction of earthen ponds ranging from 0.2 to 0.5 ha. in size, selective stocking with fast growing species at a comparatively high density ranging from 1 to 4 lakhs per hectare maintenance of water quality by exchanging 10 to 20 per cent daily, aerating the pond with air-blowers/paddle wheels and feeding the shrimps with supplementary feed. The production in this system ranges from 8 to 10 tonnes/ha/year.

Intensive System

The intensive system of shrimp involves construction of concrete ponds of 0.03 to 0.1 ha in size, selective stocking with quality prawn seeds exclusively procured from hatcheries at a density ranging from 5 to 10 lakhs per ha. maintaining water quality by exchange up to 300 per cent a day. Aerating the pond with air-blower/ agitators etc and feeding the shrimps with nutritionally balanced high energy feed. The production from this system ranges from 20 to 30 tomes per hectare per year.

Natural Beach Filter for Seawater

Suitable Beaches can be used as natural seawater filters for hatcheries. Some hatcheries draw water from perforated pipes protected by a 150 μm nylon screen buried approximately 1 m deep in the beach. However, the screens are prone to damage and it is better to develop the beach itself as a filter. This annex describes a cheap filter probe made from plastic, which is derived from a stainless steel probe developed by a zoologist, the late George Cansdale. The following notes have been extracted from Suwannatous and New (1982). The system is easy and cheap to make and is described here in spite of some scepticism that such simple systems are effective.

Careful thought should be given to the location of the beach filter. A permeable beach is needed, with a depth of 2-3 m under a minimum of 30 cm water. Beaches of a wide range of types may be used, including sand, gravel, broken coral, shell, etc. The bulk of the sand grains should be between 0.5 mm and 5.0 mm but a great advantage of this system is that, during the development of the filter, excess fine sand is pumped out, leaving the larger grains in and around the filter probe; thus a precise sand specification is not needed. Uniformly fine sand, especially of wind-blown origin, is unsuitable on its own, but it can be graded up by adding coarse sand or gravel under and around the unit. If most grains are above 2 mm diameter, it helps to add fine sand on the surface around the unit during development. 'Fine sand' is defined for the purposes of this annex as material up to 1 mm and 'coarse sand' from 2 mm to 5 mm, but these are not technical terms. A few stones in the beach of up to 50 mm do not prevent it being developed as a filter bed but larger stones will reduce the efficiency of the filter and should be cleared away (or a different site chosen). Sites with little or no sand are not suitable for natural beach wells. Those with soft mud cannot be used. Where the beach is rocky some people have found that excavating a large hole and filling it with sand from another site, into which the filter probe is inserted, is effective. However, this may be very difficult and costly to construct and maintain. If the hatchery site is not adjacent to a beach with a favourable structure for a beach well there are a number of choices, including choosing a better site, bringing seawater (or brine) from another location (this is essential for inland hatcheries anyway), or pumping raw seawater and treating it within the hatchery.

The capacity of the pump required to operate the filter probe and the jet probe and the correct pipe diameter, depends on the water requirements of the hatchery, as well as its elevation above sea level and the distance between the pump and the filter probe, and between the pump and the seawater holding tank in the hatchery. It is important to have no noticeable flow resistance at the maximum water flow required. Equipment should not be too large for the characteristics of the site and the amount of water required by the hatchery, because this will result in excessive capital costs. Conversely, buying equipment which is too small for the site is a waste of money. The choice of pipe size is discussed in detail in an FAO manual (FAO 1992b). As an example of pump capacity, a 3 HP, 1 440 RPM self-priming electric pump sucking water from a filter probe 35 m distant through a 10 cm flexible hose (adapted down to 5 cm near the pump) and delivering water through a 10 cm pipe to a hatchery above the highest high-water mark to a site 350 m distant is capable of pumping about 20 m^3/hr of seawater.

Construction of the Filter Probe

Soften the end of a 1.5 m piece of 10 cm diameter PVC pipe with heat, taper it to a point, and make sure that it is sealed. Then cut three sets of slits into it (Figure 11.1). The three sets of slits should be cut in rings. The lowest set should be 20 cm from the bottom of the pipe and the space between the three sets of rings should be 40 cm. From the upper set of slits to the top of the pipe should measure about 45 cm. Each ring of slits should be 2.5 cm long. The individual slits should be 1-2 mm wide and the spaces between them 1 cm wide. Flow can be increased by inserting more rows of slits but care must be taken not to weaken the pipe so much that it will fracture.

Maintaining the Efficiency of the Filter

There is a tendency for the flow of water passing through any filter to gradually decrease, as the spaces between the filter bed particles become blocked. In marine sites tidal and wave movements generally keep the surface of the beach filter clear. If blocking does occur, this will only be in the top 1-3 cm, usually only the top 1 cm. If

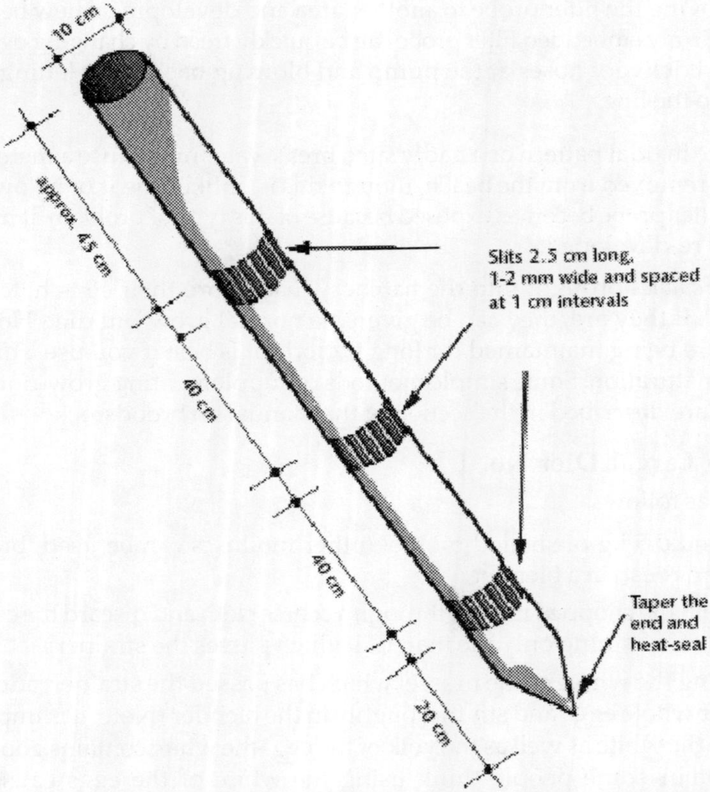

Good quality water can sometimes be obtained with simple beach filters; this figure illustrates a simple plastic beach filter probe

Figure 11.1: Simple Plastic Beach Filter

the flow from the filter becomes reduced, and this is not due to declining pump performance or other factors, this suggests that there is surface blocking. This can be cured in several ways, including:

☆ Stopping the pump and raking an area of about 5 m radius around the probe, working to a depth of about 5 cm, then pumping the water to waste and redeveloping the filter for as long as needed;

☆ Skimming off about 3 cm of sand from the surface and replacing it with new sand;

☆ Forking the area over lightly and back-washing the filter by moving the suction hose to the pump outlet and drawing water through a spare hose. Somebody needs to make sure that the probe does not become displaced during this operation;

- Moving the filter probe to another area and developing a new beach filter. A firmly embedded filter probe can be quickly freed by changing over suction and delivery hoses at the pump and blowing back, after letting some air into the line.

A change in tidal pattern or a badly sited breakwater may cause a meter or more of sand to be removed from the beach, though this is unlikely near or below low tide mark. If the filter probe becomes exposed because of this type of problem, it must be re-installed and re-developed.

Berried Females brought into the hatchery just before their eggs hatch are not normally fed. If they are, they can be given the normal grow-out diet. However, if broodstock are being maintained for long periods, it is best if you use a diet which encourages maturation. Some simple methods of supplementing grow-out feeds for this purpose are described in the section of the manual on broodstock.

Farm-made Larval Diet No. 1

Prepare as follows:

- Blend 0.5 kg of shelled mussel (other molluscs can be used, but mussel seems best) in a blender;
- Strain the chopped mussel through a coarse cloth and discard the connective tissue, retaining only the material which passes the strainer;
- Using the whole of the mussel which has passed the strainer, add three or four whole eggs and stir thoroughly in the blender (Note: it is important to use the white as well as the yolk of the egg–the white contains good quality protein–some people think using the white of the eggs causes water pollution but it does not if homogenized properly);
- Steam the mixture over water (like poaching an egg) until it solidifies into a custard;
- Screen to the correct size (see the main text of the manual) and feed directly; or
- You can refrigerate it for a few days for later use (however, the quality of frozen EC is not as good as fresh EC for feeding purposes).

Stock Estimation

Estimating of the Number of animals present under hatchery or pond conditions is difficult. The four critical times when it is important to assess the number (and sometimes the size) of prawns present in the system are:

- When postlarvae are harvested, to provide a record and an assessment of the production efficiency of each larval batch and tank;
- When postlarvae are transferred from the hatchery to the pond, to control stocking density and determine feeding rates;
- At intervals during the grow-out period, to check growth rate and survival; and

☆ When market-sized prawns are harvested, to provide a final or cumulative record (an assessment of the productivity of the pond and management system being used).

The following methods are suggested for stock estimation.

Stock Estimation when Postlarvae are Harvested

The following system is suggested:

1. Before the harvested PL are transferred to the PL holding tank, suspend them temporarily in a small container with a known volume of aerated water;
2. Agitate the water thoroughly to evenly disperse the animals;
3. Take four samples from the container in 100 ml beakers;
4. Now place the bulk of the postlarvae into the holding tank (do not wait until the sample counting process is complete);
5. Count every animal in each of the four 100 ml beakers (one way of doing this is to take quantities into a graduated pipette held at a 45° angle towards a lamp and to count the animals as they swim up towards the light);
6. Average the number of postlarvae found in each 100 ml beaker and multiply this number by the volume of water in the container mentioned in (1) above (in ml) and divide by 100.

Water Quality Management

Water quality has influence on growth and survival of shrimps as they live and grow in water. Water of good quality means all the parameters are in the optimum range of requirement of shrimps. It is necessary to know the parameters which affects on the growth and survival of shrimps. They are (1) Salinity, (2) Temperature, (3) Turbidity, (4) Dissolve Oxygen, (5) pH, (6) Alkalinity and Hardness, (7) Carbon Dioxide, (8) Ammonia. Detailed requirements are as under (Source: MPEDA, 1996).

Salinity

Salinity is the total concentration of dissolved ions in water and is expressed as parts per thousand (ppt or 0/00). Salinity influences many functional responses such as metabolism, growth, migration, osmotic behavior, reproduction etc.

When shrimp is exposed to salinity higher or lower than its optimum requirement, more energy is used for osmo-regulation and hence more feed is required to obtain optimum growth. Shrimp farming is coastal activity and the water is drawn either from the sea or estuary. The estuarine salinity can drop down even to zero ppt during monsoon and increase to 35 to 45 ppt during summer in certain area. It is therefore, desirable to select the species for culture based on salinity condition of the site during the period of culture.

Temperature

Temperature affects the metabolism. The rates of chemical and biological reactions are doubled for every 10°C increase in temperature.The internal tissues of shrimp are typically at nearly the same temperature as the water.

Increase in temperature increase metabolism and thereby growth, but as the temperature increases the demand for oxygen also increase. At higher temperatures, the demand for oxygen increases to such an extent that physiological capacity of shrimp fails to meet the demand. The capacity of water to hold oxygen also decreases with increase in temperature.

The decrease in temperature reduces metabolism and thereby growth and feed consumption. Shrimps hardly feed below 18° C Temperature below 15° C is stressful for shrimp. Surface water gets heated up during the day leaving bottom layer cool. During the night the surface layer cools down sufficiently to cause complete mixing with deep cool water layer. The difference in temperature form surface layer to bottom layer should not be more than 1°C as the same cause cramps (curved stiff) in shrimps.

To maintain optimum temperature level, water depth is to be increased during summer and decrease in winter as well as provide sufficient aerators to mix the water to main same temperature.

Turbidity

The term turbid indicates that water contains suspended materials like silt, clay phytoplankton etc. which interferes with light penetration in the water column. The suspended solids originate either from the external or within the farm (internal). External load depends on the quality of silt, clay and phytoplankton of intake water. The farm source included shrimp feces, uneaten food; algae and erosion of bunds. Since turbidity limits light penetration, it limits photosynthetic activity in the bottom layer. High turbidity can cause temperature and Dissolve Oxygen stratification in shrimp.

To observe turbidity Secchi disc visibility reading should be taken. Secchi disc visibility reading should be between 30 and 40 cm. If reading is less than 30 cm, the phytoplankton density is high. In that case water exchange carried out to flush out excess bloom. When Secchi disc reading is 40 cm or more it indicates low population of phytoplankton. In that case water should be fertilized. The reading should be taken daily at 11=00 hrs.

Dissolved Oxygen (DO)

Oxygen is most important for survival of shrimp. It affects feed consumption, metabolism, environmental conditions, solubility and availability of nutrients. Under anaerobic conditions nitrate, sulfate and carbon dioxide are reduced to ammonia, hydrogen sulfide and methane which are harmful to shrimps. It is very necessary to maintain dissolved oxygen above 3.5 ppm.

When DO concentration is less than 1 mg/lit., normally shrimp becomes lethal, if exposure persists for more than a few hours. When concentration reaches to 1–3.5 mg/lit., shrimp survive but there may be sub lethal effects with poor growth. When it reaches more than 3.5 mg/lit to saturation it considered as the best condition for good growth. However, when the concentration is increased to saturation point it can be harmful, if super saturated condition exists through out pond volume.

D.O. level increases during the day and reaches its maximum in the afternoon due to photosynthesis. During the night oxygen production decreases and is utilized by all living organisms in the pond in respiration.

pH

pH is the hydrogen ion concentration in water. pH indicates the extent of acidic or basic nature of water. Water with pH 7 is said to be neutral. Water with pH below 7 indicates water is acidic and a pH above 7 indicates water is basic.

Water pH affects metabolism and other physiological process of shrimp, exerts considerable influence on toxicity of ammonia and hydrogen sulfide and affects solubility of nutrients and thereby fertility. pH of brackish water is usually between 7 and 9. Shrimps in general are intolerant of extremes of water pH out side the range of pH 6–9. Alkaline and acidic waters reduce swimming performance of shrimp due to ammonia accumulation at high pH and to an impairment of oxygen transfer at low pH. The optimum range for shrimp culture is pH 6.8 to 8.7. In intensive culture pH is maintained between pH 7.6 to 8.5. (MPEDA, 1996)

The best way to counter water pH problem is to lime the pond to increase soil pH and water pH to greater than pH 6. total alkalinity and total hardness to greater than 20 ml/l as $CaCO_3$.

Alkalinity and Hardness

Total alkalinity in water refers to the total concentration of titrable bases in water and is expressed in milligrams per liter of equivalent calcium carbonate.

Total hardness of water is defined as the total concentration of divalent cations in water and is expressed in milligrams per liter of equivalent of calcium carbonate.

Alkalinity primarily determines the magnitude of duel fluctuation of pH of water. Water with low alkalinity (less than 15 mg/l), results in wide fluctuation in pH for value 6 or 7.5 at dawn to 10 or even higher in the afternoon. Very high alkalinity (200–250 mg/l) coupled with low hardness (less than 20 mg/l) results in the rise of afternoon pH beyond 11. Very low and very high alkalinity of water also results in poor productivity due to limitation of carbon dioxide for photosynthesis. pH of water with moderate to high alkalinity (20–150 mg/l) normally fluctuate between 7.5 to 8 at dawn to 9–10 in the afternoon. (MPEDA,1996)

Carbon Dioxide

Carbon dioxide is highly soluble in water. High level of carbon dioxide (more than 10 mg/l) reduces the capacity of blood to transport oxygen. High carbon dioxide concentration may be tolerated by shrimp, if DO level is also high. When oxygen is utilized in respiration it releases carbon dioxide. CO_2 level decreases during the day and reaches its minimum in the afternoon due to removal from water in photosynthesis. Concentration of CO_2 occurs during the night and peaks at dawn due to respiration of all living organisms in pond. High concentration of CO_2 occurs during cloudy weather and following phytoplankton die off. So it is very important to maintain optimum DO level before dawn, on a cloudy day and during plankton die off. Agricultural lime and hydrated lime are used for removal of CO_2 when high levels of CO_2 are encountered after phytoplankton die off.

Ammonia

Ammonia originates in pond water from microbial decomposition of organic matter (uneaten feed, feces, dead phytoplankton). Shrimp also execrate ammonia during protein metabolism. As the Ammonia concentration in water increases, excretion of by shrimp diminishes. The ammonia level in blood and other tissue of shrimp increases. The result is elevation in blood pH and adverse effects on enzyme catalyzed reactions and membrane stability. Ammonia increases oxygen consumption by tissues, damages gills and reduces the ability of blood to transport oxygen. Exposure of shrimp to sub lethal concentration of ammonia increases susceptibility of shrimp to diseases.

Recommended, safe level is less than 0.0025 mg/l of unionized ammonia and 1.0 mg/l of total ammonia. Adverse effects on the shrimp can occur with long term exposure the shrimp to 0.1 ppm ammonia. The higher ranges of ammonia (0.6 ppm–2.0 ppm) are tolerated for short term exposure. Beyond these, the shrimp may become more susceptible to diseases.

The toxic effect of ammonia is minimized by 1) maintaining sufficient level of DO to oxidize ammonia to nitrate by nitrifying bacteria, (2) Providing suitable slope to pond bottom for collection and removal of organic waste, (3) Periodic partial removal of cyano-bacterial bloom and algal bloom by flushing or scooping out scum to maintain optimum density of bloom. 4) Application of Zeolite (200 kg/ha/week) for adsorption of ammonia from the bottom sediment, particularly when sediment is saturated.

Hydrogen Sulfide

Under anaerobic condition, certain heterophic bacteria excrete hydrogen sulfide during decomposition of organic waste. The hydrogen sulfide is oxidized to sulfate by Thiobacullus bacteria in presence of oxygen. Thus H_2S does not occur in well oxygenated pond.

Culture of *Penaeus latisulcatus*

Western king prawns, *Penaeus latisulcatus* are a candidate species for culture in inland saline water. Western king prawn survival, growth, condition, osmo- and ions-regulation was studied when reared in potassium-fortified inland saline water for 202 days. PL40 prawns were stocked into three media in 250 L tanks: inland saline water with potassium fortified to 80 per cent (IS80) of the marine water concentration, 100 per cent (IS100) of the marine water concentration and marine water (MW). By the conclusion of the trial, survival was 53 per cent in IS80, 64 per cent in IS100 and 68 per cent in MW. Mean prawn weight, total length, carapace length, condition factor and moult interval were significantly higher ($P < 0.05$) in MW than in IS100 and IS80. Specific growth rate of prawns in MW was significantly higher ($P < 0.05$) than in IS80. There were no significant differences ($P > 0.05$) in osmoregulatory capacity of prawns between various media. Serum Na^+, K^+, $Ca^{2+}$and S concentrations were influenced by their concentration in the medium. Ca^{2+} was the only major cation hyper-regulated and tended to be accumulated, while Mg^{2+} was maintained at a much lower concentration in the serum than in the medium. Moisture contents in hepatopancreas, tail muscle and exoskeleton and organosomatic indices improved from IS80 to IS100 to MW. The lower growth rates and condition factors of prawns reared in potassium-fortified inland saline water suggests the presence of limiting factors other than potassium concentration. Western king prawns are stronger regulators of divalent cations than monovalent cations and the extra energy required to regulate ions in inland saline water may have been the major cause of lower growth rates.(David I. Prangnell and Ravi Fotedar,2006)

Survival was highest at 22 ppt and food conversion ratios were significantly lowest (P<0.0.5) in prawns cultured at 22 and 34 ppt. Haemolymph osmolality increased with the increase in medium salinity and physiological age of the prawns. Isosmotic points calculated from regression between haemolymph and medium osmolality were 28.87, 29.46 and 31.73 ppt at 0, 20 and 60 days of culture, when body weights were 2.95±0.26, 4.02±0.47, 5.79±0.64 g respectively. After 60 days of culturing, hepatopancreatic moisture levels increased with the increase in salinity, whereas tail moisture levels remained unchanged. Wet and dry hepatosomatic indices were highest at 22 ppt. Wet tail muscle index was highest at 34 ppt, whereas dry tail muscle index did not change by any salinity levels. The results suggest that the optimum salinity range for rearing western king prawns ranges from 22 to 34 ppt and 10 ppt was unsuitable for culture. (Huynh Minh Sang, Ravi Fotedar, 2004)

Penaeus japonicus (New name *Marsupenaeus japonicus*)

Once the eggs have hatched, the water is fertilized to stimulate the growth of diatoms. Predetermined amounts of fertilizer and seawater are added each day to the tank until the larval shrimp have reached the last mysis stage. Brine shrimp nauplii (*Artemia* spp.) are fed from the last mysis stage through the fourth postlarval stage. The shrimp are then fed fresh meats of clams *(Venerupis philippinarum)* and mussels *(Mytilus edulis)*, which are crushed and distributed throughout the ponds. Because it is too costly and time consuming to separate the crushed shell from the meats, the shell eventually covers the pond bottom, resulting in a substrate that hampers the burrowing of the shrimps. Thus, ponds must be drained or dredged periodically to remove the shell debris. (Cornelius, R. Mock, 2006)

Liao (1992) reported growing of *P. japonicus* juveniles in cement concrete tanks of 100 to 250 m^3 capacity, where the juveniles grew to adulthood of 20 to 25 g in a period of six months.

Final yield and survival obtained from *M.japonicus* were 655 kg/ha and 62 per cent. FCR ranged between 0.47 and 7.04 (mean = 3.50). Mean weight increased from 0.02 g to 7.05 g at the end the growth trial. Average daily growth rate was calculated as 0.046 g/day. SGR was the highest (20.29 per cent day-1) during the first 15 days and dropped dramatically to 0.46 per cent day-1 towards the end of growth period. Gürel Türkmen, (*Turkish Journal of Fisheries and Aquatic Sciences 7: 07-11 (2007)*)

Table 11.1: Economics for Extensive Prawn Farm (for Tiger Prawn)

I)	**ASSUMPTIONS**		
a)	Farm Size	:	4 ha.
b)	Culture period	:	4 ½–5 months
c)	Stocking rate	:	40,000 Nos per ha.
d)	Survival rate	:	70 per cent
e)	Harvest weight	:	35 gms
f)	Average yield	:	950 kg/ha/crop
g)	Feed conversion ratio	:	1:4 with wet compound feed costing Rs 8/- per kg
h)	Pond size	:	1 ha (4 Nos.)
II)	**DEVELOPMENT COST ESTIMATE (4 ha water area)**		
			Rs. In lakhs
a)	Earth work for 12,000 m^3	:	2.40
b)	Stone pitching on one of the four sides	:	0.30
c)	Main Inlet	:	0.20
d)	Main outlet (2 Nos)	:	0.30
e)	Individual outlets (4Nos)	:	0.32
f)	Pump (4 Nos. 6 HP each)	:	0.80
g)	Pump House cum shed	:	0.20
h)	Store cum office	:	0.60
i)	Miscellaneous	:	0.80
	Total	:	**5.20**
	Development cost per ha.	:	**Rs. 1.3 lakh**
III)	**OPERATIONAL COST FOR ONE CROP (4 ha. water area)**		
a)	Seed (1.6 lakhs)	:	0.240
b)	Feed (15200 kg @ Rs 8/- per kg)	:	1.216
c)	Eradicator (MOC @ 600 kg/ha @ Rs 3/- per kg)	:	0.072
d)	Fertilizer and lime (800 kg.lime @ Rs 1/- per kg; 20 tonnes of cowdung @ Rs. 150/- per tonne; 600 kg Urea and 600 kg Super phosphate @ Rs. 2.50 per kg each)	:	0.068
e)	Staff salary (one Farm Supervisor @ Rs 1200/- per month and one Watchman @ Rs 500/-per month)	:	0.204
f)	Diesel	:	0.090
g)	Harvest	:	0.038
h)	Causal labourers, Chemicals and other miscellaneous expenses	:	0.048
	Total	:	**1.976 lakh**
	Operational cost per crop per ha	:	Rs. 0.494 lakhs
	Operating cost per year (two crops)	:	Rs 3.952 lakhs
	Operating cost per year per ha	:	Rs 0.988 lakhs
	Production cost per ha = $\frac{\text{Rs } 197600/-}{3800 \text{ kg}}$	:	Rs 52/- per kg

Contd...

Table 11.1–Contd...

IV)	**GOSS INCOME FROM SALES**		
a)	From one crop	:	3800 kgs X 110/- per kg
		:	Rs 4.18 lakhs
b)	From one crop per ha	:	Rs. 1.045 lakhs
c)	Per year	:	Rs 8.36 lakhs
d)	Per year per ha	:	Rs 2.09 lakhs
V)	**ECONOMICS AND REPAYMENT**		
a)	Total Investment	:	Cost of development + operational cost for 1st year
		:	Rs 5.20 lakhs + Rs 3 .952 lakhs
		∴	Rs. 9.152 lakhs
b)	Gross profit before interest and repayment		
		:	Gross income per year–(Operational cost for next year + Depreciation @ 10 per cent of the cost of development.)
		:	Rs. 8.36 lakhs–(3.952 + 0.520) lakhs
		:	Rs 3.888 lakhs.

Assuming that the Total Investment is borrowed from bank at the rate of 12.5 per cent interest, a replacement schedule for 7 years is shown below.

Table 11.2: Bank Loan, Interest and Replacement Schedule for Extensive Farming (All figures are Rs in lakhs)

Year of Culture	*Principal Outstanding*	*Yearly Payment to Bank*			*Gross Yearly Income*	*Net Profit after payment to Bank*
		Annual Repayment	*Interest*	*Total Amount*		
1	9.152	1.30	1.144	2.444	3.888	1.444
2	7.852	1.30	0.982	2.282	3.888	1.606
3	6.552	1.30	0.819	2.119	3.888	1.769
4	5.252	1.30	0.657	1.957	3.888	1.931
5	3.952	1.30	0.494	1.794	3.888	2.094
6	2.652	1.30	0.332	1.632	3.888	2.256
7	1.352	1.352	0.169	1.521	3.888	2.367

After 7th year no further payment is to be done to the bank and the gross yearly income becomes the net profit.

Source: M.P.E.D.A. (Govt. of India) 1991.

Table 11.3: Economics for Semi-Intensive Prawn Farm (for Tiger Prawn)

I)	**ASSUMPTIONS**		
a)	Farm Size	:	4 ha
b)	Culture period	:	4 ½–5 months
c)	Stocking rate	:	2.5 lakhs per ha
d)	Survival rate	:	70 per cent
e)	Harvest Weight per prawn	:	30 gms.
f)	Average yield	:	5000 kh/ha/crop
g)	Feed conversion ratio	:	1:2 with formulated palletized feed costing Rs 21/- per kg.
h)	Pond size	:	0.5 ha (8 Nos)
II)	**DEVELOPMENT COST (4 ha water area)**		**(Rs. In lakhs)**
a)	Earth work (16200 m^3)	:	3.24
b)	Elevated canal with brick masonry (200m)	:	0.40
c)	Pond inlets (16 No. free fall PVC pipe)	:	0.24
d)	Pond outlet (8 No. Hume pipe Monk)	:	0.64
e)	Main outlet sluice (2 Nos. RCC open type)	:	0.30
f)	Pump House	:	0.10
g)	Generator house	:	0.10
h)	Watchman shed (2 Nos.)	:	0.16
i)	Office, Laboratory, dormitory	:	1.50
j)	Store	:	0.60
k)	Drinking water bore well	:	0.75
l)	Pump (3 No. mixed flow pump each 25 HP)	:	2.25
m)	Generator (2 No. 50 KV each)	:	3.00
n)	Aerators (32 No. 1 HP each)	:	5.44
o)	Electrical Installations	:	0.80
p)	Area lighting	:	0.30
q)	Miscellaneous Expenditure	:	0.18
	Total	:	**20.00**
	Development cost per ha	:	**Rs. 5 lakhs**
III)	**OPERATIONAL COST FOR ONE CROP (4 ha water area)**		**Rs in lakhs.**
a)	Seed (10 lakhs)	:	1.500
b)	Feed (40,000 kg @ Rs 21/- per kg)	:	8.400
c)	Eradicator (200 kg/ha @ Rs 3/-per kg)	:	0.024
d)	Fertilizer and lime (800 kg lime @ Rs 1/- per kg 20 tonnes cow dung @ Rs 150/- per tonne; 600 kg urea and 600 kg super phosphate @ Rs. 2.50 per kg each	:	0.068

Contd

Table 11.3–Contd...

e)	Staff Salary (One Farm Manager @ Rs. 2000/- per month; one Mechanic @ Rs. 900/- per month; one Farm hand @ Rs. 500/- per month Two Watchmen @ Rs 500/- per month each)	:	0.528
f)	Labour (Pond preparation, stocking etc.)	:	0.080
g)	Harvesting	:	0.200
h)	Diesel	:	1.000
i)	Maintenance of structure, machineries etc.	:	0.300
j)	Chemicals, Office expenses and other miscellaneous expenses	:	0.300
	Total	:	**12.400**
	Operational cost per crop per ha.	:	Rs 3.1 lakhs
	Operational cost per year (Two crops)	:	Rs. 24.8 lakhs
	Operational cost per year her ha.	:	Rs. 6.2 lakhs
	Production cost per kh = Rs 12,40,000/- / 20,000	:	62/- per kh.
IV)	**GROSS INCOME FROM SALES**		
a)	From one crop = 20,000 kg X Rs 100 per kg	:	Rs 20.00 lakhs
b)	From one crop per ha.	:	Rs. 5.00 lakhs
c)	Per year	:	Rs. 40.00 lakhs
d)	Per year per ha.	:	Rs. 10.00 lakhs
V)	**ECONOMICS AND REPAYMENT**		
A)	Total Investment	:	Cost of Development + Operational cost for 1st year
		:	Rs. 20.00 lakhs + Rs. 24.80 lakhs
		:	Rs. 44.80 lakhs
B)	Gross Profit before Interest and repayment		
		:	Gross Yearly Income–(Operational cost of next year + Depreciation @ 10 per cent of the Development cost)
		:	Rs. 40.00 lakhs–(24.80 + 2.00) lakhs
		:	Rs. 13.20 lakhs

Assuming that the Total Investment is borrowed from the bank at the interest rate of 12.5 per cent, a repayment schedule for 7 years is shown in Table 11.4.

Culture of *M. dobsoni*

Of various relations attributed to the crustacean eyestalks, the relation between the eyestalks and moulting is of great significance. Moulting in crustaceans was thought to be regulated by two hormones-the moult inhibiting hormone and moulting hormone. It is believed that the moult inhibiting hormone is produced in the eyestalk and stored in the sinus gland and the moulting hormone is produced in Y organ. When the eyestalks are ablated, the moult inhibiting hormone is excluded thus allowing the moulting hormone to act. Thus the removal of eyestalks shortens the intermoult period. Several investigators worked on the endocrine control of moulting and growth in decapod crustaceans. The moulting changes in the animal *Metapenaeus dobsoni* due to different levels of eyestalk ablation Prawn fishery is an important marine resource in India and accounts for about *1 per cent* of total marine fish landing and about 50 per cent of the total marine fishery export. Prawn farming in brackish water can enhance production to augment the export potential. Two important requirements in culturing prawns are the availability of prawn seeds and methodology to accelerate growth. Eyestalk ablation is a frequently adopted procedure for induced maturation of gonads and spawning.

Table 11.4: Bank Loan, Interest and Replacement Schedule for Semi- Extensive Farming

(All figures are Rs in lakhs)

Year of Culture	*Principal Outstanding*	*Yearly Payment to Bank*			*Gross Yearly Income*	*Net Profit after payment to Bank*
		Annual Repayment	*Interest*	*Total Amount*		
1	44.80	6.40	5.60	12.00	13.20	1.20
2	38.40	6.40	4.80	11.20	13.20	2.00
3	32.00	6.40	4.00	10.40	13.20	2.80
4	25.60	6.40	3.20	9.60	13.20	3.60
5	19.20	6.40	2.40	8.80	13.20	4.40
6	12.80	6.40	1.60	8.00	13.20	5.20
7	6.40	6.40	0.80	7.20	13.20	6.00

After the 7th year no further payment is to be done to the bank and the gross income becomes the net profit.

Source: M.P.E.D.A. (Govt. of India) 1991.

The crustaceans eyestalk is known to have a neurohaemat function due to the presence of the X-organ-sinus gland system. Excision of eyestalk is a classical endocrinological experiment to determine the functions of the eyestalk neurosecretory system. Besides, the established effect of the reproductory function, manipulation of the hormonal supply by eyestalk extirpation can bring about alterations in the physiology of prawns.

Eyestalk ablation is usually resorted to obtain enhanced growth. This method is also most frequently performed for inducing maturation of gonads and spawning of prawns. Extirpation of an eyestalk removes the endocrine system of prawns located in the eyestalk which influences growth, reproduction and other metabolic activities. Besides, eyestalk ablation is reported to influence lipid metabolism, protein metabolism, carbohydrate metabolism, hydromineral regulation, gonad inhibition and limb growth, Moulting in Crustacea is part of the mechanism of growth. Change in form and increase in size can only occur when the hard calcareous exoskeleton is shed and before the new cuticle is hardened. Removal of eyestalk leads to increased weight and decreased osmolality of aemolymph. Unilateral eyestalk ablation has been employed to induce both ovarian maturation and spawning with varying success in many species. Unilateral eyestalk ablation can also be used to shorten the moult interval and to stimulate gonad development in shrimps

After eyestalk ablation the animals developed crimson red pigmentation particularly at the terminal parts of the body and also on the peopods gradually becoming grayish white in colour.

Breeding

Water to be kept fully saturated with oxygen by providing continuous aeration. *pH* of the water should be maintained close to the value 8 by adding anhydrous calcium carbonate. It has been suggested that the size alone may not be taken in account in determining the age and that older individuals of smaller size might be present in the sample (Venkitraman *et al.*).

Net Growth Efficiency

The net growth efficiency of ablated and control prawns was estimated by the formula, Net growth efficiency = (Px100)/A where P is the production in calories and A is the assimilation in calories.

Food Consumption

Eyestalk ablation significantly altered the food consumption rate. There are two types of eye ablation, unilaterally (UEA) and bilaterally (BEA). In the 35- 40 mm size group the average weight of food consumed by the BEA prawns (0.101g) was 21.5 per cent less than the consumption of control prawns (0.129g). However, the UEA prawns showed not much change in the consumption rate (0.128g) compared to control prawns. In the 48-53 mm size group consumption rate of BEA prawns (0.146g) was 43.4 per cent less than that of the control prawns (0.257g) and that of UEA prawns (0.360g), 57.8 per cent more than that of the control. The production rate of UEA prawns was 84 per cent more than that of the control whereas BEA prawns indicated negative production. The same trend followed for net growth efficiency also. The production rate of UEA prawns was 84 per cent more than that of the control here as BEA prawns indicated negative production.

Feeding

Prawns were fed with fresh clam meat twice per day. For feeding, food, approximately equivalent to 20 per cent of the body weight of the animal was used. The unused food was removed after 6 to 8 hrs. The individual compartments and PVC tanks were cleaned daily. The moult, excreta and unused food were blotted and weighed separately. They were dried for sufficient length of time at 55°C until a constant weight was achieved and stored in the desiccators for further analysis.

Chapter 12
Shrimp Feed and its Management

Introduction

Feed composition of stomach content in different prawn species in natural environment have been studied by different workers. From the results of the same it is observed that generally all prawns eat different diatoms, crustaceans, detritus, mud, polycheates, fishes, algae etc. which are available from the habitat in which they exist. The percentage and variations of feed content varies even from the same species when its living habitat changes.

It should be noted that in mass production of prawn the growth has to be faster in specified time and thus artificial feeding is preferred to natural feed. Different stages of prawn need different type of feed with different dietary lipids, fatty acids and phospholipids for growth and survival rate.

Feeding and feed management is one of the most important operational functions in shrimp farming, as adequate food supply has to be ensured to attain the cultured animals at desired harvesting size within the targeted time frame. Feed is the largest operational cost of shrimp farming and every efforts should be made to ensure efficient utilization of feeds for growth. It is therefore necessary to have adequate knowledge of the feeds and feed management for a successful farm operation.

Feed is one of the important and essential inputs in shrimp farming. In scientific shrimp culture where the cultured organisms are stocked at a high density, the amount of natural food available in the pond is not sufficient to support good growth even with fertilization. As the shrimp grows, the food requirement will increase substantially and if they do not get sufficient food, growth will retard leading to poor survival and production. Hence, it is necessary to increase the production by supplementary balanced artificial feed. The role of artificial feed in shrimp farming is greatly dependant on the culture system or cropping density employed.

TYPE OF FEEDS

Feed to be Given at Nauplier Stage

During nauplier stage no feed is required. From nauplier 5 stage, *Chaetoceros sp.* should be given as feed. From the protozoea–3 or mysis–1 stage freshly hatched and frozen artemia nauplii should also be given. It is suggested that up to PL-1 artemia nauplii of San Fransisco Bay Brand may be given as the size of cyst of these varieties is 0.17 mm (227 µ). The count of cyst is 2 to 2.7 lakh per gm.

Natural Feed

Natural food grows in shrimp pond after application of predator control chemicals and fertilizers. They are mainly blue green algae which form a complex with the associated zooplankton. The natural feed thus developed are called "Lab Lab" and "Lumut" Lab Lab is benthic blue green algae, diatoms and many other forms of plants. "Lumut" is composed of filamentous green algae and many other form of life.

Wet Feed

This type of feed can support production upto 300 kgs/ha in traditional farms, this comprises of fresh fish, mussels etc and are traditionally fed to the shrimps. These feeds are suitable in extensive farming but not in semi-intensive farming because water quality is affected due to disintegration of feed and thereby creating unhealthy environment. Feed quality is inconsistent and is not nutritionally balanced.

Pellet Feed

Considering the feeding habits of shrimps and the role of feed in growth of shrimp and the economics of culture, nutritionally balanced feed in the form of pellets is used in semi–intensive and intensive farming systems.

The thrust of the general aquaculture feed industry from the earliest stages of production had its emphasis on the lowest possible cost of production and the highest possible production rate, quality and nutrition. A wide variety of techniques are being employed to manufacture complete aquaculture feeds. From flaking to wet compaction filleting, steam compaction pelleting and extrusion pelleting.

Extrusion Feed Preparation Method

The extrusion process consists in pushing a product through a small size aperture called die. It has been used in a variety of industries for many decades. The first extruder application on feeds was done to manufacture pet food in middles of 40s by the well known pet food manufacturers Relston Purinam in the United States. For preparing feed from bacteria heat and steam process was developed.

More efficient shrimp feed production were introduced heat and steam under high pressure within an enclosed chamber in a continuous manner. The results of a process utilizing high temperature short time (HTST) cooking are gelatinization of starch, denaturing of protein, inactivation of many raw food enzymes and elimination of microbial counts in the final products. In case of compaction pellets, the starch granules would form tiny particles giving rise to a sandy, grainy texture whereas with extruded pellets the starch is fully gelatinized forming an interconnected matrix composed of long and short carbohydrate chains. In case of proteins, their structure is broken down in the extruders high temperature to form a melted dough–like product and expand to form an elastic and meat-like texture when they come out of a die. Such a matrix is strong enough to contain various soluble materials which also combine with the starch matrix. The pellets prepared with such process are most long lasting. (Source: Jaaraman, R and Karl Mark, 1998)

Advantages of Palletized Feed

1. Slow leaching of nutrients helps in maintenance of of water quality.
2. Can be well balanced with amino acids, vitamins, minerals and trace.
3. Elements for better growth.
4. 3 to 4 hour water stability enables the animals to eat the feed well.
5. Available in different shapes and sizes to suit different stages of shrimp.
6. The feed with consistent nutritional level can be purchased at a time.
7. Stored for a fairly long period.

Pellet Size

Depending up on the size of shrimp pellet size should be different to adjust mouth size. The feed should be of proper size and it should be stable, should not loose its shape and not disintegrate minimum up to two hours. Shrimp will consume feed within two hours. All pellets should be of same size and shape. Recommended size of feed for respective size of shrimp is shown as under:

Size of Shrimp	*Size of Pellets*
PL–20	0.2 to 0.5 mm pellets (Pre starter)
1–3 gm	1.0 mm pellets (Starter)
5–10 gm	1.8–2.00 mm pellets (Grower)
30–35 gm	2.3–2.5 mm pellets (Finisher)

Source: Dholakia *et al.*, 2004.

Calculation of Feeding

Shrimps are fed at a particular percentage of their body weight ranging from 10 per cent at the initial stages to 2 per cent towards harvest. Generally it is recommended that during first month 10 per cent of body weight may be given, while during second month about 8 per cent, during third month about 5 per cent and during fourth month of culture it may be 3 to 2 per cent of body weight. In order to calculate the quantity of feed required as a percentage of standing crop, regular sampling and assessment of standing crop is essential. Standing crop can be assessed by wooden frame method.

Calculation of Daily Feed Requirement

Daily feed requirement = Standing crop X percentage of feed.

Example

Assume average weight	:	8 gms
Stocking density	:	1,25,000 fry per ha.
Survival rate	:	80 per cent
Per cent of feed	:	5.8
Therefore total feed required	:	1,00,000 fry x 8gm x 5.8 per cent = **46.4 kg**

Assessing Survival Rate

To assess the feeding and to save feed from wastage and further deterioration and to increase profitability in culture, assessment of feeding rate and survival rate is essential. Put 100 gm of feed in one feeding tray (2′ X 2′) and float in different locations. Check tray after 30 minutes and see that feed is consumed or not? If total feed is consumed, add 10gm more and check. By this way calculate feed requirement considering the size of pond.

Survival rate can be calculated as under:

$$\text{Survival rate} = \frac{\text{Actual feed consumption}}{\text{Calculated feed requirement}}$$

Example

$$= \frac{24 \text{ kg}}{30 \text{ kg}} = 80 \text{ per cent survival}$$

Or using following formula survival rate can be calculated.

$$\text{Survival rate} = \frac{\text{Actual feed consumed per day}}{\text{No. of animals stocked} \times \text{Av. Body weight} \times \text{Per cent feeding rate}}$$

The growth of the animals and increase in weight of the animals over a week or fortnight, the average gain in the weight can be calculated by other data available from the pod, given the FCR.

Example

Average body weight = 20 gms

Feed percentage = 3.5 according to feeding practice.

$$\text{Growth rate per day} = \frac{\text{ABW} \times \text{Feed per cent}}{\text{FCR}}$$

$$= \frac{20 \times 3.5}{1.5 \times 100} = 0.46$$

The growth per week = 0.46 x 7 = 3.22 gms.

Feed Conversion Rate (FCR)

This is an expression denoting the quantity of feed required to get one kg of flesh. The FCR is calculated by applying following formula.

$$FCR = \frac{\text{Quantity of feed consumed}}{\text{Total weight gain}}$$

E.g. If in a pond 1000 kgs feed is used for growing 500 kgs of prawn, then the

$$FCR = \frac{1000}{500} = 2.1$$

Higher the FCR, poor the feed quality, higher the cost of production, higher amount of organic load in the pond. If the FCR is lower the pond can be kept in a good condition.

One tonne of feed is needed to produce one tonne of shrimp in intensive/semi-intensive culture. It is now well known that in shrimp culture the feed conversion ratio (FCR) falls in between 1.1 to 1.8 and feed alone accumulates for more than 50 per cent of the total production cost. The FCR value for freshwater prawn is somewhat higher than that of shrimp. Increasing aquaculture practice lead to corresponding demand for feed, which in turn leads to an increase in the feed cost since the demand exceeds the supply.

Feed Efficiency (FE)

Feed efficiency is calculated to grade the efficiency of the feed. This is an expression of quantity of shrimp obtainable per kg of feed.

$$FE = \frac{\text{Total weight gain}}{\text{Quantity of feed consumed}}$$

E.g. (1) 1000 kgs of feed is required to grow 500 kgs of shrimp.

$$FE = \frac{500}{1000} = 0.5$$

(2) 100 kgs feed is required to grow 2000 kgs of shrimp

$$FE = \frac{2000 \text{ kgs}}{1000 \text{ kgs}} = 2.0$$

In these two cases, while the first example shows lower feed efficiency, the second one shows higher feed efficiency. Here, higher the number more efficient the feed the lower the number less efficiency is the feed.

Use of Antiobiotic Feeds and Withdrawal

When more and more people resort to intensive farming system without proper management there is more chance of prawns getting disease. To present disease sometimes antibiotic incorporated feed is used. Constant use of antibiotics and the presence of antibiotics above a particular limit in the shrimp is not acceptable for human consumption. Whenever antibiotic feeds are used, towards end of culture period, feed free of antibiotics is to be used. The following table gives the withdrawal for a few antibiotics.

Table 12.1: Recommended Withdrawal Period for the Administration of Various Drugs in Different Rearing Water Temperatures

Sl.No.	*Name of Drug*	*Temperature*		
		Less than 12° C	*12–22 ° C*	*Greater than 22° C*
			Days	
1.	Oxytetracycline	60	40	15
2.	Oxolinic acid	60	40	15
3.	Furazolidone	40	20	10
4.	Sulfamonomethoxine	60	30	15
5.	Sulfadomethoxine	60	30	15
6.	Neomycin	40	30	15
7.	Nalidixic acid	40	20	10
8.	Piromidic acid	40	20	10
9.	Nifurpirinol	40	20	10

Source: MPEDA, 1998.

Selection of Good Quality Feed

Selection of good quality feed is an important parameter for better production, growth and survival. It is necessary that depending of size of prawn/shrimp size of pellets/crumbs should be selected. Pellets/crumbles should be of uniform size, shape and colour. It should be properly nutritionally balanced feed for the respective stage of prawn. A proper mixing and grinding with micro pulverizer result in uniform distribution of each and every component of the feed. Fish meal; being the major ingredient the colour of the fed should almost similar to the colour of the fish meal. Change in the colour may be due to bad quality ingredients or improper drying. The feed should be free from fungus and insects and free flowing without cake formation.

Feed should not contain powder. Powder shows poor quality of pelleting. If fed with more powder feed there will be more loss of feed as the dust will be blown up in the wind. A good feed will have a fish meal smell.

Water stability of feed is another important factor to be considered for production. This can be tested by putting small quantity of feed in a glass of water. The pellets should well on putting in water and become soft and maintain its shape in water for 2–3 hours enabling the animals to eat freely.

Feed Purchase and Storage

1. Feed should be purchased as far as possible as per requirement keeping a little buffer. Purchase of feed at a time should be restricted to one crop requirement.
2. It is always advisable to check the freshness of the feed by looking at the date of manufacture. The bag should not be damaged and should not be wet.
3. Feed bags should be stored on wooden platform in clean dry and cool (temp. 24–25°C) store room with good ventilation and less than 75 per cent humidity.
4. The store room windows should provided with metal screens to prevent entry of rats, birds and other animals. Doors and wall should be free from holes to prevent entry of rats.
5. During unloading and stocking, bags should be handled gently to prevent powder formation.
6. The store room should be used exclusively for storing feeds only and not for any other purpose.
7. Stocking should be in such manner so as to enable rotation of stock that first received should be first used.

Source: MPEDA, 1996.

Feed Management

On the basis of above calculations estimated feed broadcasting may be decided. Secondly for better feed management total feed calculated may be divided in four equal parts and 25 per cent of such feed may be broadcasted to the culture farm at proper feeding time. *i.e.* at 6.00 hrs, 11.00 hrs, 16-30 hrs, and 22.00 hrs. However, as per recent trends in shrimp feeding practice, Mariappan and Chellam (1998) suggests five times feeing with different percentage of feed. According to them, at 6.00 hrs (20 per cent), 10.00 hrs (10 per cent), 14.00 hrs (10 per cent), 18.00 hrs (30 per cent) and 22.00 hrs (30 per cent). Thus 60 per cent of feed is given during night time due to the nocturnal feeding habit of shrimp. By such increased consumption of feed during night hours than during day time an enhanced growth rate has been reported for juveniles of *P. monodon* (Reymond and Lagarder, 1988).It is explained that in *P. japonicus* the activity of digestive enzymes increases in 3–4 hours after sunset (Cuzon *et al.*, 1982). Temperature also influences the intake of feed.

Feed quality is an important criterion, which directly influences the growth rate of shrimp/prawns and contributes to profitable harvest from commercial farms.

Daily adjustment in feed ration confirms an effective strategy of feed management. Observations are made frequently to know whether the given feed is consumed or not. Depending on consumption levels, the quantity of next meal is adjusted.

The number of feed trays required in the pond is based on the area of the pond. Four to six feeding trays are normally required for a one ha pond. Generally one or two additional feed trays are used to ensure better feed management. The feeding trays should be ideally placed to facilitate feeding and to enable easy access for observation. Feeding trays also serve the purpose of observing the health status of the animals, moulting and other related observations apart from the patterns in the consumption of feed.

Proper selection of feed particle size also helps top conserve the energy of shrimp by way of minimizing its expenditure for food gathering. Along with available natural food and algal blooms in the medium ensure growth and maximum survival rate.

Feed Composition

Feed is one of the important and essential in put in shrimp farming. In scientific shrimp culture where organisms are stocked at high density, the amount of natural food available in the pond is not sufficient to support good growth even with fertilization. As the shrimp grows the requirement will increase substantially. Hence to increase the growth and keep good health, it is necessary to supply proper type of feed. Many workers have studied to find out optimum composition for shrimp feed. The consolidate results of all workers are tabulated as under:

Protein	38.00 per cent
Fat	8.00 per cent
Carbohydrate	22.00 per cent
Fibers	3.50 per cent
Moisture	8.60 per cent
Minerals	15.00 per cent

Source: Dholakia, A. D. *et al.*, 2004.

One can select different ingredients for preparing feed. It is necessary to know percentage of nutrients. It is recommended in Table 12.2.

Table 12.2: Percentage of Nutrients and Requirements

Sl.No.	*Nutrients*	*Percentages*	*Requirement*
1.	Protein	30–45 per cent	It is a main ingredient
2.	Fat	6–10 per cent	Fat should contain poly unsaturated fatty acid, Polyphosphate, Lacithin and Cholesterol
3.	Carbohydrates	Dy saccharides (Sucrose)	It is necessary to have poly saccharides in shrimp feed

Contd...

Table 12.2–Contd..

Sl.No.	*Nutrients*	*Percentages*	*Requirement*
4.	Cellulose fiber	6 per cent	Necessary for good digestion
5.	Vitamin	C	Necessary for healthy growth and reduce mortality rate
6.	Vitamin	A, D, E, and B group	Necessary for good health
7.	Minerals	Ca, P, Mg, K, Cu, Zn and Selenium	Necessary for good quality of food

Source: Desai, A. Y. *et al.*, 2004.

Protein

To obtain maximum growth, an optimum level of dietary protein is needed. Generally shrimp have a gross protein requirement of 38–46 per cent. Similarly amino acids in protein also play an important role. The proper combination of different protein sources can ensure the presence of all essential amino acids in the prepared diet. Such amino acids can be obtained from plant as well as animal protein.

Average protein content of different animal and vegetable source are as under:

Animal Protein

Fish Meal	45–60 per cent
Squid Meal	60 per cent
Jawla (type of non-Penaeid shrimp)	45 per cent
Beef meat	75–80 per cent

Vegetable protein

Soya bean oil cake	45–60 per cent
Ground nut oil cake	40–45 per cent
Maize	11–15 per cent

Lipids

Similarly dietary lip has two function (1) as a source of energy (2) as a source of essential fatty acid, steroid and phospholipids. It is recommended that lipid level should range between 6.0 to 7.5 per cent. The high content essential fatty acids found in plant oils are Linoleic and Linolenic, while in marine animal oils Eicosapentaenoic and Decosahexaenoic fatty acids are more. Recommended level for phospholipids is 2 per cent. Squid, shrimp contains 35 per cent to 50 per cent of phospholipids. Carbohydrate provides a cheap source of energy for shrimps. They include starch, cellulose and chitin. Chin and cellulose are fibrous. High levels of fiber are not recommended. A minimum level of chitin in shrimp feed is recommended.

Vitamins

Vitamin content in typical shrimp diets is slightly higher because of considering degradation of leaching. They are important organic compounds. However it required to trace quantities for metabolic process and for normal growth.

Fibers

Fiber is also one of the important ingredients. It helps in digesting food properly. Fibers are available in vegetable feed only. It is therefore necessary to mix some percentage of vegetable ingredients. Rice bran contains about 4–6 per cent of fibers, in Soyabean it is 4–7 per cent, in Ground nut oil cake it ranges from 4–6 per cent. While dry maize contains about 2–4 per cent of fiber.

Considering all above aspects different workers have prepared shrimp feed with different composition. Some of them are mentioned as under:

Table 12.3: Formula No. 1

Component	*Percentage*
Shrimp meal	15.0
Fish meal	30.0
Soyabean extract	15.0
Rice bran	15.0
Wheat Flour	15.0
Starch	5.0
Fish oil	4.0
Vitamin mineral mix	0.95
Vitamin "C"	0.05

Source: Chhaya N. D. 1993.

Table 12.4: Formula No. 2

Component	*Percentage*
Fish meal	16.3
Shrimp meal	13.2
Soyabean meal	24.0
Rice bran	21.4
Wheat bran (With Glutton)	15.0
Binder	5.0
Fish oil	2.5
DaiCalcium Phosphate	1.9
Trace mineral mixture	0.2
Vitamin mixture	0.5

Source: Chhaya N. D. 1993.

Table 12.5: Formula No. 3

Component	*Percentage*
Ground Nut oil cake	71.27
Edible Protein Concentrate (from fish, Protein 84 per cent)	23.92
Shark liver oil	2.39
Starch	1.91
Vitamin/mineral mix	0.50

Source: Dholakia, A.D., 1994.

Table 12.6: Formula No. 4

Component	*Percentage*
Molluscan meal	32.68
Shrimp meal	32.68
Wheat flour	13.32
Corn waste	13.32
Molasses	2.00
Palm oil	2.00
Cod liver oil	2.00
Vitamin/mineral mix	2.00

Source: MPEDA, 1996.

Table 12.7: Formula No. 5

Component	*Percentage*
Rice bran	20.00
Soyabean bran	20.00
Coconut cake	12.00
Fish meal	25.00
Wheat flour	20.00
Aqua mix	2.00
Fish oil	1.00

Source: MPEDA, 1996.

Table 12.8: Formula No. 6

Component	*Percentage*
Shrimp meal	27.50
Fish meal	27.50
Rice bran	20.00
Bread flour	15.00
Corn starch oil	5.00
Fish liver oil	2.00
Soyabean oil	2.00
Vitamin/mineral mix	0.95
Vitamin C	0.05

Source: MPEDA, 1996.

Table 12.9: Formula No. 7

Component	*Percentage*
Fish meal	30.60
Prawn waste	30.60
Ground nut oil cake	30.60
Wheat flour	08.20

Source: Pawase, A.S. and Shakuntla Shenoy 1998.

Table 12.10: Formula No. 8

Component	*Percentage*
Fish meal	24.01
Squid	24.01
Ground nut oil cake	24.01
Wheat flour	27.97

Source: Pawase, A.S. and Shakuntla Shenoy 1998.

Table 12.11: Formula No. 9

Component	*Percentage*
Fish meal	15.0
Squid meal	15.0
Shrimp Head meal	15.0
Soyabean extract	15.0

Contd...

Table 12.11–Contd...

Component	*Percentage*
Rice bran	14.0
Wheat flour	15.0
Cod liver oil	5.0
Soyabean lecithin	3.0
Vitamin mix	2.0
Mineral mix	1.0

Source: MPEDA 1998.

Chapter 13

Important Prawn Diseases and their Treatment

The term *DISEASE* can be defined as *"Any departure from normal structure or function of the animal"*. We know that most of the organisms are healthy in their natural habitat, but when these organisms are cultured, we may not be able to maintain all the environmental factors. When organisms are susceptible to stress, leads to disease. These diseases can be grouped in to two main group (1) Infectious disease and (2) Non-infectious disease.

Infectious Diseases

In shrimp/prawn infectious disease caused by microbial pathogens and animal parasites like different type of virus, bacteria, etc.

Non-infectious Diseases

The non-infectious diseases are genetical environmental, toxic nutritional and unknown etiology disease.

Taxonomical Word Used in Disease

The initial infection in a host with a parasite is called *Primary infection* and subsequent infection by the same parasite is termed as *re-infection*. When a host's resistance is lowered by a pre-existing disease or parasite infection, an additional disease is called *secondary infection*. If the infection is only in a localized area in a host (e.g. gill rot), it is known as *local infection* or *local sepsis*, but if the microbes multiply internally in the vital organ, then it is described as *systemic infection* (*e.g.* furunculosis). In the host, if signs of infection are not apparent, but the disease agent can be detected, then it may be denoted as *in apparent* or *latent infection*. Continuous infection with a potential pathogen is known as *persistent infection*. The presence of a known or potential bacterial pathogen in the blood is called *bacteremia*. Previously in the older tests this condition was called as *septicaemia*.

In all animals the occurrence of infectious disease is unpredictable. It is often considered that occurrence of disease in the culture system is unavoidable. However, this need not be so, for many diseases can be prevented by adopting prophylactic measures or certain management techniques. Culture systems are a variety of water sources which are classified as under:

Culture Ecosystem

- Salt water
  - Open Sea
  - Saline lagoons
  - Estuaries
  - Brackish water
- Freshwater
  - Rivers
  - Ponds (Canals)
  - Tanks
  - Lakes Reservoirs

Each water sources has their own ecological and biological characters. Depending upon the system type and occurrence of disease may change. The signs produced by a disease may be of considerable help to provisionally diagnose a case. However, detection of the a etiological agent of the disease will be necessary to confirm diagnosis. Additionally, the latter may be especially important in order to ensure correct treatment.

Details of disease, Identification symptoms, causing agents, Preventive method of control of such disease and treatment is given hereunder:

INFECTIOUS DISEASES

1. Black Spot or Brown Spot or Burnt Spot

Identification

Necrotic lesion of the exoskeleton and brown to black spot appears. Erosion of carapace, abdominal segments, rostrum, tail, gills and appendages were observed.

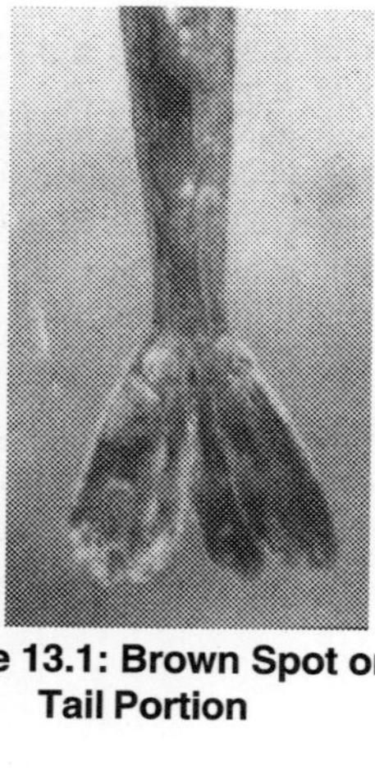

Figure 13.1: Brown Spot on Tail Portion

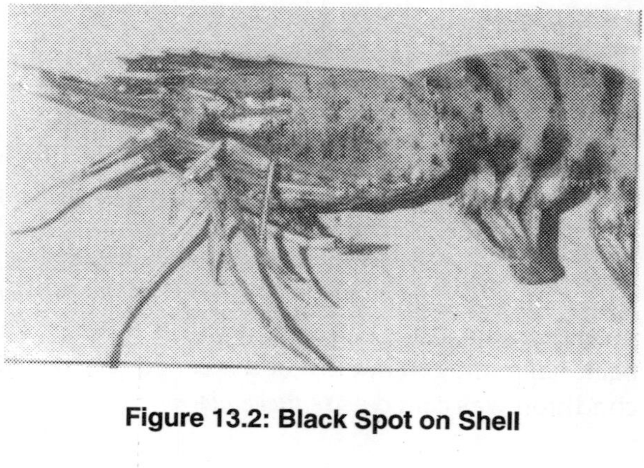

Figure 13.2: Black Spot on Shell

Figure10.3: Overhead Air and Water Distribution Systems are Used to Supply these Indoor Nursery Tanks (USA) (Page 263)

Figure 10.4: Nursery Tank (Page 264)

Figure 10.8: Pond Banks Over Laid with Turf (Page 270)

Figure 10.9: Pond Banks Over Laid with Turf (Page 270)

Figure 10.16: Using Paddlewheel Aerators keeps the Dissolved Oxygen Level High Enough to Increase Stocking Levels (Page 276)

Figure 10.18: Substrates have been Placed Vertically in this Temperate Zone Rearing Pond for *Macrobrachium rosenbergii* Culture (USA) *Source*: Charles Weibel) (Page 282)

Figure 10.19: *Macrobrachium rosenbergii* (Page 284)

Figure 10.20: View of a Freshwater Prawn Culture Pond (Page 286)

Figure 10.22: Gravel Filter to Protect Predators (Page 296)

Figure 10.21: A Haul of Giant Freshwater Prawn (Page 288)

Figure 10.23: Large Size Pond (Page 297)

Figure 10.24: Small Size Pond (Page 297)

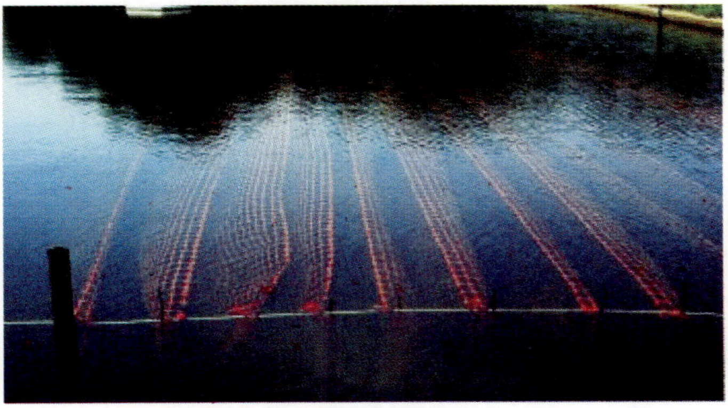

Figure 10.25: Substrates have been Placed Horizontally (Page 299)

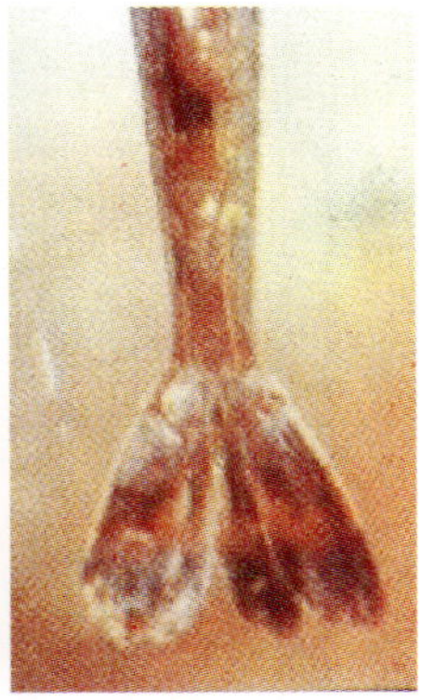

Figure 13.1: Brown Spot on Tail Portion (Page 334)

Figure 13.2: Black Spot on Shell (Page 334)

Figure 13.3: Brown Gill Due to Phytoplankton Crash (Page 335)

Figure 13.4: Swollen Gill Due to Acid Sulphate Soil Pond (Page 336)

Figure 13.5: Red Gill Due to Low Dissolve Oxygen Level (Page 336)

Figure 13.6: Black Gill (Page 337)

Figure 13.6: Cross Section of Body (Page 338)

Figure 13.7: Cramped Tail (Page 339)

Figure 13.8: Soft Shell Prawn (on Top) Normal Shell (Below) (Page 340)

Figure 13.9: Blue Disease (Page 340)

Figure 13.10: Ectocommensal Fouling Disease (Page 341)

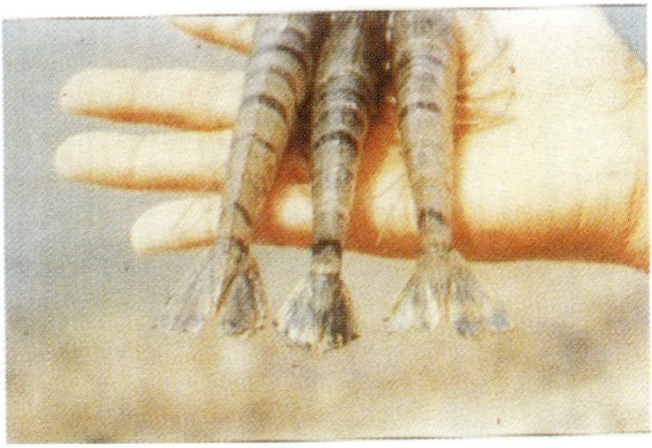

Figure 13.11: Tail Rot (Page 342)

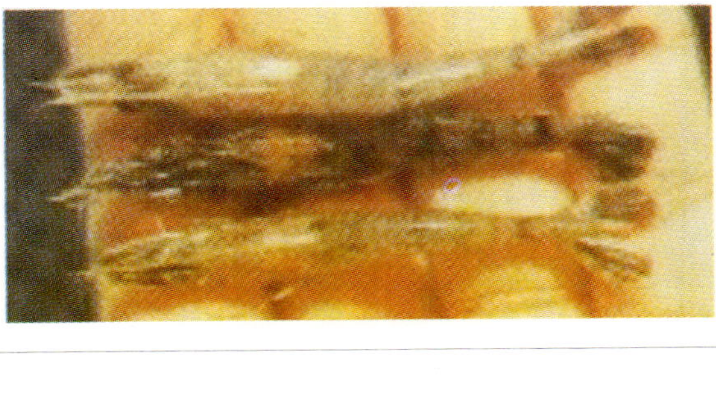

Figure 13.12: Yellow Head Disease in 25–35 Day Old Shrimp (Page 344)

Figure 13.13: Yellow Head Disease in 50–70 Day Old Shrimp (Page 344)

Figure 13.14: Normal Shrimp (Above); Affected Shrimp (Below) (Page 345)

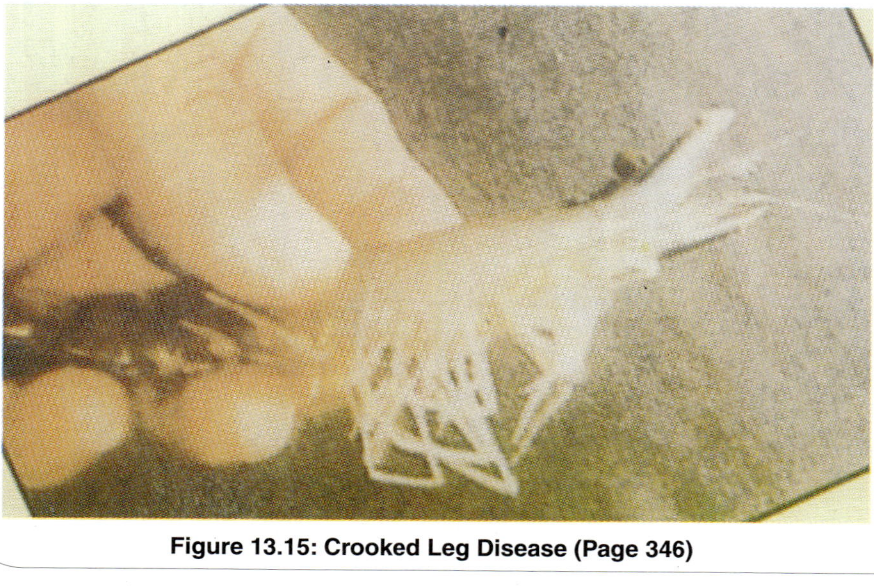

Figure 13.15: Crooked Leg Disease (Page 346)

Figure 13.16: Black Splinter in Muscle (with shell on) (Page 347)

Figure 13.17: Brown Muscle Syndrome (Page 348)

Figure 13.18: Black Death (Page 349)

Figure 13.19: Shrimp with Septic Hepatoprancreas (Page 350)

Figure 13.20: Blisters on Operculum (Page 353)

Figure 13.21: Blister in Branchial Chamber (Page 353)

Prevention

Maintenance of adequate water quality, removal of infected or dead prawns, adequate nutrition, prevention of mechanical injury, reduction of stocking density.

Treatment

A mixture of Malachite green 0.5–1 ppm and Formalin 20 to 70 ppm in water. The following were found to be effective treatments, when incorporated in food Terramycin 0.5 to 1 ppm.

Black Gill Disease

Identification

This is the most common disease problem to be observed among penaeid shrimp. The gill will be unusually dark, turning to such colours as brown, tan, red, orange and black.

Causing Agents

The main causes of gill diseases are due to particles or sediments accumulated on the pond bottom and heavy phytoplankton mortality.

2. Brown Gill

Figure 13.3: Brown Gill Due to Phytoplankton Crash

When large amount of phytoplankton die and the water becomes transparent due to dead phytoplankton gathered on the bottom, the shrimp gill gets obstructed the dead phytoplankton and becomes a brown colour.

3. Orange Gill

Orange gill colour will be found in areas with acid sulphate soil conditions. When these particles obstruct the gills, the shrimp decrease their food consumption, the carapace will become swollen and the head will be larger than normal.

Figure 13.4: Swollen Gill Due to Acid Sulphate Soil Pond

4. Red Gill

Figure 13.5: Red Gill Due to Low Dissolve Oxygen Level

Red colour gills are usually observed at night when the dissolved oxygen is low (less than 3 ppm), an indication of shrimp stressed by low oxygen. Increased aeration or more water exchange will solve the red gill colour problems.

5. Black Gill

There are many causes for black gill disease. Pond bottoms with various accumulated waste will encourage fungal growth and shrimp gills which are covered with this type of fungus will rot and turn black in colour. Bacterial infections can cause black gill but there are usually other. Clinical signs present such as black dots around the gill coverings and the abdominal region. The tail may be eroded and blackened at the edges. Black gill can also occur due to polluted pond bottom and water quality deterioration. This occurs with a phytoplankton die off. Dead shrimp with black gills will be found in the morning along the edges of the pond. The legs

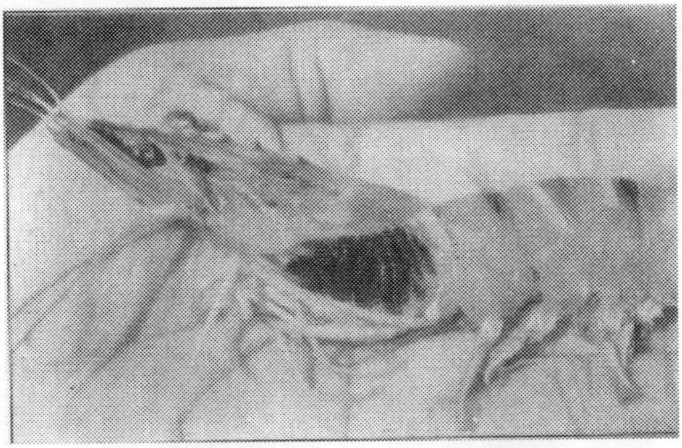

Figure 13.6: Black Gill

and appendages will be dirty; the tail may be normal or torn. The death of shrimp from this disease occurs when particles in the water clog the gills.

Prevention

Gill disease may be caused by unclean pond bottom and excess sediment, which may result from soil erosion of the pond, phytoplankton die off, over feeding or improper positioning of aerators which spread the particles over the pond rather than pushing them towards the centre of the pond. When the particles are spread out all over the pond, there is a high chance that the particles will get into the shrimp gills and obstruct them; especially black tiger shrimp prefer to stay on the pond bottom. To prevent these aerators must be positioned properly. In the early stages when the shrimp are still small, the aerators should be about 5 meters from the pond edge because during that time the feed will be scattered only along the edges and most particles will come from the erosion of the pond dike by the wind and waves as well as water currents. When the shrimp are larger, from about two and a half months old, the aerators should be moved further out to about 8 meters from the edges in order to allow them to push the particles towards the pond center. All the aerators must be in operation simultaneously to achieve a constant current revolving around the pond. If a single aerator is turned on separately, it will only move the particles around. When the particles accumulate at the pond center, there will be little chance of them clogging the gills.

Treatment

1. Give treatment with Furazolidone 2–3 ppm, for 2 to 4 nights, to affected shrimps.
2. The best way to solve this type of black gill disease is to exchange water as much as possible, maintaining the water colour with proper transparency (25–35 cm) and minimize the feeding so that, shrimp will consume all the feed. The fertilizer must be dissolved and distributed all over the pond. This will quickly add the desired colour to the water, the shrimp will feed

normally and the gill condition will improve. Aerators should be run for maximum time, until the bottom becomes clean.

6. Milk or Cotton Disease

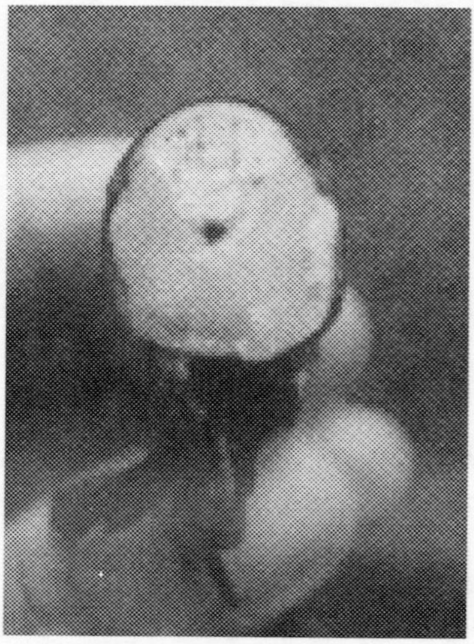

Figure 13.6: Cross Section of Body

Identification

Dorsal side of the shrimp will have group of spores under the exoskeleton and having a milky white appearance. This condition will affect the normal functioning of the various organs. The shrimp will be slower than usual, consume less feed and gradually weaken until they become prey for other shrimp or simply die off.

Causing Agents

It is affected with *Nozema nelsoni, Pleistophora* sp. mainly infect the tail and abdomen of the hosts, making them chalky white in appearance (Similar to cooked prawn) and cottony in texture. Masses of spores may be present in haemocoel, gut wall, intestine muscle etc.

Prevention

Prevention measures include destroying infected individuals and avoiding contact of infected brood stock with offspring. Any shrimp exhibiting cotton disease should be removed from the pond and destroyed. After harvesting the bottom of the pond should be scarped out and left to complete dryness to destroy spores or other infectious agents.

Treatment

No treatment.

7. Cramped Tail Disease

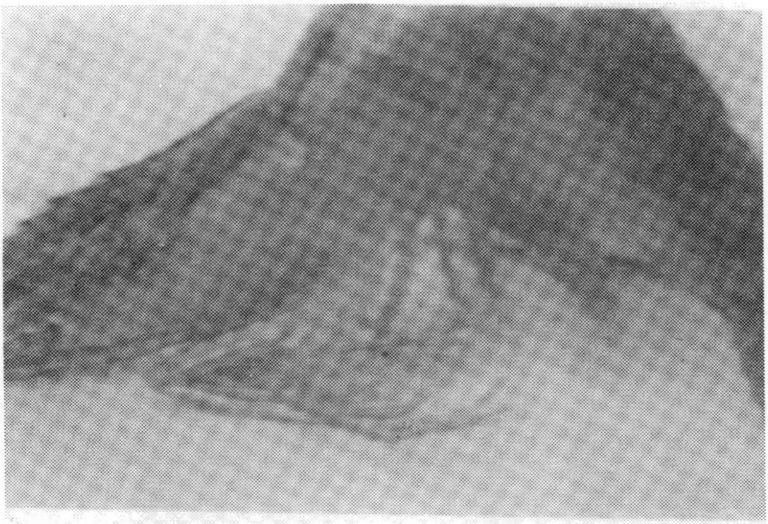

Figure 13.7: Cramped Tail

Identification

Symptom of the diseased shrimp is the flexure of the tail which is rigid and cannot be straightened. Some time muscle bulges out between the 3rd and 4th abdominal segments. Partially cramped animals swim about with a humped abdomen while those in advanced stages stop moving and lie on their sides.

Prevention

Even though exact reason is unknown, mineral imbalance, low temperature, handling stress etc have been recorded as possible causes. Hence take preventive care.

8. Chronic Soft Shelling

Identification

The affected shrimp has thin shell and persistently soft for several days, shell surface is often dark, rough and wrinkled. The shrimp are very lethargic and weak.

Causing Agent

This disease appears due to pesticide contamination, nutritional deficiency and poor pond management. It can be predicted under the conditions of high soil pH, low water phosphate and low organic matter of the soil. In adequate feeding practice like improper storage of feeds and use of old or low quality feeds are also found to be the reason.

Figure 13.8: Soft Shell Prawn (on Top)
Normal Shell (Below)

Diseased shrimp become more susceptible to wounding cannibalism and fouling by *Zoothaminium* and other organisms.

Prevention

Only good quality of feed should be used with adequate feeding. Provide nutritious food. Exchange water, maintain proper salinity, higher stocking to be avoided, maintain pH. Pond should be flushed thoroughly if any pesticide is used.

Treatment

Maximum water exchange is recommended as 20–25 per cent daily. Supplementary feeding should be provided such as mussel, clam meat at 8–14 per cent of the body weight daily for 2–4 weeks.

9. Blue Disease

Identification

The effected shrimp shows sky-blue colour instead of normal brown-black.

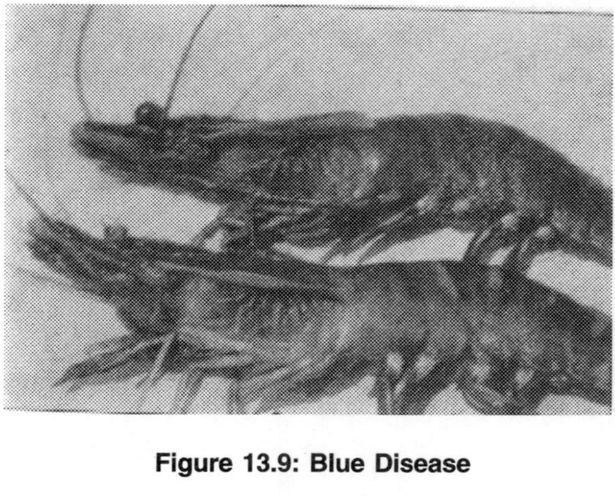

Figure 13.9: Blue Disease

Histological changes in the hepato-pancreas *e.g* disruption of the tubules also noticed. The shells become soft and thin with the rough surface and infected shrimp are found to be very lethargic.

Causing Agents

Usage of bad quality feed is the main reason for the disease. Low level of the carotenoid in diet, and acid sulphate soil, higher organic wastes and low dissolved oxygen levels may cause the disease. Apart from this, acid sulphate soil with high organic wastes and low level of dissolved oxygen in the pond water are also could be reason for this disease.

Prevention

Stocking density should not be very high. The feed should be new and a recommended bran. Water exchange should be adequate. Use diet with carotenoid sources may reduce the intensity of this disease.

Treatment

1. Use of Vitamin-B complex @ 1 kg/ha for 3 days is reported to give a good results.
2. At the first sign of this disease, feeding should be reduced by 50 per cent and shrimp density should be reduced by transferring the animals to other ponds. 15 to 20 per cent of the water should be exchanged daily depends upon the stocking density. Use only recommended feed.

10. Ectocommensal Fouling Disease

Causing Agents

Owing to the weakened condition of diseased shrimp there is more opportunity for algae, *Zoothamnium, barancles* and other organisms to attach organisms to the shrimp bodies. The damage caused by such organisms depends on the stage of infection in the shrimp.

Figure 13.10: Ectocommensal Fouling Disease

Identification

Ectocommensal fouling disease is a complication found in almost every area where shrimp have disease problems, especially in the last few months before harvesting. If shrimp with such parasitic materials on its body are found along the edge of the pond appear lethargic, dirty looking and not feeding they will soon die. It is impossible to return them to their normal condition. But if they will still eat normally even if the body not as firm as normal and they have not come up to the edge of the pond, there is a good chance these shrimp will moult and return to health.

Prevention

Zoothamnium and algae are usually found on the pond bottom in ponds with too clear water. Since shrimp stay on the bottom for most of the time, they are likely to have these organisms attached to their bodies. Control of good phytoplankton bloom around 25–35 cm transparency is recommended to prevent this problem, especially during the early weeks after stocking.

Treatment

Zoothamnium can be treated with formalin at a concentration of 20–25 ml per 1000 litres of water, by reducing the water level to 50 cm. The paddlewheels must be in operation constantly when formalin has been applied. After about 4 hours the water level must be increased back to normal. The treatment may be repeated again after 2 to 3 days if there are still large amounts to *Zoothamnium* present. For algae and other organisms, the water exchange should be increased and phytoplankton bloom should be improved with a water transparency of 25–35 cm.

11. Tail Rot Disease

Causing Agents

This is because of quantity of the quantity of excess feed and other materials which collect at the pond bottom and lead to more bacterial growth. Many type of

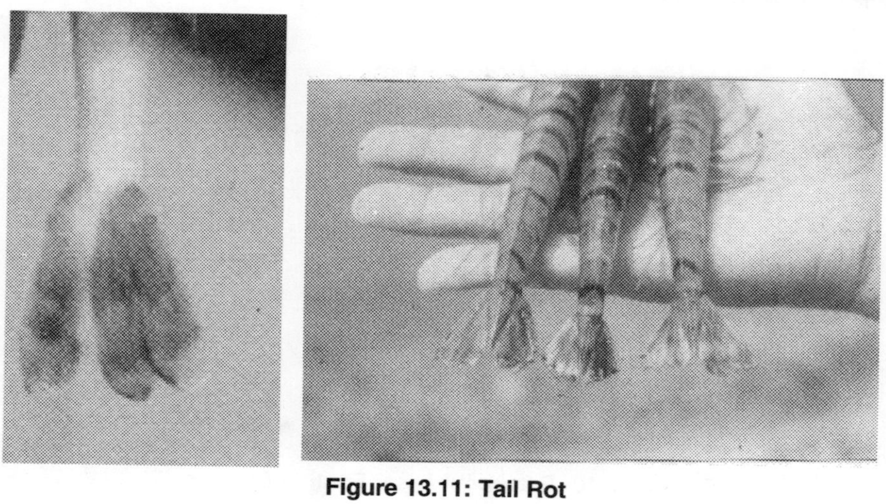

Figure 13.11: Tail Rot

bacteria are capable of producing the enzyme *chitinase*, which can break down the chitin in the shrimp exoskeleton.

Identification

Shrimp with marked, nicked, worn out or blackened tails are a common sight in semi-intensive culture shrimp farms. Ponds with high survival rates and high density will have a high chance of experiencing this disease. The eggs will turn black and look as if they were burnt. These black areas contain large amounts of bacteria, but it cannot spread into other parts of the body. If the shrimp moult before the bacteria can reach the inner part of the tail, they will be sloughed off with the old shell. However, the infection is chronic; the muscle tissue of the tail will rot and die. This will appear red-dish.

Prevention

Control of excess feed and proper placement of aerators are the best prevention measures. However, feed should not be limited too much during the time the shrimp are moulting, as hungry shrimp will cannibalise the fails of those which moult first, making them more susceptible to infection.

Aerators should be placed in the corners approximately 3 to 5 meters away from the dike and angled to achieve maximum water movement. If distance between the two aerators is greater than 50 m, additional aerators should be installed after the 6th week of culture. If the distance between the aerators is 50 to 70 m, the additional one should be installed nearer to the existing aerators, 15 to 20 m from the dike. If the distance is 70 to 100 m, then the additional ones can be installed in line with the existing aerators, 3 to 5 m from the dike. For the first 20 days of culture, aerators should be run for 2 to 6 hrs as night on every third day. From 20th day onwards a minimum of 6 hrs should be run every night.

Treatment

Firstly the bottom of the pond must be kept clean. This can be accomplished by using sufficient aerators and properly positioning them to keep the sediments from spreading over the pond bottom and the feeding rate must be carefully controlled. Treatment with antibiotics is not necessary. Although antibiotics added to the feed will cure the eroded tail, but the disease will be re-occurred if the pond bottom is not kept clean.

12. Yellow Head Disease

Identification

This disease derives it names from the clinical sign of pale yellow body and head (cephalothorax) due to the colour change of the hepatopancreas. This disease affects shrimp approximately 35 days after stocking and shrimp size 5–15 gms. Before infection, shrimp show excellent feeding behaviour and good growth. Infected shrimp exhibit yellow cephalothorax and stop eating. Many more shrimp than usual are found in the check trays but most of them are not eating. Once mortalities start, the entire population dies within 2–3 days. Yellow head disease in fries 25–35 days after

Figure 13.12: Yellow Head Disease in 25–35 Day Old Shrimp

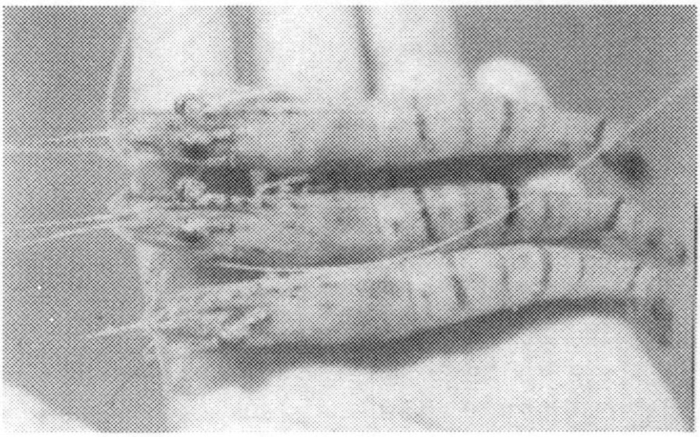

Figure 13.13: Yellow Head Disease in 50–70 Day Old Shrimp

stocking are found usually in areas where the farmers cannot maintain a consistent phytoplankton bloom during the season when salinity and temperature of the water are higher than normal.

Prevention

The yellow head disease does not appear in ponds which have a good and consistent phytoplankton bloom with a water transparency of 25–35 cm. In order to solve the problem or prevent the disease the phytoplankton must be cultured both in reservoirs and in grow out ponds. Application of agricultural lime at a rate of 100–200 kg/ha, for every five days from one month onwards is the best prevention method.

Treatment

Chemicals such as chlorine and iodine have proved to be effective disinfectants. However, these chemicals are not suitable for use during the culture period. They can only be used for pond preparation or disinfection of the pond after the disease outbreak. To date, there is no effective treatment for this disease.

13. Red Disease

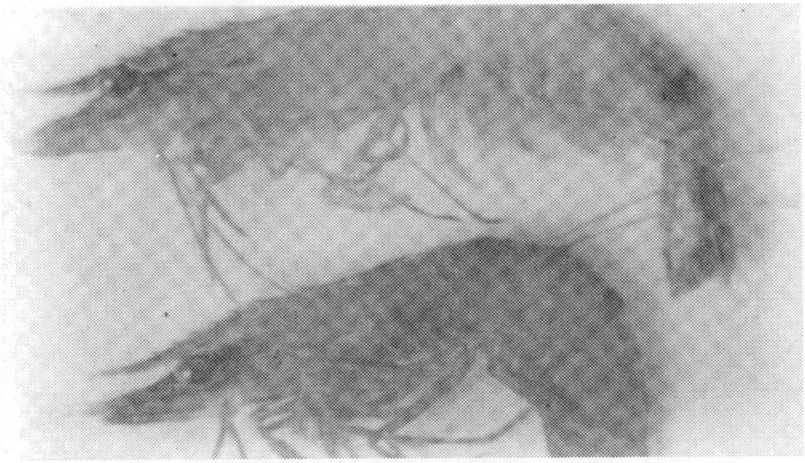

Figure 13.14
Normal Shrimp (Above)
Affected Shrimp (Below)

Identification

One of the first signs of red disease is the greatly reduced size of the hepatopancreas. Severely affected shrimp will have a white area under the exoskeleton in the head region, the body will feel soft and limp, the muscle tissue will be whitish and cloudy and the hepatopancreas will be largely destroyed. Shrimp displaying such clinical signs certainly die.

Causing Agent

Red disease is a condition caused by bacterial infection in which affected shrimp exhibit red coloured tails and legs.

Prevention

30 days of culture onwards water transparency should be strongly maintained between 25–35 cm by adding organic (Cow dung or chicken manure) and inorganic (Urea and super phosphate) fertilizers at a regular interval. Excessive feeding should be avoided and water exchange must be adequate.

Treatment

Oxytetracycline is effective for red disease. It should be given at the rate of 3–5 gms per 1 kg of pellet feed per day for 5–7 days. Constant phytoplankton bloom and water quality management are important in solving this problem. If conditions in the pond are not improved after antibiotics treatment, the shrimp may begin to die. Antibiotics should be used on shrimp weighing 20 gms or more, but instead the water quality and pond bottom conditions should be improved.

14. Crooked Leg Disease

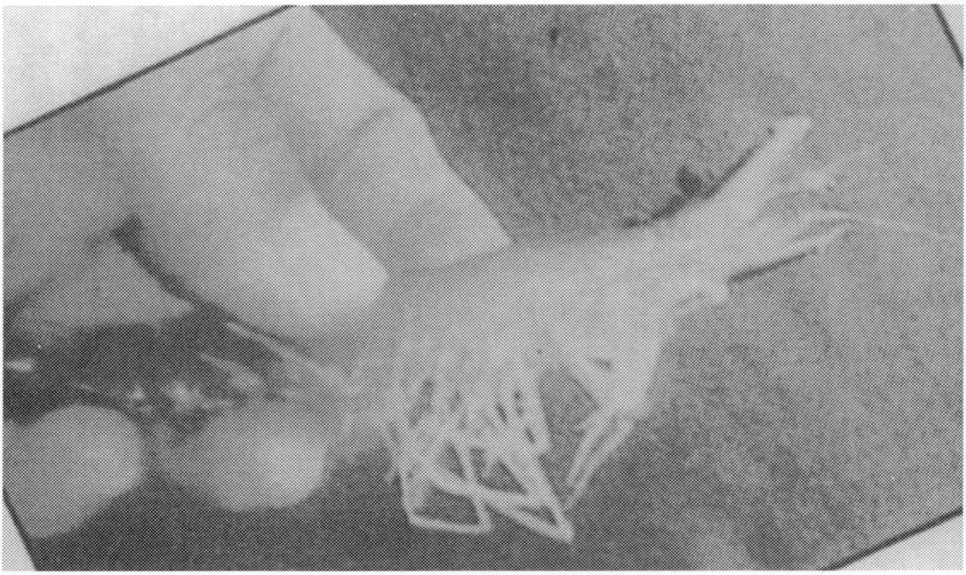

Figure 13.15: Crooked Leg Disease

Identification

Shortly after stocking postlarvae, many will die during moulting and those which survive will be left with deformed legs.

Causing Agents

This disease is found in ponds which suffer a drop in salinity (may be due to heavy rainfall) and low pH. This disease is very prevalent in area with acid sulphate soil condition. The problem will occur when the insufficient calcium is available in the water; calcium will be utilized from other parts of the shrimp's body resulting in crooked legs.

Prevention

The best prevention measure is to prepare the pond well and make sure that the pH and alkalinity of the water are suitable before stocking the shrimp. The water should be monitored and maintained at a pH level not lower than 7.5 and alkalinity not less than 70 ppm. This can be maintained by the addition of agricultural lime or dolomite to the water which will raise the pH and alkalinity. After heavy rainfall when the salinity of the water is reduced to less than 10 ppt, lime should also be added to bring the pH and alkalinity to appropriate level. The addition of lime to the water will increase its alkalinity and make more calcium available, both of which will avoid this problem.

Treatment

Generally this disease occurs only after a rainy season. The surface water should be drained away completely from the ponds, whenever heavy rain fall occurs so the ponds can maintain normal salinity. It has been observed that in ponds where agricultural lime or dolomite were added to raise the pH and alkalinity, most of the shrimp with crooked legs returned to normal after several moults.

15. Black Splinter Disease

Figure 13.16: Black Splinter in Muscle (with shell on)

Identification

The disease is characterized by change in the tissue and the affected shrimp will have black spots or splinters around the joints between each segment of the shell or under the tail. When the shell is removed a thin flat black marks will be seen under the shell. The growth of shrimp with black splinter disease is not noticeably different from that of normal shrimp. Normally black spots will appear when the shrimp is two months old.

Causing Agent

Black splinter disease results from an infection of *Vibrio vulnificus* bacteria. The disease is thus often experienced in the rainy season when the salinity of water is very low for long periods.

Prevention

The bacteria found in black splinter disease are found mostly in water of low salinity, the farmer should avoid the stocking schedule in a period of high rainfall and low salinity. Surface water should be drained off at the time of heavy rain fall. Big reservoir with high saline water can help to reduce the severity of the problem. Another prevention method is to reduce the bacterial population by keeping the pond bottom clean and avoiding over feeding. Having a lower population of shrimp in the pond at time of low salinity will reduce the seriousness of the condition.

Treatment

In the first stages when only a few shrimp have died, the shrimp should be fed less because they do not feed as much as healthy shrimp. If the water is too clear usually because the phytoplanktons die off and the transparency is more than 45 cm, fertilizer should be added to increase the phytoplankton bloom. Aerators should be installed to increase the dissolved oxygen availability and to clean the pond bottom. These measures should make the shrimp healthy but if they continue to die antibiotics should be added to the food. One to two gm of Oxytetracycline should be added per one kg of pellet per day for 3–5 days where disease is not severe. When disease is more severe, the Oxytetracycline can be increased to 3–4 gms per one kg of pellet feed for another 3–7 days. This treatment proves to be very effective if the farmer can increase the phytoplankton bloom and maintain transparency to 25–35 cm.

16. Brown Muscle Syndrome

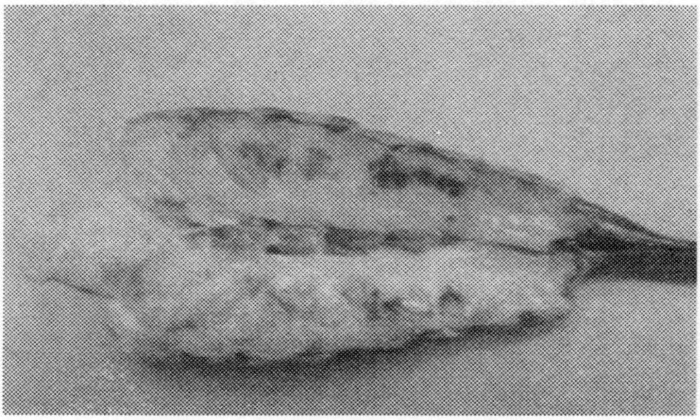

Figure 13.17: Brown Muscle Syndrome

Identification

Brown muscle disease or syndrome was named due to brownish colouration along the abdominal muscle. Some affected shrimp showed only small brown spots along the two muscle bundles which join the abdomen and cephalothorax. Whereas some showed diffuse brown spots or patches throughout the abdominal muscle. Generally affected shrimp showed no abdominal external signs, but if the shrimp were examined against the sunlight, brown to black spots or patches were easily observed underneath the shell.

Prevention

Over feeding should be avoided and water exchange should be followed as recommended.

17. Black Death Disease

Identification

The symptoms include blackened areas of tissue under the exoskeleton of the carapace, in the abdomen, in the foregut and hind gut. Affected animal goes off feed and death follows within 24 to 72 hrs.

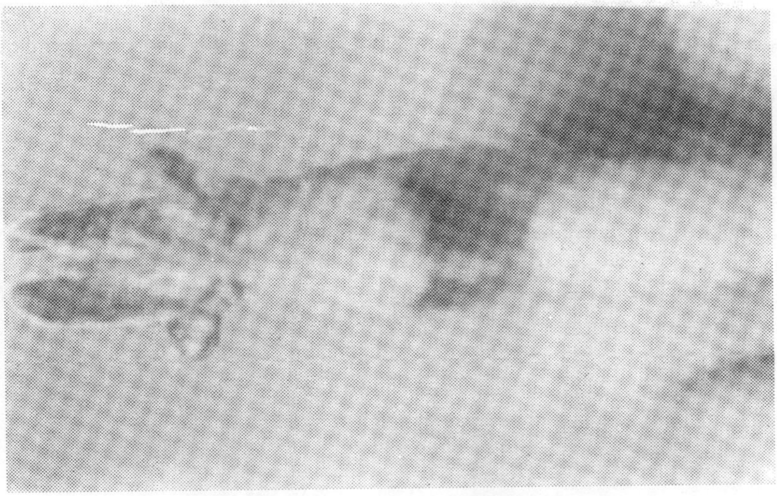

Figure 13.18: Black Death

Causing Agents

This disease has been observed repeatedly in prawns feed with artificial diets low in ascorbic acid for several weeks in tanks that contain no plant material. So its considered as nutritional disease.

Prevention

Give good quality feed sufficient quantity of ascorbic acid and plant material.

18. Change in Eye Colour

Identification

At night the farmer should shine a flash light around the edges of the ponds. Shrimp that are becoming sick will often coming up and float around the edges of the ponds. When the light is shone down to look at the shrimp eye and if they appear red, then the shrimp is still healthy. Normal shrimp will swim away when the light comes closer to them. Diseased shrimp will take longer time to escape the light and will have paler eye colour. If the eyes are almost white, then they are already severely affected.

19. Diseases Affected by Virus

Viral disease is the most infectious in the shrimp and there is no treatment available at this point time. Viruses multiply within the host cell and infect. Some of the virus and their effect are shown as under.

19.1. *Baculovirus penaei* (BVP)

This virus is characterized by large polyhedral shape inclusion bodies having 20 micron size located at nuclei of hepatopancreatic cells. Larvae, juveniles and adults are susceptible. Infected nuclei enlarge with degeneration of nucleoli. Infections in natural populations may reach up to 80 per cent, with subsequent mortalities.

19.2. *Monodon Baculovirus* (MBV)

In hepatopancreas and mid gut epithelium intra-nuclear spherical inclusion bodies formed. It causes destruction of hepatopancreas and lining of digestive tract. Affected shrimps exhibit pale bluish-gray to dark blue colouration, sluggish and inactive swimming movements, loss of appetite, retarded growth. Necrosis of hepatopancreastic is found in tubules, followed by haemocytic encapsulation. Heavily infected individuals become lethargic. Mortality occurs in high density culture. This virus is found in *Penaeus monodon* and *P. merguiensis.*

Figure 13.19: Shrimp with Septic Hepatoprancreas

19.3. Baculoviral Midgut Necrosis Virus (BMNV)

This disease is diagnosed by histological demonstration of hypertrophied nuclei of hepatopancreatic epithelial cells. It is typical that the virus completely fills the infected nucleus. It causes severe mortalities in cultured larvae and post larvae resulting in hepatopancreatic necrosis. This virus is found in *Penaeus japonicus.*

19.4. Infections Hypodermal and Hematopoietic Necrosis Virus (IHHNV)

When the shrimp attains 0.5–1 gm, the virus shows its effect. The first sign of disease is erratic swimming behaviour, rising to surface and sinking. Mortalities may occur usually within 4–12 hrs. In acute stage animal cuticle acquires a whitish appearance. Peak mortalities is observed within 2–3 weeks after the infection of the disease. In the nuclei of ectodermally and mesodermally derived tissues, large inclusion bodies are found. Larval stages (Zoea and Mysis) of black tiger shrimp are presumed to be latently infected. Causes diffuse tissue necrosis with generalized inflammation. Death of the cells in theses tissue can cause abnormal metabolism which eventually leads to mortalities. *P. monodon*, *P. vannamaei* and *P. stylirostris* are main hosts of this virus.

19.5. White Spot Virus (WSV) or White Spot Baculovirus (WSBV)

Serious disease in grows out ponds and much more serious threat to the shrimp industry. The post larvae are implicated as a possible route of WSV transmission to ponds. The moribund shrimp swim to the surface of the water and gather at the pond dikes. Typical signs included broken antennae; circumscribed whitish spots pinpoint 1 mm in diameter in the cuticle or shell that first started in the carapace and fifth-sixth abdominal segments and later the entire body shell and reddish boy discolouration, empty guts and sometimes body shell/gills epibiont fouling and sometimes lymphoid organ swelling. The white spots can be most easily seen by removing the carapace and holding up to the light. The virus effects ectodermal and mesodermal origin tissue. The virus may absent in midgut. Infected cell are with vacant cytoplasom. It is DNA virus.

Prevention

For hatchery: Do not give crabs or other crustaceans as feed to brood stock.

For Farms: It is possible that the latex of the medicinal plant, *Calotropis gigantean*, nutrient supplements and bacterial amendments are found to improve rearing success.

20. Disease Affected by Bacteria

20.1. Filamentous Bacterial Disease

Filamentous growth of the bacteria can be seen on appendages and body surface even under 100 times magnification. In adult the filamentous growth can be seen on setae of uropods and pleopods and on gill filaments, as fine coloured and thread like growth. Heavily infected animal often show discolouration of gills that range from yellow to brown to green. Such heavy infestations prevent gas diffusion across gill which may result in death of the animal. Filamentous growth on appendages and body surface may interfere with normal locomotory process.

Prevention

By avoiding crowded population of prawn and organic rich seawater and maintaining good water quality with optimum dissolved oxygen level.

21. Disease Affected by Protozoa

21.1. Gregarines

Gregarines are not known to cause significant disease in penaeids. About eight species of genera *Nempatopsis* eg. *Nematopsis duorari, N, penaeus, N. vannameri, N. brasiliensis etc., are* the causing agents. Their presence in the digestive tract in large number may be demonstrated in fresh squashes of hind and midgut contacts. Gregarines require a mollusk for completion of their life-cycle nad hence may be excluded from pond. Cotton shrimp disease is one of the result of effect of protozoa.

21.2. Ciliates

Unique and relatively common, these are typically commedals rather than true parasites. Stalked ciliates especially those of genus Zoothamnium can be a problem in high density shrimp culture. Extensive colourization in gill may interfere with gas exchange and result in mortalities.

22. Disease Affected by Worms

Worms that have been found in shrimp are trematodes (flukes), Cestodes (tapeworms) and Nematodes (round worms). Some species are more common than others ans as yet none have been known to cause widespread mortality. Worms may be found in various species of penaeid shrimp.

23. Non-Infectious Disease

23.1. Blister (Hemolymphoma)

The condition occur primary in the carapace covering the gills (branchiostegal region) and less on the ventrolateral portions of the abdominal pleural plates. The water blisters normally contain some squid which varies from clear to brown or black colour. Cause unknown. Almost all penaeids may be susceptible (Figures 13.20 and 13.21).

23.2. Muscle Necrosis

When shrimps are exposed to certain stressors, such as low oxygen levels, temperature or salinity shock, over crowding etc, the tissues of the body will quickly undergo necrosis.

The most common manifestation is the necrosis of musculature which shows tself as a general opacity. Although a detriment, it gives aqua culturist a sign that conditions are not suitable. Focal point for this action in penaeid shrimp is the posterior portion of the tail.

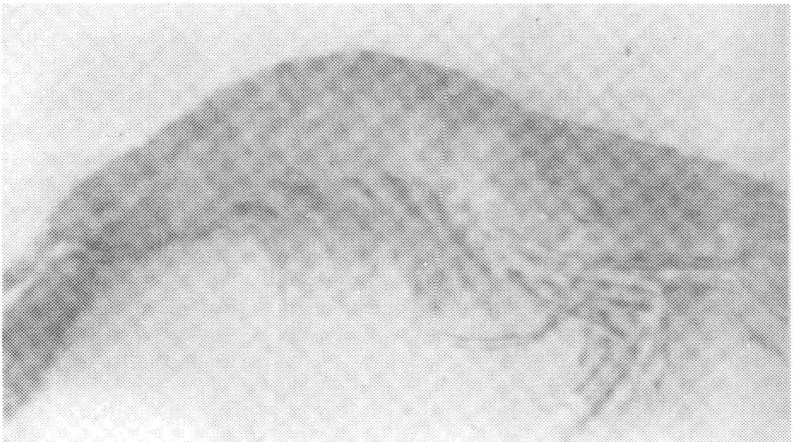

Figure 13.20: Blisters on Operculum

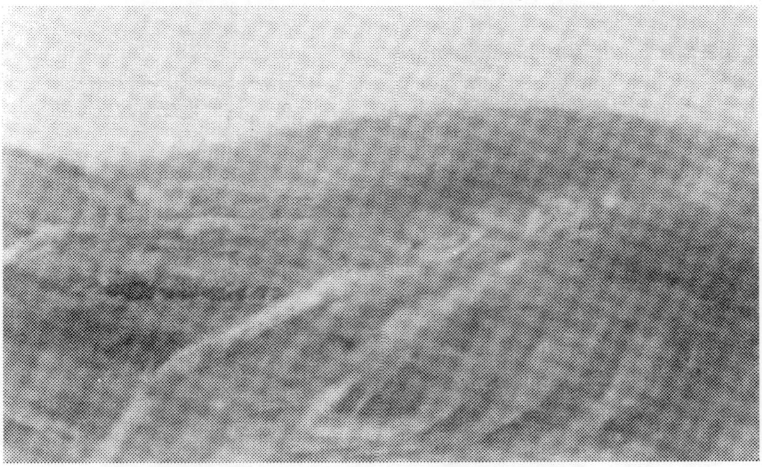

Figure 13.21: Blister in Branchial Chamber

23.3. Gas Bubble Disease

Mostly juveniles are known to be affected by this disease. Many small and large bubbles of gas develop in gills and other tissues, which may be due to supersaturated water conditions. This disease is not common and is considered to be serious only in culture situation utilizing heated water.

23.4. Hemocytic Enteritis (Disease Related to Toxic Algae)

Blooms of certain filamentous green algae, all belonging to the family *Oscillatoriaceae* have been implicated as causing disease syndrome primarily in young and juvenile penaeids. The principal lesion of this disease is the result of algal endotoxins release in the gut from ingested algae. The cause of death of shrimp may be due to osmotic imbalances poor absorption of nutrients or secondary bacterial infections. Use of algaecides or antibiotic therapy, controls the disease.

24. Change in Eye Colour

Identification

At night the farmer should shine a flash light around the edges of the ponds. Shrimp that are becoming sick will often coming up and float around the edges of the ponds. When the light is shone down to look at the shrimp eye and if they appear red, then the shrimp is still healthy. Normal shrimp will swim away when the light comes closer to them. Diseased shrimp will take longer time to escape the light and will have paler eye colour. If the eyes are almost white, then they are already severely affected.

In Table 13.1 some recommended medicines and application for bacterial disease are given. Similarly medicines recommended for fungus disease are given in Table 13.2. Recommended medicines and application for protozoan disease are given in Table 13.3.

Table 13.1: Recommended Medicines and Application for Bacterial Disease

Sl.No	*Name of Medicines*	*Recommended Dose*
1.	Erythromycin	1-2 ppm
2.	Oxy tetracycline	0.5–2 ppm
3.	Kannamycin	0.1–0.25 ppm
4.	Diametin	25 ppm
5.	E.F. Mycin	2.5 ppm
6.	Chloromycetin	2–10 ppm
7.	Doxycline	0.5–1 ppm
8.	F.G.C. Mycin	1 kg/ha
9.	K-3 Mycin	1 Kg/ha for 2–3 days continuous
10.	G.P.N/Mycin	1 Kg/ha for 2 days continuous
11.	O.K. Mycin	2 Kg/ha, for 7 days continuous
12.	T Mysin	2 Kg/ha for 2–3 days
13.	Reco Mycin	2 Kg/ha for 2–3 days
14 .	Tea seed powder	10 ppm

Source: MPEDA, 1996: 138.

Table 13.2: Recommended Medicines and Application for Some Fungal Disease

Sl.No	*Name of Medicines*	*Recommended Dose*
1.	Parazin (Oxolinic acid)	0.1–0.5 ppm for 3–5 days
2.	Furasol	1.0 to 2.5 ppm
3.	Ampicillin	0.5–1 ppm

Contd...

Table 13.2–Contd...

Sl.No	*Name of Medicines*	*Recommended Dose*
4.	Chloramphanicol	2.0–10 ppm
5.	Gentan violet	0.1–0.2 ppm
6.	Formalin	25–50 ppm
7.	Potassium permanganate	0.1–0.3 ppm
8.	F.G.C. Mycin	1 Kg/ha for 2-3 days continuous
9.	K-3 Mycin	2 Kg/ha for 2 days.
10.	G.P.N. Nycin	2 Kg/ha for 2 days continuous
11.	O.K.Mycin	1 Kg./ha 7 days continuous
12.	Y- Mycin	2 Kg/ha 2–3 days
13.	Reco Mycin	2 Kg/ha 2–3 days
14.	Lufutalon	1 Kg/ha 3–5 days

Source: MPEDA, 1996:139.

Table 13.3: Recommended Medicines and Application for Some Protozoan Disease

Sl.No	*Name of Medicines*	*Recommended Dose*
1.	Formalin	25 ppm
2.	Chloramin T	5 ppm
3.	Quinine bisulphate	5 ppm
4.	Cupric sulfate	1 ppm
5.	Methylene blue	0.01–1 ppm
6.	Malachite green	0.02–0.25 ppm
7.	Nitrofurazone	0.05–2.5 ppm
8.	E.F. Mycin	0.05–2.5 ppm
9.	Pyrimethamine	10.1–0.6 ppm
10.	Duocoxinsy	0.1–0.5 ppm
11.	Furasol	1–2.5 ppm
12.	Parasin (Oxolinic acid)	0.1–0.5 ppm
13.	G.F.C. Mycin	1 kg.ha for 2–3 days continuous
14.	K-3 Mycin	2 Kg/ha 2–3 days
15.	G.P.N. Mycin	2 Kg/ha 2–3 days continuous
16.	O.K. Mycin	2 Kg/ha 7 days continuous
17.	Y-Mycin	2 Kg/ha 2–3 days
18.	Reco Mycin	2 Kg/ha 2–3 days
19.	Lujuralon	1 kg/ha 3–5 days
20.	Pysolpre P.S.	2.5 Kg/ha for 3–5 days

Source: MPEDA, 1996:140.

References

Alston, D.E. and Sampaio, C.M.S. 2000. Nursery systems and management. In M.B. New and W.C. Valenti, eds. Freshwater prawn culture: the farming of *Macrobrachium rosenbergii*, pp. 112-125. Oxford, England, Blackwell Science.

Anon 1883.

FAO SPECIES IDENTIFICATION SHEETS PEN Pen 26 1983.

FAO SPECIES IDENTIFICATION SHEETS PEN Pen 27 1983.

FAO SPECIES IDENTIFICATION SHEETS PEN Pen 20 1983.

FAO SPECIES IDENTIFICATION SHEETS PEN Pen 15 1983.

FAO SPECIES IDENTIFICATION SHEETS PEN Pen 14 1983.

FAO SPECIES IDENTIFICATION SHEETS PEN Parap 6 1983.

FAO SPECIES IDENTIFICATION SHEETS PEN Pe 3 1983.

FAO SPECIES IDENTIFICATION SHEETS PEN Para 14 1983.

FAO SPECIES IDENTIFICATION SHEETS PEN Para 12 1983.

FAO SPECIES IDENTIFICATION SHEETS PEN Para 8 1983.

FAO SPECIES IDENTIFICATION SHEETS PEN Para 4 1983.

FAO SPECIES IDENTIFICATION SHEETS PEN Metap 8 1983.

FAO SPECIES IDENTIFICATION SHEETS PEN Metap 20 1983.

FAO SPECIES IDENTIFICATION SHEETS PEN Meta 16 1983.

FAO SPECIES IDENTIFICATION SHEETS PEN Metap 17 1983.

FAO SPECIES IDENTIFICATION SHEETS PEN Metap 12 1983.

FAO SPECIES IDENTIFICATION SHEETS PEN Metap 22 1983.

Anon 2003. A report on the International Symposium on Freshwater Prawns, 2003 held at the College of Fisheries, Panangad from 20 to 23 August 2003. Current Science, Vol 85 (10), November 2003.

Anon 2005. FAO Fisheries Global Aquaculture Production Database for freshwater crustaceans. The most recent data sets are for 2003 and sometimes contain estimates. Accessed June 28, 2005.

Anon, 2006. Museum Victoria, Information sheet No 10295 August 2006.

Anon, 2007. Turkish Journal of Fisheries and Aquatic Sciences 7: 07-11.

Achuthan Kutty C. T., Anil Chatterjee, R.A.Sripada, and Ulhas M.Desai, 2003. Production of Giant Freshwater Prawn Postlarvael in Penaeid Prawn *(shrimp)* Hatchery: An Experience. National Institute of Oceanography, Dona Paula, Goa–403 004.

Alekhnovich AV and Kulesh VF (2001). Variation in the parameters of the life cycle in prawns of the genus *Macrobrachium* Bate (Crustacea, Palaemonidae). Russian Journal of Ecology 32: 420-424.

Armstrong KF and BallLSL (2005). DNA barcodes for biosecurity: invasive species identification. Philosophical Transactions of the Royal Society B 360: 1813-1823.

Bal, D.V. and K. Rao 1990. *Marine Fisheries of India*. Tata Mac. Graw-Hill Publishing Co. Ltd.

Boyd, C.E. 1990. Water quality in ponds for aquaculture. Auburn, Alabama, USA, Alabama Agricultural Experiment Station.

Boyd, C. E. and Tucker, C.S. 1998. Aquaculture water quality management. Boston, USA, Kluwer Academic Publishers.

Bojan, J. 2003. A report on the International Symposium on Freshwater Prawns, 2003 held at the College of Fisheries, Panangad from 20 to 23 August 2003. Current Science, Vol 85 (10), November 2003.

Chhaya, N. D. 1993. Marine prawn culture (Edited), Workshop, organized by College of Fisheries, Guj. Agri. Uni. Veraval. pp15.

Chhaya, N, D, and D. V. Nandasana. 1990. Some observation on availability of culturable penaeid prawns juveniles in Navibandar creek in Junagadh district–Gujarat. *Fishing Chimes*, 10 (2): 45–46.

Cornelius, R. Mock, 2006. Crustacean culture, Shrimp, *Penaeus japonicus* Nansei Regional Fisheries Research Laboratory.

Cuzon, G., M. Hew, Cognie, D. and Soletchnik, P. 1982. The lag of effect of feeding on growth of juvenile shrimp *Penaeus japonicus* Bate. *Aquaculture*, 29: 33–44.

David I. Prangnell, Ravi Fotedar, 2006. Effect of sudden salinity change on *Penaeus latisulcatus* Kishinouye osmoregulation, ionoregulation and condition in inland saline water and potassium-fortified inland saline water. *Comparative Biochemistry and Physiology–Part A: Molecular and Integrative Physiology*, Volume 145, Issue 4, December 2006, Pages 449-457.

Devanand Kavlekar 1998. Distinctive characters of species *Atypopenaeus stenodactylus* (Stimpson, 1860). Bioinformatics Centre, National Institute of Oceanography, Dona Paula, Goa, India.

Devarajan, K., J.Sunny Nayagam, V.Selvaraj and N. N. Pillai. Larval development *Penaeus semisulcatus* de Haan 1978. CMFRI Bull 28: 22–29.

Dimmock A, Williamson L and Mather PB (2004). The influence of environment on the morphology of *Macrobrachium australiense* (Decapoda: Palaemonidae). Aquaculture International 12: 435-456.

Desai, A. Y., A. D. Dholakia, A. A. Vyas, 2004. Nutritious feed for shrimp and it's technology. Extension Booklet, Junagadh Agricultural University, Veraval.

Dholakia, A. D. 1986. List of important fishes and crustaceans available in Gujarat with scientific, English and local names. Fish Processing Preservation and Quality control pp 133–149.

Dholakia, A. D., M. I. Patel and N. D. Chhaya 1991. Observation on growth of banana prawn *Penaeus merguiensis* (de-Man) using different feed. National Seminar on Shrimp production and farming, 12–13 Feb. Held at Berhampur, Orissa.

Dholakia, A. D., N. G. Akolkar, M. I. Patel and N. D. Chhaya 1991. Constrains and solution of transportation of prawn seeds of *Penaeus merguiensis* (de-Man) in tropical areas. National Seminar on Productivity, constraints in coastal areas, held at Calcutta.

Dholakia, A. D. and N. D. Chhaya, 1993. Some information on artemia which plays an important role in prawn culture. "Krushi go Vidya", Gujarat Agricultural University, 46 (4): 26–28.

Dholakia, A. D. 1993 (i). Availability of prawn seed from different stages of growth in prawn culture. Workshop organized by College of Fisheries, Gujarat Agricultural University, Veraval on 20 June.

Dholakia, A. D. 1993 (ii). Details of feeds to be given during different stages of growth in prawn culture. Workshop organized by College of Fisheries, Gujarat Agricultural University, Veraval on 20 June.

Dholakia, A. D. 1994. Marine prawn Fishery and culture in Saurashtra with special reference to *Penaeus merguiensis* (de-Man)–Ph. D. Thesis, with Saurashtra University, Rajkot.

Dholakia, A. D. and N. D. Chhaya 1995. Model size group of banana prawn *Penaeus merguiensis* (de-Man) GAU, RES. J., 20 (2): 94–97.

Dholakia, A. D. and D. V. Nandasana 1996. Information on Commercial production of *Metapenaeus kutchensis* and its availability in Kutch and Gujarat Marine waters. Sovenior at Guj, Agri. University Junagadh on 16th January. Pp 146.

Dholakia, A. D. 1996. Study on maturity of *Penaeus merguiensis* (de-Man) banana prawn in Gujarat–Saurashtra coast. *Fisheries World*, 4 (4): 13–15.

Dholakia, A. D. 2004. Prawn Fishery in "Fisheries and Aquatic Resources of India". pp 97–125. Daya Publishing House, New Delhi.

Dholakia, A. D., A. A. Vyas, A. Y. Desai, 2004. Prawn feed and it's management Extension Booklet, Junagadh Agricultural University, Veraval.

Dimmock A, Williamson L and Mather PB (2004). The influence of environment on the morphology of *Macrobrachium australiense* (Decapoda: Palaemonidae). Aquaculture International 12: 435-456.

Doyle J. J. and Doyle J. L. 1987. Procedure for small quantities of leaf tissue. Phytochemistry Bulletin 19: 11-15.

FAO. 1992 b. Pond construction for freshwater fish culture: pond farm structures and layouts.

FAO Training Series No. 20/2. Rome.

FAO. 1995. Pond construction for freshwater fish culture: building earthen ponds. FAO Training Series No. 20/1. Rome.

Gürel Türkmen, 2006. Pond Culture of *Penaeus semisulcatus* and *Marsupenaeus japonicus* (Decapoda, Penaeidae) on the West coast of Turkey Ege University, Faculty of Fisheries, Department of Aquaculture, 35100, Bornovaizmir, Turkey.

George, M. J. 1999. Synopsis of Biological data on the penaeid prawn *Metapenaeus monoceros* (Fabricius, 1798). FAO Fisheries Synopsis No. 104.

George, M. J. 2000. Synopsis of Biological data on the penaeid prawn *Metapenaeus brevicornis* (H. Milne Edward, 1837). FAO Fisheries Synopsis No.105.

George, P.C., M. J. George and P. Vedvuasa Rao. 1963. *Metapenaeus kutchensis* sp. Nov., A Penaeid prawn from the Gulf of Kutch. J. Mar. Boil. Ass. India, 5 (2); 284–288.

Hebert PDN, Cywinska A, Ball SL and De Waard JR (2003). Biological identifications through DNA barcodes. Proceedings of the Royal Society B 270: 313-321.

Holthuis L. B. 1980. FAO species catalogue. Shrimps and prawn of the world. An annotated catalogue of species of interest to fisheries. FAO Fisheries Synopsis, (125) Vol. 1: pp. 261.

Huynh Minh Sang, Ravi Fotedar, 2004. Growth, survival, haemolymph osmolality and organosomatic indices of the western king prawn (*Penaeus latisulcatus* Kishinouye, 1896) reared at different salinities. *Aquaculture, Volume 234, Issues 1-4, 3 May 2004, Pages 601-614.*

Ismael, D. and New, M.B. 2000. Biology. *In* M.B. New and W.C. Valenti, eds. *Freshwater prawn culture: the farming of Macrobrachium rosenbergii*, pp. 18-40. Oxford, England, Blackwell Science.

Jayachandran, K.V.; Joseph, N.I.1985. A new species of Macrobrachium from the south-west coast of India (Decapoda: Palaemonidae). Journal of Natural History, Volume 19, Number 1, January-February 1985 pp. 185-190(6).

Jayachandran K. V., R.S. Lal Mohan and A.V. Raji. 2007. A new species of *Macrobrachium* Bate, 1868 (Decapoda, Palaemonidae) from the dolphin trenches of Kulsi River, N. India, possibly under threat Zoologischer Anzeiger–A Journal of Comparative Zoology Volume 246, Issue 1, 22 May 2007, pp. 43-48.

Jay Chandran, K.V. and N.I. Joseph 1992. A Key for the commercially important *Macrobrachium* spp. of India-with a review of their bionomics. *In: Freshwater prawn proceeding of National symposium on Freshwater Prawns Macrobrachium spp.*, 12-14 December, 1990, Kochi,.

Jayaraman, R. and K. Karl Mark. 1998. Extruded feeds for successful aquaculture. Fisheries World, December: 23-24.

Joshi, V. P., S. M. Wasave and J. M. Koli 2005. Growing Banana shrimp into Adult in Masonry Tank in Cement Concrete Tank. www.Fishing Chimes.com.

Kunju, M. M. 2005. Some aspects of the Biology of *Solenocera indica* Nataraj. FAO.Org/ decapod/05/AC740t/Ac740T.27.

Kurian C. V. and V. O. Sebastian 1976. Prawn and Prawn Fisheries of India. Hindustan Publishing Co. New Delhi.

Mali Boonyartpalin, Thailand (2003). A report on the International Symposium on Freshwater Prawns, 2003 held at the College of Fisheries, Panangad from 20 to 23 August 2003. Current Science, Vol 85 (10), November 2003.

Mashiko K and Numachi K (2000). Derivation of populations with different-sized eggs in the palaemonid prawn *Macrobrachium nipponense*. Journal of Crustacean Biology 20: 118-127.

Michael New, 2003. A report on the International Symposium on Freshwater Prawns, 2003 held at the College of Fisheries, Panangad from 20 to 23 August 2003. Key note address European Aquaculture Society, UK. Current Science, Vol 85 (10), November 2003.

Mohamad K. H., M. S. Muthu, N. N. Pillai and K. V. George. 1978. Larval development *Metapenaeus monoceros* (Fabricius). CMFRI Bull 28: 50–59.

Mohmad K.H, P. Vednyas Rao, and M. J. George, 2006. Post larvae of penaeid prawns of southwest coast of India with a Key to other identification Ingenta connect, 2007.

MPEDA, 1996. A manual on shrimp farming.: Species selection for culture, 58-62.

MPEDA, 1998. Shrimp feed Management. Manual on shrimp farming, 108–117s.

Muthu, M. S. and G. Sudhakar Rao 1973. On the distinction between *Penaeus indicaus* and *P. merguiensis* with special reference to juveniles. Indian Journal of Fisheries Vol–20 No 1: 61–69.

Muthu, M, S., N. N. Pillai and K. V. George 1978. Larval development *M. dobsoni* (miers) CMFRI Bull, 28: 30–40.

Muthu M. S., N. N. Pillai and K. V. George 1978. Larval development *Metapenaeus affinis* (H. Milne Edward). CMFRI Bull 28:50–59.

Muthu M. S., N. N. Pillai and K. V. George 1978. Larval development *Parapenaeopsis stylifera* (H. Milne Edward). CMFRI Bull 28: 65–74.

New, M.B., 2005. Freshwater prawn farming: global status, recent research and a glance at the future. Aquaculture research 36: 210-230.

Pawase, A. S. and Shakuntala Shenoy 1998. Growth Experiments on *Penaeus merguiensis* and *Metapenaeus monoceros* from Ratnagiri using different pelleted feeds. Current and Emerging Trends in Aquaculture. Ed. P. C. Thomas. 244–260.

Pitchaimuthu Mariappan and Chellam Balasundaram 2004. Studies on the Morphometry of *Macrobrachium nobilii* (Decapoda, Palaemonidae). Brazilian Archives of Biology and Technology Vol.47, n. 3: pp. 441-449, July 2004.

Raje, P. C. Ranade, M. R. 1972. Larval Developments of Indian Penaeid shrimp *Penaeus merguiensis* De Man. *Journal of the Indian Fisheries Association*. Vol. No. 1 and 2. pp. 1–16.

Rajyalakshmi, T., 1961. Observations on the biology and fishery of *Metapenaeus brevicornis* (M. Edw.) in the Hooghly estuarine system. Indian J. Fish., 8(2): 383–402.

Rasheed, M. A. and CM Bull, 2000. Behaviour of the western king prawn, *Penaeus latisulcatus* Kishinouye: effect of food dispersion and crowding. *Australian Journal of Marine and Freshwater Research* 43(4):745–751.

Reymond, H. and Lagarder, J. P. 1988. Rythme alimentaire de *Penaeus japonicus* (Bate) en marais maritime. *Compte Rendu Hebdomadaire des Seances del'Academie des Science de Paris*, 307: 407–417.

Richardson L.R. and J.C. Yaldwyn 1958. A Guide to the Natant Decapod Crustacea (Shrimps and Prawns) of New Zealand Tuatara: Volume 7, Issue 1, September.

Salman D. Salman, Timothy J., Murtada D. Naser and Ama'al G. Yasser 2006. The invasion of *Macrobrachium nipponense* (De Haan, 1849) (Caridea: Palaemonidae) into the Southern Iraqi Marshes Aquatic Invasions (2006) Volume 1, Issue 3: 109-115.

Satyal Nadlal and Timothy Pickering. Freshwater prawn *Macrobrachum rosenbergii* Hatchery Operation Vol 1.

Sharma, A. and B. R. Subba 2005. General Biology of Freshwater prawn *Macrobrachium lamarrei* (H. Milne-Edward 1937) at Biratnagar Nepal. Our Nature, (3) 31–41.

Shadiq Ahamed, M. 1998. Shrimp diseases and treatment. Fisheries worlds, Dec.: 15.

Subrahmanyam, C.B., 1965. On the unusual occurrence of penaeid eggs in the inshore waters of Madras. *J.mar.biol.Ass.India.*, 7(1):83–8.

Tidwell, J.H. and D'Abramo, L.R. 2000. Grow-out systems–culture in temperate zones. *In:* M.B. New and W.C. Valenti, eds. *Freshwater prawn culture: the farming of Macrobrachium rosenbergii*. Blackwell Science, Oxford, England, pp. 177-176.

Ujaini Halim, 2004. Shrimp Monoculture in India Impact on the livelihood of coastal poor people.

Valenti, W.C. and Daniels, W. 2000. Recirculation hatchery systems and management *In:* M.B. New and W.C. Valenti, eds. *Freshwater prawn culture: the farming of Macrobrachium rosenbergii*. Blackwell Science, Oxford, England, pp. 69-90.

Valenti, W.C. and New, M.B. 2000. Grow-out systems–monoculture. *In:* M.B. New and W.C. Valenti, eds. *Freshwater prawn culture: the farming of Macrobrachium rosenbergii.* Blackwell Science, Oxford, England, pp. 157-176.

Vedvyas Rao P. 1973. Larval development of penaeid prawns. J. mar. boil. Asso India Vol 15 (1) 95–124.

Venkitraman P. R., K V Jayalakshmy, T Balasubramanian, Maheswari Nair and K K C Nair, 2003. Effects of eyestalk ablation on moulting and growth of penaeid prawn *Metapenaeus dobsoni* (de Man) National Institute of Oceanography, Regional Centre, Kochi 14, India.

Vyas, A.A., Desai, A.Y., Dholakia, A.D. and Chaya, N.D., 1993. Unavoidable live feed: Chaetoceros and artemia in prawn hatchery. Workshop on "Marine Prawn Culture" organised by College of Fisheries, Gujarat Agricultural University, Veraval on 20 June.

Wynne, F. 2005. Grow-out Culture of Freshwater Prawns in Kentucky, 2000. Last accessed July 4, 2005. Wikimedia Foundation, Inc.

Glossary

DO: Dissolved Oxygen.

Feed Efficiency (FE): Feed efficiency is calculated to grade the efficiency of the feed. This is an expression of quantity of shrimp obtainable per kg of feed.

Food Conversion Ratio (FCR): Food conversion ratio was determined by the formula of Sedgwick. FCR = Total dry weight of food fed (g)/total wet weight gain (g) over the experimental period.

Net Growth Efficiency: The net growth efficiency of ablated and control prawns was estimated by the formula, Net growth efficiency = (Px100)/A where P is the production in calories and A is the assimilation in calories.

pH: pH is the hydrogen ion concentration in water. pH indicates the extent of acidic or basic nature of water.

Protein Efficiency Ratio (PER): Protein efficiency ratio = Wet weight gain/dry weight of protein.

Salinity: Salinity is the total concentration of dissolved ions in water and is expressed as parts per thousand. (ppt or 0/00)

Telson: Last abdominal segment consists of an elongated sharp spine known as Telson.

Total Alkalinity: Total alkalinity in water refers to the total concentration of titrable bases in water and is expressed in milligrams per liter of equivalent calcium carbonate.

Total Hardness: Total hardness of water is defined as the total concentration of divalent cations in water and is expressed in milligrams per liter of equivalent of calcium carbonate.

MTL = Mean total length

MCL = Mean carapace length

MW = Mean width

MSF = Mean length of longest pair of furcal setae

A1 = Antennual

A2 = Antenna

Md = Mandible

Mx1 = Maxillute

Mx2 = Maxilla

Mxp1 = First maxilliped

Mxp2 = Second maxilliped

Mxp3 = Third maxilliped

P1 = First pereopod

P2 = Second pereopod

P3 = Third pereopod

P4 = Fourth pereopod

P5 = Fifth pereopod.

Seawater salinity = generally 35 ppt.

ppt = Parts per thousand

ppm = Parts per million